Principles of
Acoustic Devices

Principles of Acoustic Devices

VELIMIR M. RISTIC
Professor of Electrical Engineering
University of Toronto

A Wiley-Interscience Publication
JOHN WILEY & SONS
New York · Chichester · Brisbane · Toronto · Singapore

Library of Congress Cataloging in Publication Data:

Ristic, Velimir M., 1936–
    Principles of acoustic devices.

    "A Wiley-Interscience publication."
    Includes index.
    1. Ultrasonic transducers. I. Title.
TK5982.R57   1983   621.38′0412   82-20278
ISBN 0-471-09153-7

Printed in the United States of America

10 9 8 7 6 5 4 3 2 1

To

Jelica
for her patience and understanding

Foreword

For many decades, ultrasonic acoustic waves in liquids and solids have been used in a variety of engineering applications—sonar, electrical filters, nondestructive testing, and medical instrumentation. Since the 1960s, with the rapid growth of a sophisticated Rayleigh wave, or surface acoustic wave (SAW), technology, this subject has come more and more into the domain of the electrical engineer. Acoustic devices realized during this period are largely based on ideas conceived in the development of radar and of laser optics. For this reason the field theory of acoustic waves and vibrations is now closely related to applied electromagnetic theory. This facilitates the transfer to acoustics of applied electromagnetics formalism concerning a wide diversity of topics—frequency and pulse compression filters, phased arrays, waveguide propagation and components, slow wave structures, distributed feedback, etc.—and provides conceptual guidance in devising new types of devices.

With acoustic device technology now reaching maturity, the need has arisen for a simple textbook presentation of basic physical principles and design procedures both for the student and for the practising engineer. This book provides a systematic development of the necessary concepts, relating them at every step to simple models of actual devices and questions arising in practical design. To the degree required in achieving his chosen goal, the author covers bulk and surface acoustic waves, electromagnetic and piezoelectric transducer mechanisms and transducer design, elastic and piezoelectric properties of crystalline materials, reflection and refraction, radiation and diffraction. A unique feature is the first textbook treatment of electromagnetic transducers (EMATs), for which the principles are established with a simple one-dimensional bulk-wave model and then applied to practical geometries for the noncontacting plate and surface wave now commonly used in nondestructive testing. The chapter on acoustic wave radiation and diffraction is also a new contribution at this textbook level.

The author developed this book during the course of his research and teaching in the Department of Electrical Engineering at the University of Toronto, where he is active in advancing the design and application of acoustic devices.

B. A. AULD

Stanford, California
October 1982

vii

Preface

This book is designed primarily for graduate use. However, it can also be used in an advanced undergraduate course.

The book presents the principles of the basic acoustic devices most commonly used in engineering practice (nondestructive testing, communications, sonar, radar), and medical ultrasonics. Some of these principles are also directly applicable to other fields such as acousto-optics and laser optics. The book is intended to quickly introduce the reader to design aspects of acoustic devices while providing the underlying physics. To this end a few basic concepts such as the solution of the wave equation, acoustic impedance, impulse response, acoustic Poynting vector, etc., are developed and used consistently throughout, without resort to any theories that cannot be linked directly to basic principles. The design equations are developed, whenever possible, through one-dimensional modeling of the devices. The generalization of equations is done inductively. In many cases the mathematical proof is omitted, and the reader is directed to more advanced texts such as *Acoustic Waves in Solids* by B. A. Auld (two volumes, Wiley, 1973). With the exception of Chapters 6 and 7, which can be omitted in teaching the undergraduate course, the text emphasizes the real problems of acoustic device design.

The manuscript was tested in the classroom over a number of years in a basic graduate course on acoustic devices, and was used as an introductory text for more advanced design books such as *Surface Wave Filters* (H. Matthews, Ed., Wiley, 1977) and *Acoustic Surface Waves* (A. A. Oliner, Ed., Springer-Verlag, 1978), and more advanced textbooks such as *Acoustic Waves in Solids* (B. A. Auld, two volumes, Wiley, 1973).

I have many people to thank for their suggestions and their specific help. In particular, B. A. Auld from Stanford University and R. M. White from the University of California at Berkeley gave some general and specific advice that contributed considerably to the improvement of the manuscript.

I am also indebted to Mrs. Linda Espeut for her skillful typing of the manuscript.

VELIMIR M. RISTIC

Toronto, Ontario
January 1983

Contents

Chapter 5.   Piezoelectric Bulk Wave Transducers, 117

Chapter 6.   Piezoelectric Materials, 174

Chapter 7.   Wave Propagation in Anisotropic Materials, 212

Chapter 8.   Surface Acoustic Wave Transducers, 238

Chapter 9.   Reflection and Refraction of Bulk Waves and the Coupling Prism, 277

Chapter 10.   Radiation and Diffraction of Acoustic Waves, Planar Radiators, and Acoustic Lenses, 294

Tables

Principles of
Acoustic Devices

Chapter 1

Basic Concepts
of Acoustic Devices

The basic concepts of acoustic devices employing acoustic waves comprise definitions of engineering design quantities such as acoustic impedance and impulse response, synthetic acoustic materials for impedance matching purposes, and the properties of various acoustic modes in specific materials. A skillful usage of these properties for conversion of electromagnetic into acoustic energy and vice versa is the object of the design. The conversion processes can be phenomena inherent to materials, such as piezoelectricity or magnetostriction, or they may be mechanisms such as the Lorentz force mechanism. These phenomena and mechanisms form the operational basis of acoustic transducers. The finite size of transducers cause various boundary phenomena such as reflection, refraction, and diffraction. These phenomena are taken into account in the design of either one-dimensional devices, such as surface acoustic wave transducers, or two-dimensional devices such as lenses, coupling prisms, and bulk transducers. More advanced device design usually requires the implementation of a specific signal-processing function such as the generation of dispersive signals for use in radar or sonar, the imaging capability of a phased acoustic array for use in medical ultrasonics or nondestructive testing, or the Q-factor of an acoustic resonator for communication signal processing. In fact, the design of acoustic devices makes use of physics and chemistry, wave engineering and optics, and communication signals, coupled to mathematical skills.

The treatment of general acoustic wave propagation[1,2] is complicated by the fact that parameters of acoustic waves such as strain, stress, and material parameters are tensor quantities. This requires a mathematical formalism that often masks the simple physical meaning of equations and alienates a reader from the theory. In the first few sections of this volume, the approach is to present the simplest types of waves, which do not require the tensor formalism. However, the presentation is rigorous, and the quantities and equations derived here can be applied to the design of various acoustical components and systems.

Basically, it may be said that acoustic waves are propagated in the bulk of material in either the longitudinal or transverse (shear) mode. In the *longitudi-*

nal mode, the motion of the acoustic medium is only in the direction of propagation. This is shown in Fig. 1.1(a), where the material is expanding and contracting in the z direction, which is the direction in which acoustic waves are propagating. A type of longitudinal motion may be visualized by imagining the stretching and compression of a spring. The longitudinal mode requires a change in the volume or mass density of the material. In the *shear mode*, the motion of the medium is transverse (perpendicular) to the direction of acoustic wave propagation, as illustrated in Fig. 1.1(b), where the motion of the medium is in the y direction, and the resulting mode is called the *y-polarized shear mode*. Rotational torsion and flexing or bending of the material about an axis perpendicular to the direction of propagation are associated with the shear mode. The longitudinal and shear modes so defined are termed *pure modes*.

An acoustic wave propagating through an anisotropic material in an arbitrary direction (which does not coincide with axes or planes of symmetry of the material) is in a combination of longitudinal and shear modes. This type of wave is called either *quasi-longitudinal* or *quasi-shear*, depending on which component is predominant. In anisotropic media there are so-called *pure mode directions*, which coincide with axes of crystal symmetry. Along these axes, waves can propagate in either the pure shear or pure longitudinal mode or both. In most acoustic devices using anisotropic material, pure mode directions are chosen for propagating directions to facilitate the wave excitation and to maximize the electrical-to-mechanical conversion efficiency.

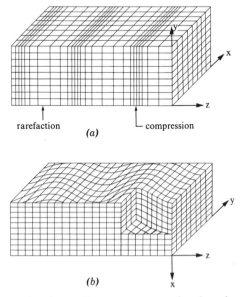

Figure 1.1.   Grid diagrams for plane uniform waves propagating along the z axis in a material of infinite extent: (a) longitudinal mode; (b) y-polarized shear mode.

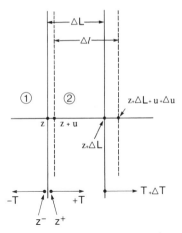

Figure 1.2. Deformation of a small slab of cross-sectional area A and thickness ΔL subject to time-varying longitudinal stress propagating in the $+z$ direction. By convention, $-T$ is the traction force exerted by medium 1 on medium 2.

On the microscopic scale,[2] atoms in the lattice of a solid are coupled to each other by binding forces, and any local disturbance of the lattice can propagate to other parts of the solid. The spectrum of frequencies and wavelengths of mechanical vibrations that may be allowed in a given solid depends on the detailed nature of the lattice and binding forces. The energy associated with the vibration of an atom about its rest position in the lattice is called a *phonon*. This energy is proportional to the discrete frequency of atomic vibration, that is, the phonon is quantized. An acoustic wave in a solid is considered to be a group of coherent phonons. The smallest volume of material that is considered in device design, an *acoustic particle*, is large enough compared with the spacing between atoms that a piece of material is like a continuum.[1] Within the volume of an acoustic particle, the magnitudes of macroscopic physical quantities are constants, that is, they are independent of spatial coordinates.

1.1 PROPAGATION OF LONGITUDINAL WAVES

Consider a slab of material of thickness ΔL, small compared to the wavelength but large compared to the spacing between atoms, so the medium is like a continuum, subject to force $-T$ per unit area at the plane z^-, as shown in Fig. 1.2. The force per unit area is termed *stress*[3] (*traction force* or *force density*) and is measured in units of N/m^2. The traction force is transmitted through the material. For instance, the stress at the plane z^+ will be $+T$ since the material at plane z is in equilibrium.[†] Time-varying stress can be used to excite vibrations in the material. The stress waves propagating through the materials in the z direction will depend upon both time and coordinate z, that is,

[†] The force per unit area exerted from the left across the plane $z = 0$ is $-T$ and that from the right is $+T$. This is a convention.

$T = T(t, z)$. Due to the action of time-varying stress, the boundaries of the slab will be displaced with respect to the center of mass of the body. Thus, the plane z will move to $z + u$, where u is the displacement, measured in meters (m). The plane[†] $z + \Delta L$ will be displaced to $z + \Delta L + u + \Delta u$, where Δu is the change of displacement over the thickness ΔL. Since the stress wave is a function of z, its intensity will be different at the plane $z + \Delta L$; that is, at this plane the stress is $T + \Delta T = T + (\Delta T/\Delta Z)\Delta L$, thus producing a net force density of $(\partial T/\partial z)\Delta L$ that acts on the slab relative to the center of mass of the body. The displaced boundaries of the slab in the figure are an indication of the deformation inside the body of the material. The *deformation*, $\mathcal{D}$, measured in square meters is usually defined[1] (see Fig. 1.2) as

$$\mathcal{D} = (\Delta l)^2 - (\Delta L)^2 \equiv 2S(\Delta L)^2 \qquad (1.1)$$

where S, the relative deformation, is termed *strain* and is a purely numerical quantity. From Fig. 1.2 it is seen that $\Delta l = \Delta L + \Delta u = \Delta L + (\partial u/\partial z)\Delta L$, and therefore

$$S = \frac{1}{2} \frac{(\Delta l)^2 - (\Delta L)^2}{(\Delta L)^2} \simeq \frac{\partial u}{\partial L} \simeq \frac{\partial u}{\partial z} \qquad (1.2)$$

where the quadratic terms are neglected. This is the *linearized strain–displacement relation*. Thus, the strain S is a change of length per unit length and is positive for extension and negative for compression.

If ρ_m is the mass density [measured in kilograms per cubic meter (kg/m³)] of the material in the stationary state, and A is the cross-sectional area of the slab, then, according to Newton's second law, the force acting on the slab will be

$$F = A\frac{\partial T}{\partial z}\Delta L = \rho_m A \,\Delta L \frac{\partial^2 u}{\partial t^2} \qquad (1.3)$$

or

$$\frac{\partial T}{\partial z} = \rho_m \ddot{u} = \rho_m \dot{v} \qquad (1.4)$$

where u and v are the displacement and particle velocity, respectively, at the plane z of the slab, and the dots over variables represent partial differentiation with respect to time. The change of particle velocity over the thickness ΔL of the slab is given by

$$\Delta v = \frac{\partial v}{\partial z}\Delta L \qquad (1.5)$$

[†]Plane z^- or plane $z + \Delta L$ means a plane orthogonal to the z axis and intersecting the z axis at the points z^- and $z + \Delta L$, respectively.

or, using Eq. (1.2), by

$$\Delta v = \frac{\partial(\Delta u)}{\partial t} = \frac{\partial S}{\partial t} \Delta L \tag{1.6}$$

Combining Eqs. (1.5) and (1.6) it follows that

$$\frac{\partial S}{\partial t} = \frac{\partial v}{\partial z}. \tag{1.7}$$

This equation can be further specified by invoking the *linear Hooke's law* for small strains $S < 10^{-4}$ given by

$$T = cS \tag{1.8}$$

where c is the *elastic stiffness constant* of the material measured in newtons per square meter (N/m^2) or pascals (Pa). If we combine the last two equations, it follows that

$$\frac{1}{c} \frac{\partial T}{\partial t} = \frac{\partial v}{\partial z}. \tag{1.9}$$

The wave equation, obtained from Eqs. (1.4) and (1.9), is

$$\frac{\partial^2 T}{\partial z^2} - \frac{\rho_m}{c} \frac{\partial^2 T}{\partial t^2} = 0. \tag{1.10}$$

In the one-dimensional case, all the field quantities such as S, T, u, and v, are functions of z and t only. Since the linearity of the phenomenon is assumed, each of the physical quantities can be expressed as

$$Q = Q(z)e^{j\omega t} \tag{1.11}$$

that is, a single-frequency harmonic time dependence is assumed. Because the material is homogeneous, it is possible to seek solutions of Eq. (1.10) in the form $\exp[j(\omega t - \beta_a z)]$ where β_a is the *acoustic propagation constant* (*wave number*). Using Eq. (1.10), it is easy to show that

$$\beta_a^2 = \frac{\omega^2 \rho_m}{c} \tag{1.12}$$

From the general formula for phase velocity, $v_{\text{phase}} = \omega/\beta$, the *acoustic phase velocity* v_a is found as

$$v_a = \frac{\omega}{\beta_a} = \left(\frac{c}{\rho_m}\right)^{1/2} \tag{1.13}$$

Using Eqs. (1.7) and (1.8) it follows that

$$u(z) = \frac{v_a S(z)}{j\omega}.$$ (1.14)

Thus, for a given strain and a given acoustic phase velocity, the displacement $u(z)$ is inversely proportional to the frequency.

By analogy with plane electromagnetic waves, it is possible to define an *acoustic impedance* of a plane acoustic wave traveling in the $+z$ direction as

$$Z_a = \frac{-T}{v} > 0$$ (1.15)

which is measured in units of kilograms per square meter per second ($\text{kg/m}^2 \cdot \text{s}$). The reason for the negative sign is the convention used for orienting the stress. By convention, $-T$ is the traction force exerted *by* medium 1 *on* medium 2 as shown in Fig. 1.2. Using Eqs. (1.15) and (1.14), it follows that the acoustic impedance is given by

$$Z_a = \rho_m v_a = \sqrt{\rho_m c}$$ (1.16)

In the derived expressions the acoustic phase velocity v_a is a constant for a given homogeneous and isotropic material. The particle velocity v is a function of both time and coordinate z, for instance, $v = \dot{u}$.

1.2   MATERIAL PARAMETERS

In engineering design, an important material property, *Young's modulus of elasticity*, is sometimes experimentally determined from extensional waves on a thin rod. A rod of cross-sectional dimensions much smaller than the wavelength of the wave is subject to periodic force along the rod. The only stress component, T_z, is along the axis of the rod, but there are two displacements. The one along the axis of the rod results in the strain $S_z = \partial u / \partial z$. The other strain component is along the radius of the rod, since the rod's diameter is free to expand and to contract. The strain component S_z is related to the applied stress T_z by Hooke's law, Eq. (1.8), as

$$T_z = E_Y S_z,$$ (1.17)

where E_Y is Young's modulus of elasticity. In all cases $E_Y < c$, since a larger force is necessary to obtain the same strain S when the bulk solid is constrained than when it is free to expand and contract transversely.

In classical elasticity design the properties of the material such as density and Young's modulus E_Y are determined experimentally.[†]

[†]Young's modulus of elasticity, E_Y, is usually measured by static extensometer testing.

TABLE 1 Characteristics of Some Nonpiezoelectric Materials[a]

Material	Mass Density, kg/m^3	Longitudinal Velocity, m/s	Acoustic Impedance, $\times 10^6$ kg/m$^2 \cdot$ s
Air (20°C)	1.21	340	411 $\times 10^{-6}$
Water (20°C)	1,000	1,480	1.5
Glycerine	1,260	1,920	2.5
Castor oil	950	1,540	1.46
Motor car oil (SAE 20)	870	1,740	1.5
Polyalkyleneglycol (PAG), 20°C	973	1,330	1.29
Choloroform	1,480	1,010	1.5
Carbon tetrachloride, 20°C	1,590	940	1.5
Paraffin	860	1,420	1.22
Ice	900	3,980	3.6
Lithium	5,330	3,000	16
Teflon	2,160	1,340	3
Acrylic resin (Lucite, Perspex, Plexiglass)	1,182	2,680	3.17
Polyethylene	900	1,940	1.75
Polystyrene	1,056	2,340	2.47
Epoxy resin	1,100–1,250	2,400–2,900	2.7–3.6
Nylon, Perlon	1,100–1,200	2,200–2,600	2.4–3.1
Silicon rubber (GE, RTV 615)	1,010	1,030	1.04
Aluminum	2,695	6,350	17.1
Indium	7,310	2,500	18.3
Beryllium	1,850	12,500	23.1
Brass	8,100	4,430	35.9
Lead	11,400	1,960	22.3
Copper	8,900	4,700	42
Gold	19,300	3,210	62
Iron (steel)	7,830	5,950	46.6
Mercury	13,600	1,450	20
Mylar	1,180	2,540	3
Platinum	21,400	4,152	89
Silver	10,490	3,440	36.1
Tungsten	19,100	5,500	105.1
Zinc	7,100	4,200	30
Titanium	4,500	6.071	27.3
Nickel	8,905	6,030	53.7
Glass T-40	3,390	4,310	14.6
Sapphire (z axis)	3,990	11,100	44.3
Glass, Pyrex	2,320	5,610	13
Glass, crown	2,243	5,090	11.4
Glass, flint	3,879	4,030	15.6

TABLE 1   (continued)

Material	Mass Density, kg/m^3	Longitudinal Velocity, m/s	Acoustic Impedance, $\times 10^6$ kg/m$^2 \cdot$ s
Fused quartz (silica)	2,200	5,970	13.1
Rutile (z axis)	4,260	7,900	33.7
Methanol, 30°C	796	1,088	0.866
Acetone, 30°C	791	1,158	0.916
Ethanol, 30°C	789	1,127	0.890
Zircaloy (Zr–2.5Nb)	6,500	4,686	30.5
Magnesium	1,720	5,800	10
Graphite	2,170	4,210	9.14
Magnesia (MgO)	3,842	8,850	34
Silicon, along [100]	2,337	8,430	19.7
Flutec PP3 ($C_7F_{14} + C_8F_{16}$)	1,900	612	1.12
Stainless steel	8,090	5,610	45.4
Porcelain	2,400	5,600	13.4

[a] The parameters listed for some materials should be considered as an average. The parameters of these vary depending on the production method used. Ice and metals are assumed to be polycrystalline.

The speed of sound v_e (*extensional wave velocity*) and the acoustic impedance Z_Y can be calculated using Eqs. (1.13) and (1.16) with E_Y instead of c. The numerical values for Z_Y and v_e are not of direct interest to us, since in the wave propagation studies considered here the wavelength of the propagating waves will be, in most cases, much smaller than the transverse dimensions of the sample. Thus, the computation of the acoustic velocity v_a and acoustic impedance Z_a will be based on the elastic constant c. In Table 1, the longitudinal wave velocity v_a and the acoustic impedance Z_a for longitudinal waves are given for a number of nonpiezoelectric materials of interest. The characteristics for some piezoelectric materials are given in Table 2. The computation of longitudinal velocity for specific directions in some piezoelectric materials requires that the piezoelectric properties of the material be taken into account (see Chapter 5). In gases and nonviscous liquids (excluding liquid crystals[4]), only longitudinal waves can be propagated, since those media cannot support shear stress.

1.3   BOUNDARY CONDITIONS

The general statement for boundary conditions between two acoustic materials that are intimately in contact (the so-called *acoustic bond*) is that stress and

TABLE 2   Characteristics of Some Piezoelectric Materials

Material	Mass Density, kg/m³	Longitudinal Velocity, m/s	Acoustic Impedance, $\times 10^6$ kg/m² · s
Quartz (X cut)	2650	5740	15.2
Lithium niobate (Y + 30° cut)	4640	7400	34
Lithium niobate (Z cut)	4640	7330	34
Lithium tantalate (Y + 47° cut)	7460	6300	47
Zinc oxide (Z cut)	5680	6330	36
Cadmium sulfide (Z cut)	4840	4460	21.6
Paratellurite			
(along ⟨001⟩)	6000	4260	25.56
(along ⟨110⟩)	6000	4660	27.96
(along ⟨100⟩)	6000	3050	18.3
Potassium sodium[a] niobate ceramics	4460	6940	31
PZT-5A[a]	7750	4350	33.7
PZT-5H[a]	7500	4560	34.2
PZT-7A[a]	7600	4800	36.5
PVF$_2$[a]	1777	2150	3.82
Lead metaniobate ceramics[a] (K-81)	6200	3323	20.6

[a]Along poling axis.

particle velocity must be continuous across the interface. In what follows, this statement will be proven for longitudinal waves with propagation direction normal (at right angles) to the interface. Note that if a longitudinal wave is incident at an oblique angle ($\alpha_i < 90°$) to the interface, two reflected waves are generated. One is a longitudinal wave reflected at an angle equal to the angle of incidence, and the other is a shear wave reflected at a smaller angle α_r, where $\sin \alpha_r / \sin \alpha_i = v_S / v_L$, and where v_S and v_L are the shear and longitudinal wave velocities, respectively, with $v_L > v_S$. Furthermore, two refracted waves are generated. Shear waves are always generated if the angle of incidence $\alpha_i \neq 90°$.

Consider a sinusoidal longitudinal wave propagating across an interface, as shown in Fig. 1.3, assuming that the stress and the displacement are different on the two sides of the boundary. Integration of Eq. (1.4) over the prismatic

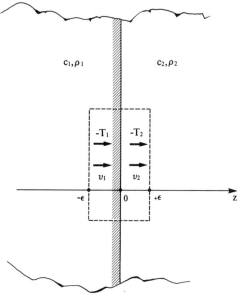

Figure 1.3. Longitudinal wave incident from the left, normal to the interface between two intimately bonded acoustic media. The shaded area is the region of the bond, assumed much smaller than the acoustic wavelength.

surface consisting of the cross section A and a side 2ϵ yields[†]

$$A \lim_{\epsilon \to 0} \int_{-\epsilon}^{+\epsilon} \frac{\partial T}{\partial z} \, dz = A \lim_{\epsilon \to 0} \int_{-\epsilon}^{+\epsilon} \rho_m \dot{v} \, dz \qquad (1.18)$$

or

$$-T_2 + T_1 = (\rho_2 \dot{v}_2 - \rho_1 \dot{v}_1)\epsilon, \qquad \epsilon \to 0 \qquad (1.19)$$

Since $\epsilon \to 0$, $T_2 = T_1$. Note that if T_1 were different from T_2 this would mean physically that there is a finite force density on an arbitrarily small slab of material. Again, we must conclude that the particle displacement is continuous, since otherwise the two materials would not be in contact, which is contrary to our initial assumption of an intimate bond. This is an important point since, for instance, in the production of epoxy-bonded piezoelectric transducers, extreme care must be taken to drive out all the air bubbles from the bond. The presence of air with its low impedance (see Table 1) means a short circuit (i.e., total reflection) for the incident wave. Since $\dot{u} = v$, the continuity of the displacement implies the continuity of particle velocity. Note that the strain, in

---

[†]Note that ϵ is used in the text as an auxiliary variable and sometimes as an index, while ε is used to denote permittivity.

general, is not continuous. Using Eqs. (1.19) and (1.8), it can be shown readily that $S_1/S_2 = c_2/c_1$, and, since c_2 and c_1 are in general different, the continuity of the displacement does not imply the continuity of its gradient (strain). An analysis of bulk shear waves incident normal to the boundary shows that the tangential components of stress and displacement must be continuous as well.

To summarize, at the interface between two intimately bonded acoustic media, the appropriate boundary conditions are that both the normal and tangential components of the stress and displacement (or particle velocity) must be continuous across the interface.

From the analogy between electric field and stress, and magnetic field and particle velocity Eq. (1.15), it is possible to define an acoustic counterpart of an electromagnetic transmission line (slab geometry).[5] The wave equations for acoustic and electromagnetic waves (TEM mode) can be written as

$$\frac{\partial^2}{\partial z^2}\left\{ \begin{array}{c} T \\ v \end{array} \right\} - \frac{\rho_m}{c}\frac{\partial^2}{\partial t^2}\left\{ \begin{array}{c} T \\ v \end{array} \right\} = 0 \qquad (1.20)$$

$$\frac{\partial^2}{\partial z^2}\left\{ \begin{array}{c} E \\ H \end{array} \right\} - \varepsilon\mu\frac{\partial^2}{\partial t^2}\left\{ \begin{array}{c} E \\ H \end{array} \right\} = 0 \qquad (1.21)$$

where μ and ε are the permeability and the permittivity of the slab line, respectively. Furthermore, it follows from Eqs. (1.4) and (1.9) that

$$\frac{\partial(-T)}{\partial z} = -j\omega\rho_m v \quad \text{corresponding to} \quad \frac{\partial E}{\partial z} = -j\omega\mu H \qquad (1.22a)$$

and

$$\frac{\partial v}{\partial z} = -j\frac{\omega}{c}(-T) \quad \text{corresponding to} \quad \frac{\partial H}{\partial z} = -j\omega\varepsilon E \qquad (1.22b)$$

The mathematical equivalence in form of the electromagnetic and acoustic equations (1.21) and (1.22) implies that the definition of acoustic impedance given by Eq. (1.15) is consistent with the definition of the electric impedance $Z = E/H$, with μ analogous to ρ_m and the permittivity ε analogous to $1/c$. In the same way, the *stress reflection* and *transmission coefficients* can be defined by analogy with electric field reflection and transmission coefficients, for the case of normal incidence, as shown in Fig. 1.4. It is readily computed from the boundary conditions and Eq. (1.21) that the stress reflection and transmission coefficients for the wave incident from the left of the boundary are, respectively,

$$r = \frac{Z_2 - Z_1}{Z_2 + Z_1} \quad \text{and} \quad \tau = \frac{2Z_2}{Z_2 + Z_1} \qquad (1.23)$$

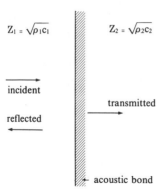

$Z_1 = \sqrt{\rho_1 c_1}$ $Z_2 = \sqrt{\rho_2 c_2}$

incident

reflected

transmitted

Figure 1.4. Normal incidence of a plane longitudinal wave on the interface between two materials. The shaded area is the region of the bond, assumed much smaller than the acoustic bond acoustic wavelength.

Slabs (plates) of transverse dimensions much larger than acoustic wavelength and of prescribed thickness, and made of material with a prescribed acoustic impedance, are used for matching purposes (Exercise 4). In this application, such a plate, often called an *acoustic transformer*, is bonded between the body of an acoustic generator and an acoustic load. Acoustic transformers are analogous to matching transformers[5] (radomes, plates) used in optical and electromagnetic engineering.[6] In practice it is often difficult to find materials with the prescribed acoustic properties that can be bonded conveniently.

1.4 ACOUSTIC LOSS MECHANISMS

Acoustic wave propagation losses (attenuation) are caused by a variety of physical mechanisms.[7] For instance, in materials with finite-size grains, such as piezoelectric ceramics and polymers, *scattering losses* exist.[8] Various imperfections in materials, such as dislocations in a crystal lattice, will also cause scattering losses. These losses become increasingly important at higher frequencies, so that at microwave frequencies only crystalline materials are used, and piezoelectric ceramics are rarely used above 20 MHz. One exception is potassium sodium niobate ($K_{0.5}Na_{0.5}NbO_3$), which can be operated in the 100-MHz region. Another type of loss is the *thermal conduction loss*[7,9] or *thermoelastic loss*. The temperature of an ideal acoustic material deformed by the passage of an acoustic wave will increase during compression and decrease during rarefaction, without loss of heat energy to its surroundings (adiabatic process). In a real acoustic material, however, there is a loss of energy due to heat conduction. This type of attenuation is proportional to the square of the frequency and occurs with longitudinal but not shear waves. A property of some liquids, glasses, and polymers is that they possess large molecular chains (short-range order). Under the action of acoustic waves, some of these chains reorient themselves irreversibly, causing losses. This type of loss is called *hysteresis absorption*[7] or the *change-of-state loss mechanism*[10,11] and is proportional to the frequency.

The most important cause of acoustic attenuation in liquids and solids, however, is the *viscous loss mechanism*[7] (*relaxation absorption*). In this case, as the wave propagates through the material, the neighboring particles intercepted by a given acoustic wavefront are not moving at the same speed. The viscous losses can be quantified in the following way. For a plane wave of acoustic power P propagating through the material in the $+z$ direction, the expression for real power is given by

$$P_a = \text{Re}\{P\} = P_0 e^{-2\alpha z} \tag{1.24}$$

where α is the *attenuation constant* measured in nepers per meter (Np/m). The attenuation constant due to viscous loss is given by[†]

$$\alpha = \frac{\eta \omega^2}{2 v_a^3 \rho_m} \tag{1.25}$$

where ω is the angular frequency and η is the *coefficient of viscosity* measured in newton-seconds per square meter ($\text{N} \cdot \text{s/m}^2$) or pascal-seconds ($\text{Pa} \cdot \text{s}$). It is seen that the attenuation constant is inversely proportional to the cube of the velocity and directly proportional to the square of the frequency. Since it is known that in most materials shear-wave velocities are typically about half as large as longitudinal velocity, the shear wave attenuation constant is about eight times the longitudinal wave attenuation constant.

The viscous-loss mechanism can be taken into account in Eq. (1.8) by assuming an additional force density acting on the particle during acoustic plane wave propagation. This force density can be defined consistently with Eqs. (1.24) and (1.25) as

$$T_\eta = \eta \frac{\partial v}{\partial z} = \eta \frac{\partial S}{\partial t}. \tag{1.26}$$

From Eqs. (1.8) and (1.26) the total force acting on the particle is then

$$T = cS + \eta \frac{\partial S}{\partial t} \tag{1.27}$$

and for sinusoidal excitation this can be written as

$$T = (c + j\omega\eta)S = \underline{c}S \tag{1.28}$$

[†]Equation (1.25) can be written as $\alpha = Bf^2$, where B is the classical absorption parameter. In highly viscous nonmetallic liquids, the measured and predicted values of B are within a factor of 3. In other cases the difference may be orders of magnitude. Also, in many materials, due to other loss mechanisms, the square frequency dependence is not obeyed.

where $\underset{\sim}{c}$ is the *complex elastic stiffness*, given by $\underset{\sim}{c} = c + j\omega\eta$. Equation (1.22) can now be written as

$$\frac{\partial(-T)}{\partial z} = -j\omega\rho_m v$$

$$\frac{\partial v}{\partial z} = -j\frac{\omega}{\underset{\sim}{c}}(-T) \qquad (1.29)$$

leading to the definition of a lossy acoustic transmission line and complex acoustic impedance.

In experimental practice the acoustic loss properties of a material are found by measuring the transmission loss of shear and longitudinal waves over the distance $z = l$ for a range of frequencies. *Attenuation* is then defined, using Eq. (1.24), as

$$\text{Attenuation [dB]} = 10 \log e^{-2\alpha l} = 10 \log \frac{\text{Re}\{P\}}{P_0} \qquad (1.30)$$

The attenuation constant is evaluated in either nepers per meter (Np/m) or decibels per meter (dB/m), using the well-known relation

$$\alpha \, [\text{dB/m}] = 20 \, \alpha \, [\text{Np/m}] \log e \approx 8.686\alpha \, [\text{Np/m}] \qquad (1.30a)$$

From these data and Eq. (1.25), the viscosity constant η is then found. Such measurements indicate that the attenuation, in general, is smallest for high quality crystal materials. For instance, for sapphire and rutile, the longitudinal attenuation constants at 3 GHz are 6.5 and 3.5 dB/cm, respectively. Although these figures appear high, it must be remembered that what matters is the attenuation per wavelength. The wavelength at these frequencies is typically 1 μm, leading to an attenuation on the order of 10^{-4} dB per wavelength. Another way of representing the attenuation data is to compute the attenuation for length l corresponding to 1 μs delay time, in which case the attenuation is given in dB/μs. For instance, at 10 MHz in water, the attenuation is 0.032 dB/μs. Typical values for η for high quality single crystals are on the order of 20×10^{-5} N · s/m². The tabulation of attenuation coefficients for various materials (except for crystals) would be of doubtful value, since the attenuation is strongly dependent on the manufacturing process used.[12] In general, at 2 MHz, materials such as steel, aluminum, silver, and tungsten exhibit losses of up to 10 dB/m; cast steel, copper, brass, bronze, and lead between 10 and 100 dB/m; cast copper, cast iron, porous ceramics, plastics with fillers, rubber, and wood, above 100 dB/m; while for plastics such as polystyrene, Lucite, PVC, and synthetic resins, the attenuation constant varies from 10 to greater than 100 dB/m. Materials such as rubber and various waxes are therefore used for acoustic absorbers.

For instance, the attenuation constant of water is given by

$$\alpha\,[\text{dB/m}] = 0.217 f^2 \qquad (1.31)$$

where f is the frequency in MHz ($\eta \simeq 10^{-3}$ Pa $\cdot$ s). On the other hand, silicon rubber compound (GE, RTV 615) has an attenuation constant[13]

$$\alpha\,[\text{dB/m}] = 130 f^{1.4} \qquad (1.31a)$$

where f is in MHz.

Room-temperature losses in Lucite and polyethylene, materials exhibiting hysteresis absorption, are linearly dependent on frequency. For Lucite,[14]

$$\alpha_L \left[\frac{\text{dB}}{\text{m}}\right] = 70 f \qquad \text{and} \qquad \alpha_S \left[\frac{\text{dB}}{\text{m}}\right] \simeq 210 f$$

where f is in MHz and α_L and α_S are the attenuation constants of longitudinal and shear waves, respectively.

Combining Eqs. (1.27), (1.4), and (1.2) results in the wave equation

$$\rho_m \frac{\partial^2 u}{\partial t^2} = c\frac{\partial^2 u}{\partial z^2} + \eta\frac{\partial^3 u}{\partial t\,\partial z^2}$$

Using $u = u_0 \exp j(\omega t - kz)$, $k = k_r - j\alpha$, and $\alpha \ll k_r = \omega/v_a$, and neglecting the small terms, the value of α, as given by Eq. (1.25), is found.

1.5 ENERGY CONSERVATION THEOREM

In order to formulate an energy conservation theorem, it is necessary to include external forces in Eq. (1.4). In addition to external *surface forces*, such as traction force (measured in units of N/m^2), there may also be *body forces*. For instance, the gravitational field produces a body force $F^b = \rho_m g$, where g is the gravitational acceleration measured in m/s^2. Another type of body force is the Lorentz force, which for a current-carrying conductor of current density J immersed in magnetic flux density B is given by

$$\vec{F}^b = \frac{d\vec{F}}{dV} = \vec{J} \times \vec{B} \qquad (1.32)$$

Body forces are measured in units of N/m^3. In the presence of body forces Eq. (1.4) becomes

$$\frac{\partial T}{\partial z} = \rho_m \frac{\partial v}{\partial t} - F^b \qquad (1.33)$$

To obtain the acoustical counterpart of Poynting's theorem, it is necessary to multiply Eq. (1.33) by v, and Eq. (1.7) by T. Then, by addition, it follows that

$$-\frac{\partial(Tv)}{\partial z} = -\left[\frac{\partial}{\partial t}\left(\tfrac{1}{2}\rho_m v^2\right) + T\frac{\partial S}{\partial t}\right] + F^b v \qquad (1.34)$$

Applying Eq. (1.27), this becomes

$$-\frac{\partial(Tv)}{\partial z} = -\left[\frac{\partial}{\partial t}\left(\tfrac{1}{2}\rho_m v^2\right) + \frac{\partial}{\partial t}\left(\tfrac{1}{2}cS^2\right) + \eta\left(\frac{\partial S}{\partial t}\right)^2\right] + F^b v \qquad (1.35)$$

Consider now a region of material as shown in Fig. 1.5(a), traversed by an acoustic wave propagating along the z direction. Integration of Eq. (1.35) over

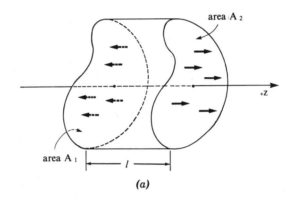

(a)

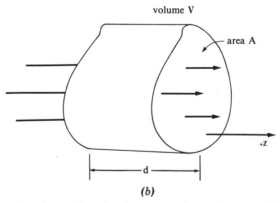

(b)

Figure 1.5. (a) A slice of material used to demonstrate the one-dimensional Poynting's theorem. (b) Acoustic beam propagating through a plate of face area A and thickness d. Arrows show the direction of the Poynting vector.

the volume V in Fig. 1.5(a), using a volume element $dV = dA\, dz$, gives

$$\iiint_V -\frac{\partial(Tv)}{\partial z}\, dz\, dA = \iint_{A_2} (-Tv)\, dA - \iint_{A_1} (-Tv)\, dA$$

$$= -\iiint_V \frac{\partial}{\partial t}\left(\tfrac{1}{2}\rho_m v^2\right) dV - \iiint_V \frac{\partial}{\partial t}\left(\tfrac{1}{2}cS^2\right) dV$$

$$- \iiint_V \eta\left(\frac{\partial S}{\partial t}\right)^2 dV + \iiint_V F^b v\, dV \qquad (1.36)$$

The integral on the left-hand side can be identified as the net power emitted through the surface $A_1 + A_2$ and, by analogy with electromagnetics,

$$\Pi = -Tv \qquad (1.37)$$

can be defined as the *time-varying acoustical Poynting vector* measured in units of W/m². Furthermore, $u_v = \tfrac{1}{2}\rho_m v^2$ is the *kinetic energy density* and $u_s = \tfrac{1}{2}cS^2$ is the *strain energy density* (*potential energy density*), both measured in units of J/m³ = N/m²,

$$P_d = \iiint_V \eta\left(\frac{\partial S}{\partial t}\right)^2 dV \qquad (1.38)$$

is the *total viscous power loss* in the material, measured in watts (W), and

$$P_s = \iiint_V F^b v\, dV \qquad (1.39)$$

is the *power supplied by sources*. Denoting the sum of the total kinetic and strain energies by U, Eq. (1.36) can be written as

$$\iint_{A_2} \Pi\, dA - \iint_{A_1} \Pi\, dA + \frac{\partial U}{\partial t} + P_d = P_s \qquad (1.40)$$

According to the energy conservation theorem, Eq. (1.40), the power supplied by sources inside the volume of Fig. 1.5 is partly radiated into the surroundings, partly stored as field energy, and partly dissipated in the material. The instantaneous radiated acoustic power is given by

$$P_a = \iint_{A_2} \Pi\, dA - \iint_{A_1} \Pi\, dA \qquad (1.41)$$

For time-harmonic excitations, where $\exp(j\omega t)$ dependence is assumed, a complex form of the energy conservation theorem can also be derived. Multiplying Eq. (1.33) by v^*, and the complex conjugate of Eq. (1.7) by T, it follows, after addition, that

$$\frac{\partial}{\partial z}\left(-\frac{Tv^*}{2}\right) = \tfrac{1}{2}\left[j\omega(-\rho_m vv^* + cSS^*) - \omega^2\eta SS^* + F^b v^*\right] \quad (1.42)$$

which after integration yields

$$\iint_{A_2}\left(\frac{-v^*T}{2}\right)dA - \iint_{A_1}\left(-\frac{v^*T}{2}\right)dA - j\omega[U_s - U_v] + \langle P_d\rangle = P_s \quad (1.43)$$

In Eq. (1.43),

$$\Pi = -\frac{v^*T}{2} \quad (1.44)$$

is the *complex Poynting vector*,

$$U_s = \iiint_V \frac{cSS^*}{2}dV \quad \text{and} \quad U_v = \iiint_V \rho_m\frac{vv^*}{2}dV \quad (1.45)$$

are the *peak strain* and *kinetic energies*, respectively,

$$\langle P_d\rangle = \omega^2\eta\iiint_V \frac{SS^*}{2}dV \quad (1.46)$$

is the *time-averaged power loss* due to viscous damping, and

$$P_s = \tfrac{1}{2}\iiint_V F^b v^*\,dV \quad (1.47)$$

is the *complex power* supplied by sources. The *average radiated acoustic power* is then given by

$$\langle P\rangle = P_a = \text{Re}\left[\iint_{A_2}\left(-\frac{v^*T}{2}\right)dA - \iint_{A_1}\left(-\frac{v^*T}{2}\right)dA\right] \quad (1.48)$$

and is measured in watts.

In a lossless medium $\langle P_d\rangle = 0$, and for wave propagation excited by sources external to V, $P_s = 0$. In that case, consider a slice of material of area A and volume $V = Ad$ as shown in Fig. 1.5(b). Since $-T/v = \sqrt{\rho_m c}$, it follows that $U_s = U_v$, resulting in cancellation of the surface integrals in Eq. (1.43). In other words, the wave energy is conserved, as expected. In a slightly lossy medium ($\eta \neq 0$), the *Q factor* of the medium is defined as

$$Q = \frac{\omega \langle U \rangle}{\langle P_d \rangle}, \tag{1.49}$$

where $\langle U \rangle$ is the *average stored energy* in the material,

$$\langle U \rangle = \frac{U_s + U_v}{2} \approx \frac{cSS^*V}{2} \tag{1.50}$$

where $U_s \approx U_v$ has been used since η is small. From

$$\langle P_d \rangle = \omega^2 \eta \frac{SS^*V}{2} \tag{1.51}$$

and Eq. (1.25), the Q factor is given by

$$Q = \frac{c}{\omega \eta} = \frac{\beta_a}{2\alpha} = \frac{\pi}{\alpha \lambda_a} = \frac{v_a^2 \rho_m}{\omega \eta} \tag{1.52}$$

where $\lambda_a = 2\pi/\beta_a$ is the acoustic wavelength and $\alpha \lambda_a$ is termed the *attenuation per wavelength*. The Q factor for the best crystals may be on the order of 10^6 around 100 MHz and can be close to 1 around 1 MHz in materials with high attenuation per wavelength, such as rubber and wax.

In what follows it is shown that the attenuation constant due to viscous loss, Eq. (1.25), is consistent with the energy conservation theorem. For a wave normally incident on a surface A in a slightly lossy medium ($U_s \approx U_v$), Eq. (1.42) can be written as

$$\frac{\partial}{\partial z}\left(-\frac{v^*T}{2}\right) = \tfrac{1}{2}\omega^2 \eta SS^* \tag{1.53}$$

or using Eqs. (1.47), (1.13), and (1.14) it follows that

$$\frac{\partial P_a}{\partial z} = \tfrac{1}{2}\omega^2 \eta SS^*A = -\frac{\omega^2 \eta}{v_a^3 \rho_m} P_a \tag{1.54}$$

The solution of Eq. (1.54) is Eq. (1.24), with α given by Eq. (1.25).

1.6 SYNTHETIC ACOUSTIC MATERIALS

From the preceding sections it is seen that in many design problems, such as in acoustic impedance matching, there is a need for acoustic materials of a specified acoustic impedance. In other situations a lossy acoustic material may be required for attenuation of acoustic waves. Experimentally, materials of required acoustic properties (velocity, impedance, attenuation) are synthesized by using multicomponent mixtures of selected primary materials.[15-18] For instance, from Table 1, one would expect by mixing different proportions of tungsten powder with epoxy to obtain a material with a wide range of acoustic impedances. Tungsten is used because of its very high acoustic impedance. In practice,[15] this particular mixture yields materials with impedances in the range 2.5×10^6 to 40×10^6 kg/m² · s and corresponding velocities of 2 to 2.6 km/s, Fig. 1.6. The attenuation characteristic of such materials can also be controlled by the processing procedure and is generally in the very useful region of 25 dB/cm at 1 MHz. Rubber-based mixtures with glass microballoons[17] (approximately 20 μm in diameter) or lead can also be used for these purposes. Materials such as boron nitride, silicon carbide, magnesium oxide, and many others are also used.

Certain predictions can be made about the properties of the composite materials if it is assumed that there is no shrinkage or expansion on mixing (ideal mixture model).[17] Denoting by V_j, ρ_{mj}, and κ_j the volume, mass density, and compressibility, respectively, of the jth component in the mixture, it can be shown that

$$\frac{V}{\rho_m} = \sum_j \frac{V_j}{\rho_{mj}} \tag{1.55}$$

and

$$V = \sum_j V_j \kappa_j, \tag{1.56}$$

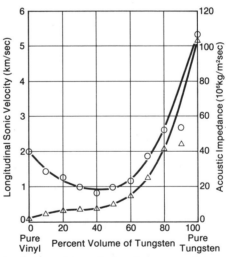

Figure 1.6. Longitudinal velocity and acoustic impedance versus percent volume of tungsten. (O) Velocity and (△) impedance for tungsten-vinyl composites. (*From Lees et al.,* [15] *reproduced with permission.*)

where V, ρ_m, and κ are the volume, mass density, and compressibility, respectively, of the composite material. In such materials, the velocity and impedance for longitudinal waves are found from

$$v_a = \frac{1}{\sqrt{\rho_m \kappa}} \tag{1.57}$$

$$Z_a = \rho_m v_a \tag{1.58}$$

Figure 1.6 is a graph of measured longitudinal velocity and acoustic impedance[15] versus percent volume of tungsten for tungsten-vinyl composites. Methods for measuring velocity and attentuation are well established.[19,20]

REFERENCES

1. B. A. Auld, *Acoustic Fields and Waves in Solids*, Vol. I, Wiley, New York, 1973, Chap. 1.

2. H. J. Pain, *The Physics of Vibrations and Waves*, 2nd ed., Wiley, 1976, Chap. 8; H. F. Pollard, *Sound Waves in Solids*, Methuen, New York, 1977, Chap. 7.

3. See Ref. 1, Chap. 2.

4. K. Miyano and J. B. Ketterson, "Sound Propagation in Liquid Crystals," in *Physical Acoustics* (W. P. Mason and R. N. Thurston, Eds.), Vol. XIV, Academic, New York, 1979, Chap. 2.

5. S. Ramo, J. R. Whinnery, and T. Van Duzer, *Fields and Waves in Communication Electronics*, Wiley, New York, 1965, Chap. 6.

6. R. Collin, *Foundations for Microwave Engineering*, McGraw-Hill, New York, 1966, Chap. 5.

7. H. F. Pollard, *Sound Waves in Solids*, Methuen, New York, 1977, Chap. 6.

8. E. P. Papadakis, "Scattering in Polycrystalline Media," in *Methods of Experimental Physics* (P. D. Edmonds, Ed.), Vol. 19, Academic, New York, 1981, Chap. 5.

9. G. Harrison and A. J. Barlow, "Dynamic Viscosity Measurement," in *Methods of Experimental Physics* (P. D. Edmonds, Ed.), Vol. 19, Academic, New York, 1981, Chap. 3.

10. E. Freedman, *J. Chem. Phys.*, **21**, 1784 (1953).

11. L. J. Slutsky, "Ultrasonic Chemical Relaxation Spectroscopy," in *Methods of Experimental Physics* (P. D. Edmonds, Ed.), Vol. 19, Academic, New York, 1981, Chap. 4.

12. J. and H. Krautkrämer, *Ultrasonic Testing of Materials*, Springer-Verlag, New York, 1977, Chap. 6.

13. J. Fraser, B. T. Khuri-Yakub, and G. S. Kino, "The Design of Efficient Broadband Wedge Transducers," *Appl. Phys. Lett.*, **32** (11), 698–700 (1978).

14. B. Hartmann and J. Jarzynski, "Ultrasonic Hysteresis Absorption in Polymers," *J. Appl. Phys.*, **43**, 4304–4312 (1972).

15. S. Lees, R. S. Gilmore, and P. R. Kranz, "Acoustic Properties of Tungsten-Vinyl Composites," *IEEE Trans. Sonics Ultrason.*, **SU-20** (1), 1–2 (1973).

16. J. D. Larson and J. G. Leach, "Tungsten-Polyvinyl Chloride Composite Materials—Fabrication and Performance," *1979 IEEE Ultrason. Symp. Proc.*, 342–345 (1979).

17. R. D. Corsaro and J. D. Klunder, "A Filled Silicone Rubber Materials System with Selectable Acoustic Properties for Molding and Coating Applications at Ultrasonic Frequencies," *NRL Rep.* 8301, ADA070461, 1979.

18. K. F. Bainton and M. G. Silk, "Some Factors Which Affect the Performance of Ultrasonic Transducers," *Brit. J. NDT*, **22**, (Jan), 15–20 (1980).

19. M. A. Breazeale, J. H. Cantrell, Jr., and J. S. Heyman, "Ultrasonic Wave Velocity and Attenuation Measurements," in *Methods of Experimental Physics* (P. D. Edmonds, Ed.), Vol. 19, Academic, New York, 1981, Chap. 2.

20. H. F. Pollard, *Sound Waves in Solids*, Methuen, New York, 1977.

EXERCISES

1. (a) Show that the solution for the wave propagation in an acoustic material of infinite extent given by

$$T_2 = T_2^+ \exp[\,j(\omega t - \beta_2 z)] \quad \text{and} \quad v_2 = \frac{-T_2}{Z_2} \quad \text{(E1.1)}$$

where T_2^+ is a constant and Z_2 and β_2 are the acoustic impedance and wave number, respectively, indeed satisfies the wave equation.

(b) Consider an acoustic wave incident from the left on the interface between two materials of infinite extent as shown in Fig. 1.7. Representing the acoustic wave in region 1 with

$$T_1 = T_1^+ \exp[\,j(\omega t - \beta_1 z)] + T_1^- \exp[\,j(\omega t + \beta_1 z)]$$

and

$$v_1 = -\frac{1}{Z_1}\{T_1^+ \exp[\,j(\omega t - \beta_1 z)] - T_1^- \exp[\,j(\omega t + \beta_1 z)]\} \quad \text{(E1.2)}$$

and the acoustic wave in region 2 with Eq. (E1.1), find the stress reflection and transmission coefficients, Eq. (1.23), defined as

$$r = \frac{T_1^-}{T_1^+} \quad \text{and} \quad \tau = \frac{T_2}{T_1^+} \quad \text{(E1.3)}$$

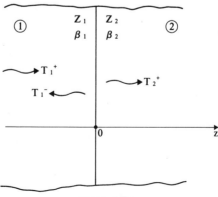

Figure 1.7.

(c) Show that the ratio of reflected to incident power in region 1 is given by
$P_a^- / P_a^+ = |r|^2$.

(d) Show that for the wave traveling in the $-z$ direction,

$$T = T_1^- e^{j(\omega t + \beta z)} \quad \text{and} \quad v = \frac{T}{Z_1} \tag{E1.4}$$

(e) Find the ratio of transmitted to incident power, in region 2, given by P_a^T / P_a^+.

2. A plane wave of 2 MHz is incident on an aluminum plate immersed in water. How thick must the plate be to retard the wave in phase by 90° behind a wave that passes through a large hole in the plate? (Neglect edge effects.)

3. Consider a plane wave incident from the left on the interface of two materials, as shown in Fig. 1.8. Show that the impedance, looking to the right at distance z from the interface located at $z = 0$, is given by

$$Z = Z_1 \frac{Z_2 + jZ_1 \tan \beta_1 z}{Z_1 + jZ_2 \tan \beta_1 z} \tag{E1.5}$$

4. Consider two materials, 1 and 3 in Fig. 1.9, separated by a layer of a third material, 2. Using Eq. (E1.5) show that the impedance looking to the right (i.e., the incident wave is from the left) at $z = 0$ is given by

$$Z = Z_2 \frac{Z_3 + jZ_2 \tan \beta_2 l}{Z_2 + jZ_3 \tan \beta_2 l} \tag{E1.6}$$

resulting in the reflection coefficient at $z = 0$ of

$$r = \frac{Z - Z_1}{Z + Z_1}. \tag{E1.7}$$

Show that the reflection coefficient, r will be zero in the following cases:
(a) $\beta_2 l = \pi$ (half-wave transformer)
(b) $\beta_2 l = \pi/2$ and $Z_2 = \sqrt{Z_1 Z_3}$ (quarter-wave transformer)

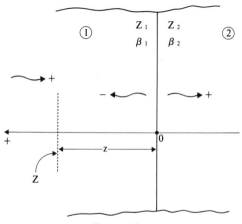

Figure 1.8.

5. Consider an acoustically thin plate, one for which $\tan x \simeq x$ in Eq. (E1.6).
(a) if $Z_1 = Z_3$ show that

$$Z \approx Z_1\left[1 + j\beta_2 l\left(\frac{Z_2}{Z_1} - \frac{Z_1}{Z_2}\right)\right] \qquad \text{(E1.8)}$$

and

$$r \approx j\frac{\beta_2 l}{2}\left(\frac{Z_2}{Z_1} - \frac{Z_1}{Z_2}\right) \qquad \text{(E1.9)}$$

(b) Take Z_1 to be water and Z_2 to be polyethylene. At a frequency of 5 MHz, find the thickness of polyethylene that results in $|r| \simeq 0.1$.
(c) Find the power transmission coefficient for this case.

6. Assume that the material represented by Z_1 in Fig. 1.9 is PZT-5A and that Z_3 corresponds to water.
(a) Compute the parameters of a quarter-wave transformer required for matching at a frequency of 2 MHz. Which material would you choose for that purpose from Table 1?
(b) The fractional bandwidth of the quarter-wave transformer is given by[6]

$$\frac{\Delta f}{f_0} = 2 - \frac{4}{\pi}\arccos\left|\frac{2r_m\sqrt{Z_1 Z_3}}{(Z_3 - Z_1)\sqrt{1 - r_m^2}}\right| \qquad \text{(E1.10)}$$

where r_m is the maximum tolerable reflection coefficient in the passband and f_0 is the center frequency. For $r_m = 0.1$, find the fractional bandwidth.

7. A two-section binomial transformer[6] consisting of two quarter-wave transformers denoted by Z_1 and Z_2, is shown in Fig. 1.10. Matching is achieved if

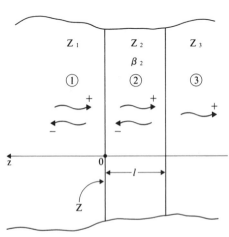

Figure 1.9.

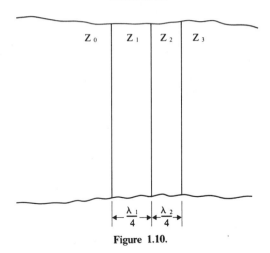

Figure 1.10.

$Z_1 = Z_3^{1/4} Z_0^{3/4}$ and $Z_2 = Z_3^{3/4} Z_0^{1/4}$. In Fig. 1.10, λ_1 and λ_2 are the acoustic wavelengths in the materials at the center frequency.
(a) Take Z_0 to be PZT-5A and Z_3 to be water. Find Z_1 and Z_2. Choose two materials closest in impedance values to the computed results. For $f = 2$ MHz, find λ_1 and λ_2. *Note*: For accurate results, the range of variables is usually restricted to $0.5Z_0 \le Z_3 \le 2Z_0$.
(b) The fractional bandwidth of the two-section binomial transformer is given by

$$\frac{\Delta f}{f_0} = 2 - \frac{4}{\pi} \arccos \left| \frac{2 r_m}{\ln(Z_3/Z_0)} \right|^{1/2} \tag{E1.11}$$

where r_m is the maximum tolerable reflection coefficient in the passband. For $r_m = 0.1$, find the fractional bandwidth and compare the result with the corresponding bandwidth obtained in Exercise 6.

8. An acoustic beam with a power density of 1 W/cm² is propagating through water. Find (a) the peak strain and kinetic energy densities if $S = 10^{-8}$, (b) time-average power loss in 1 cm³ at a frequency of 2 MHz, and (c) the acoustic Q factor of water at a frequency of 2 MHz.

9. Using Eqs. (1.55) and (1.56), compute Z_a and v_a for the tungsten-vinyl composite and compare your results with Fig. 1.6. For vinyl, take $v_a = 2$ km/s and $\rho_m = 1260$ kg/m³. (Express compressibility as $\kappa_j = 1/v_{aj}^2 \rho_{mj}$, where v_{aj} is the acoustic velocity of the jth component.)

10. Compute the impedance of water under the assumption of viscous losses ($\eta \simeq 10^{-3}$ Pa · s) at frequencies of 1, 10, and 50 MHz.

11. Using the computed data from Exercise 6, determine the characteristics of the composite materials needed to achieve matching. Use data from Fig. 1.6.

12. Repeat Exercise 11, using computed data from Exercise 7.

13. Show that any complex load impedance can be transformed into a real impedance by a $\lambda/8$ transformer having an acoustic impedance equal to the magnitude of the load impedance.

Chapter 2

Bulk Acoustic Waves and Exponential Horn

When a solid body is subjected to external deformational forces, translation and rotation of the body can occur. We are interested here only in the various deformations obeying the linear Hooke's law, Eq. (1.8), in which the strain is linearly related to stress. Therefore, in what follows, it is assumed that the body is in static equilibrium at all times so that no translation or rotation of the body can occur.

The smallest volume of material considered here is the acoustic particle, defined in Chapter 1, which restricts our approach to *large-scale acoustic phenomena*. For such phenomena the material subjected to forces is assumed to be continuous. Furthermore, it is assumed to be *elastic*; that is, if the forces are removed it will regain its original shape. The cases where applied forces exceed the elastic limit and cause permanent deformation and fracture are not of interest here.

In Chapter 1, two types of forces have been described, the surface force (stress) where the force is proportional to the surface area, and the body force where the force is proportional to the volume. When stress is applied, the force acting on the acoustic particles at the surface of the material is transmitted to the acoustic particles inside the material. The stress, defined as force per unit surface area is, in general, dependent on the orientation of the surface.

The simplest case of stress occurs in liquids where the hydrostatic pressure (stress) is independent of the orientation of the surface. If an acoustic (liquid) particle in the form of a cube is considered, and a rectangular coordinate system is oriented so that each of its axes is perpendicular to one of the three faces of the cube, it is obvious that the pressure on each of the three faces will be identical.[†] If the stress (pressure) T acting on a face is denoted by a double subscript ij, where i denotes the direction of the stress and j the orientation of the face, it follows that

$$T_{ij} = -p\delta_{ij}$$

where p is the hydrostatic pressure, δ_{ij} is the Kronecker δ symbol,[‡] and the negative

---

[†] The pressure in this case is due to gravitational body force acting on liquid particles.
[‡] $\delta_{ij} = 1$ for $i = j$, $\delta_{ij} = 0$ for $i \neq j$.

sign is due to the convention that compression is a negative stress. This can be written in matrix form as

$$[T_{ij}] = \begin{bmatrix} -p & 0 & 0 \\ 0 & -p & 0 \\ 0 & 0 & -p \end{bmatrix}$$

The concept of stress in solids was introduced in Section 1.1, where uniaxial homogeneous stress (along the z axis) was considered. In general, the force does not need to be normal to the surface of a specimen, as tangential components of force may also exist across the surface. These are termed *shear forces*. Consider, for instance, a unidirectional force $d\vec{F}$ acting on a side of a small cube, Fig. 2.1. Gravitational forces acting on the body are neglected, and the cube is assumed to be in static equilibrium. If the force is resolved into its components, dF_x, dF_y, and dF_z, and the area upon which $d\vec{F}$ acts is denoted by $dA_x = dy\,dz$, where the subscript x denotes the direction perpendicular to the surface, then the *stress components* can be defined as

$$T_{xx} = \frac{dF_x}{dA_x}, \qquad T_{yx} = \frac{dF_y}{dA_x}, \qquad T_{zx} = \frac{dF_z}{dA_x} \qquad (2.1)$$

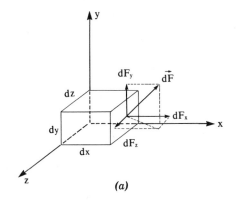

(a)

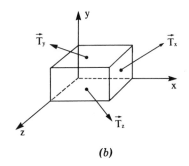

(b)

Figure 2.1. (*a*) The force $\overrightarrow{dF}$ acting on a face of a small cube. (*b*) Traction forces acting on surface elements of a cube with normals in the x, y, and z direction, respectively.

In the notation used for the stresses T_{ij}, i denotes the component of force and j the direction normal to the surface. Similarly, one can define the remaining six components of stress acting upon surfaces $dA_y = dx\,dz$ and $dA_z = dx\,dy$ as T_{xy}, T_{yy}, T_{zy} and T_{xz}, T_{yz}, T_{zz}, respectively. Since the cube is in equilibrium, the stress components on the other three faces of the cube must be opposite and equal to the ones shown. Having nine components, stress is a second-rank tensor, which can be conveniently tabulated in a 3×3 matrix as

$$[T_{ij}] = \begin{bmatrix} T_{xx} & T_{xy} & T_{xz} \\ T_{yx} & T_{yy} & T_{yz} \\ T_{zx} & T_{zy} & T_{zz} \end{bmatrix} \tag{2.2}$$

Since

$$d\vec{F} = dF_x \hat{x} + dF_y \hat{y} + dF_z \hat{z} \tag{2.3}$$

where $\hat{x}$, $\hat{y}$, and $\hat{z}$ are unit vectors, it follows from Eq. (2.1) that

$$\vec{T}_x = T_{xx} \hat{x} + T_{yx} \hat{y} + T_{zx} \hat{z} \tag{2.4}$$

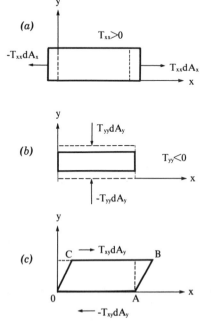

Figure 2.2. Illustration of (a) extension, (b) compression, and (c) shear stress.

where $\vec{T}_x$ is the traction force acting on a surface element with the normal in the x direction. Similar expressions can be written for $\vec{T}_y$ and $\vec{T}_z$, acting on the other two faces of a cube, as shown in Fig. 2.1(b). In Eq. (2.2), T_{jj} are normal components of stress representing either extension ($T_{jj} > 0$) or compression ($T_{jj} < 0$), while T_{ij}, $i \neq j$ are shear components of stress. The three cases are illustrated in Fig. 2.2. In order to define the traction force, it is always necessary to specify the orientation of the surface. In general, the convention shown in Fig. 2.3(a) is used in evaluating the traction force $\vec{T}_a$ on a surface element $d\vec{A} = \vec{a}\,dA$, that is, $\vec{F}_a = \vec{T}_a\,dA$ is the force acting from medium 1 on medium 2. Conversely, $-\vec{T}_a\,dA$ is the force acting from medium 2 on medium 1.

In all cases to be considered here, the *stress tensor* denoted symbolically by $\overset{\leftrightarrow}{T}$ will be symmetric, that is, $T_{ji} = T_{ij}$, in which case only six independent components exist. Instead of using the full tensor notation given by T_{ij}, it is often convenient to use the *abbreviated subscript* or *engineering notation*[1] T_J,

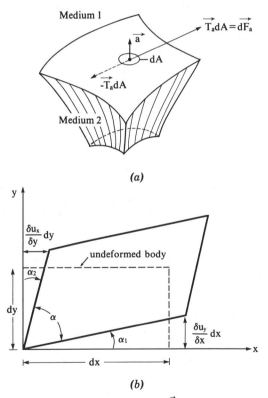

(a)

(b)

Figure 2.3. (a) Convention for using stress. The force $\vec{F}_a$ is acting from medium 1 on medium 2. (b) Illustration of shear strain $S_{xy} = \frac{1}{2}[\partial u_x/\partial y + \partial u_y/\partial x]$. Dashed lines denote the original shape of the body of cross section $dx\,dy$.

where the subscripts are derived from the following conversion table:

ij or kl	I or J
xx	1
yy	2
zz	3
$yz = zy$	4
$xz = zx$	5
$xy = yx$	6

$$(2.5)$$

so that

$$
\begin{bmatrix} T_{xx} & T_{xy} & T_{xz} \\ T_{xy} & T_{yy} & T_{yz} \\ T_{xz} & T_{yz} & T_{zz} \end{bmatrix} = \begin{bmatrix} T_1 & T_6 & T_5 \\ T_6 & T_2 & T_4 \\ T_5 & T_4 & T_3 \end{bmatrix}
$$

In this case it is possible to reduce the matrix given by Eq. (2.2) to a column vector

$$
[T_J] = \begin{bmatrix} T_1 \\ T_2 \\ T_3 \\ T_4 \\ T_5 \\ T_6 \end{bmatrix}. \tag{2.6}
$$

In Chapter 1 an elementary concept of deformation was introduced by defining the longitudinal strain. In general, deformation can cause a change in either volume or shape or both. Three types of changes can occur in a specimen subjected to forces: change of volume with no change in shape, change of length, and relative displacement of parallel layers. They correspond to *bulk compressional*, *longitudinal*, and *shear* (*tangential*) *strains*, respectively. Bulk compression strain, as explained in the example at the beginning of the chapter, occurs in liquids.

For solids, the defining equation for deformation and strain, Eq. (1.1), can also be generalized to three dimensions.[1] In three dimensions it is written as

$$
\mathcal{D} = d\vec{l} \cdot d\vec{l} - d\vec{L} \cdot d\vec{L} \equiv 2S_{ij}\, dL_i\, dL_j \tag{2.7}
$$

where the Einstein summation convention is assumed. For small strains,

$$
S_{ij} = \frac{1}{2}\left[\frac{\partial u_i}{\partial r_j} + \frac{\partial u_j}{\partial r_i} \right], \qquad i, j = x, y, z \tag{2.8}
$$

and

$$\vec{r} = x\hat{x} + y\hat{y} + z\hat{z}$$

Equation (2.8) is the defining equation for the second-rank *strain tensor* which, like stress, is also symmetric; $S_{ji} = S_{ij}$. A (longitudinal) component of the strain tensor obtained for $i = j = z$ has already been derived in Section 1.1, Eq. (1.2), which can now be written as

$$S_{zz} = \frac{\partial u_z}{\partial z} \tag{2.9}$$

Of the six strain components, three—S_{xx}, S_{yy}, and S_{zz}—represent either pure extension or pure compression of the material along the x, y, and z axis, respectively, while the remaining three, S_{yz}, S_{xz}, S_{xy}, represent pure shear strains. Consider, for instance,

$$S_{xy} = \frac{1}{2}\left[\frac{\partial u_x}{\partial y} + \frac{\partial u_y}{\partial x}\right] \tag{2.10}$$

as illustrated in Fig. 2.3(b). The angle α, which before deformation was equal to $\pi/2$, is now

$$\alpha = \frac{\pi}{2} - (\alpha_1 + \alpha_2) \simeq \frac{\pi}{2} - \left(\frac{\partial u_y}{\partial x} + \frac{\partial u_x}{\partial y}\right) = \frac{\pi}{2} - 2S_{xy} \tag{2.11}$$

In general, the variation, due to shear strain S_{ij}, of an angle between two initially perpendicular elements of length dr_i and dr_j is equal to $-2S_{ij}$.

Equations (2.1) and (2.8) defining the stress and strain components have been developed on the assumption that the material is continuous. Solids, such as crystals, possess crystal lattices, and stress and strain at this (atomic) level must be redefined since it cannot be assumed that either of them is a continuous function of position. The discontinuities in stress and strain at the atomic level occur whenever a defect in the crystal lattice is encountered.[2]

Another tensor, useful in defining gradient operation on tensors, is the *gradient displacement tensor* $\overleftrightarrow{\xi}$ defined in matrix form as[1]

$$[\xi_{ij}] = \begin{bmatrix} \dfrac{\partial u_x}{\partial x} & \dfrac{\partial u_x}{\partial y} & \dfrac{\partial u_x}{\partial z} \\[2mm] \dfrac{\partial u_y}{\partial x} & \dfrac{\partial u_y}{\partial y} & \dfrac{\partial u_y}{\partial z} \\[2mm] \dfrac{\partial u_z}{\partial x} & \dfrac{\partial u_z}{\partial y} & \dfrac{\partial u_z}{\partial z} \end{bmatrix} = \left[\frac{\partial u_i}{\partial r_j}\right] \tag{2.12}$$

and written symbolically as

$$\overset{\leftrightarrow}{\xi} = \nabla \vec{u} \tag{2.13}$$

where $\vec{u}$ is the displacement vector and ∇ is the gradient operator. The displacement tensor, in general, is asymmetric and can be decomposed into symmetric and antisymmetric parts as

$$\overset{\leftrightarrow}{\xi} = \tfrac{1}{2}\left[\overset{\leftrightarrow}{\xi} + \overset{\tilde{\leftrightarrow}}{\xi}\right] + \tfrac{1}{2}\left[\overset{\leftrightarrow}{\xi} - \overset{\tilde{\leftrightarrow}}{\xi}\right] \tag{2.14}$$

where the tilde ($\tilde{\ }$) indicates the transpose. The symmetric part can be written as

$$\xi_s = \frac{1}{2}[\xi_{ij} + \xi_{ji}] = \frac{1}{2}\left[\frac{\partial u_i}{\partial r_j} + \frac{\partial u_j}{\partial r_i}\right] = S_{ij} \tag{2.15}$$

and is equal to the strain tensor, while the antisymmetric part describes local and rigid rotations that are of no interest in acoustic wave propagation studies.[†] On combining Eqs. (2.13) and (2.15) it can be seen that

$$\overset{\leftrightarrow}{S} = \tfrac{1}{2}[\nabla \vec{u} + \tilde{\nabla}\vec{u}] = \nabla_s \vec{u} \tag{2.16}$$

where ∇_s is the symmetric part of the gradient operator. In terms of abbreviated subscripts, Eq. (2.5), this can be written as

$$S_I = \begin{cases} S_{ij}, & i = j \\ 2S_{ij}, & i \neq j \end{cases} \tag{2.17}$$

where $I = 1$ to 6 and i, $j = x$, y, z. Making use of Eqs. (2.17), (2.10), and (2.5), the strain column vector $[S_I]$ is defined as

$$\begin{bmatrix} S_1 \\ S_2 \\ S_3 \\ S_4 \\ S_5 \\ S_6 \end{bmatrix} = \begin{bmatrix} S_{xx} \\ S_{yy} \\ S_{zz} \\ 2S_{yz} \\ 2S_{xz} \\ 2S_{xy} \end{bmatrix} = \begin{bmatrix} \partial/\partial x & 0 & 0 \\ 0 & \partial/\partial y & 0 \\ 0 & 0 & \partial/\partial z \\ 0 & \partial/\partial z & \partial/\partial y \\ \partial/\partial z & 0 & \partial/\partial x \\ \partial/\partial y & \partial/\partial x & 0 \end{bmatrix} \begin{bmatrix} u_x \\ u_y \\ u_z \end{bmatrix} \tag{2.18}$$

where the matrix containing differentials is denoted by $[\nabla_{Ij}]$ and is termed the

---

[†]The antisymmetric part of $\partial u_i/\partial r_j$, defined as $R_{ij} = \frac{1}{2}[(\partial u_i/\partial r_j) - (\partial u_j/\partial r_i)] = -R_{ji}$, plays an important role in acousto-optic interaction. The reader can verify that R_{ij} has only three independent components.

symmetric gradient operator matrix.[1] Equation (2.8), given symbolically by Eq. (2.16), can now be written in terms of abbreviated subscripts as

$$S_I = \nabla_{Ij} u_j, \qquad I = 1 \text{ to } 6; j = x, y, z \qquad (2.19)$$

The gradient operation on a vector (first-rank tensor) in Eq. (2.16) increases the rank of tensor by one. This is not an unexpected result, since mathematical homogeneity requires that both sides of a tensorial equation must contain tensors of the same rank.

The translational equation of motion, given by Eq. (1.33), in the three-dimensional case is now written as[1]

$$\nabla \cdot \overleftrightarrow{T} = \rho_m \frac{\partial^2 \vec{u}}{\partial t^2} - \vec{F}^b \qquad (2.20)$$

where ∇ is a vector (gradient) operator. Scalar multiplication of a vector ∇ and a tensor $\overleftrightarrow{T}$ in Eq. (2.20) reduces the rank of tensor by one. This divergence operation results in

$$\nabla \cdot \overleftrightarrow{T} = \frac{\partial \vec{T}_x}{\partial x} + \frac{\partial \vec{T}_y}{\partial y} + \frac{\partial \vec{T}_z}{\partial z} \qquad (2.21)$$

where, for instance, $\vec{T}_x$ is given by Eq. (2.4). It can be shown[1] that the divergence operator $\nabla \cdot$ is the transpose of the symmetric gradient operator ∇_s, the latter being defined by Eqs. (2.16) and (2.18), that is,

$$\nabla \cdot = \tilde{\nabla}_s \quad \text{or} \quad [\nabla_{iJ}] = [\tilde{\nabla}_{Ij}] \qquad (2.22)$$

resulting in

$$[\nabla_{iJ}] = \begin{bmatrix} \partial/\partial x & 0 & 0 & 0 & \partial/\partial z & \partial/\partial y \\ 0 & \partial/\partial y & 0 & \partial/\partial z & 0 & \partial/\partial x \\ 0 & 0 & \partial/\partial z & \partial/\partial y & \partial/\partial x & 0 \end{bmatrix} \qquad (2.23)$$

The translation equation of motion given symbolically by Eq. (2.20) can now be written in terms of abbreviated subscripts as

$$\nabla_{iJ} T_J = \rho_m \frac{\partial^2 u_i}{\partial t^2} - F_i^b, \qquad i = x, y, z; J = 1 \text{ to } 6 \qquad (2.24)$$

or in full tensor notation as

$$(\nabla \cdot \overleftrightarrow{T})_i = \rho_m \frac{\partial^2 u_i}{\partial t^2} - F_i^b \qquad (2.25)$$

or

$$\frac{\partial T_{ij}}{\partial r_j} = \rho_m \frac{\partial^2 u_i}{\partial t^2} - F_i^b \tag{2.26}$$

From the preceding considerations, it is easy to show that the divergence operation on tensors can be represented in matrix form as

$$[\nabla \cdot T] = \begin{bmatrix} \dfrac{\partial}{\partial x} & \dfrac{\partial}{\partial y} & \dfrac{\partial}{\partial z} \end{bmatrix} \begin{bmatrix} T_{xx} & T_{xy} & T_{xz} \\ T_{yx} & T_{yy} & T_{yz} \\ T_{zx} & T_{zy} & T_{zz} \end{bmatrix} = [\nabla_{iJ}][T_J] \tag{2.27}$$

where the column vector $[T_J]$ is given by Eq. (2.6).

2.1 THE TENSORS OF COMPLIANCE AND STIFFNESS

With no external forces applied, static equilibrium at the atomic level in the solid can be explained in terms of the minimum potential energy obtained as a balance between attractive (Coulomb) forces and repulsive (atomic) forces.[3] This balance, for instance, determines the separation r_0 between neighboring ions in the crystal lattice. When external forces are applied, for the body to stay elastic, only small deviations ($< 10^{-4}$) around r_0 are allowed. The magnitude of r_0 is on the order of 10^{-10} m. In other words, the relation between the strain and the stress is linear. Any larger excursions result in nonlinearities. For continuous material, taking strain as an independent variable, this can be expressed for small deformations in terms of the series[4,5]

$$T_{kl}(S_{ij}) = T_{kl}(0) + \left(\frac{\partial T_{kl}}{\partial S_{ij}}\right)_{S_{ij}=0} S_{ij} + \frac{1}{2}\left(\frac{\partial^2 T_{kl}}{\partial S_{ij} \partial S_{mn}}\right)_{S_{ij}=0; S_{mn}=0} S_{ij}S_{mn} + \dots$$

$$\tag{2.28}$$

By definition, a body is elastic if it returns to its initial state when the outside agents (forces) return to zero. Therefore $T_{kl}(0) = 0$, since the internal restoring forces and the strain (deformation) converge simultaneously to zero. Neglecting the nonlinear term, this can be written as

$$T_{kl} = \left(\frac{\partial T_{kl}}{\partial S_{ij}}\right)_{S_{ij}=0} S_{ij} = c_{klij}S_{ij} \tag{2.29}$$

which can be written symbolically[1] as

$$\overset{\leftrightarrow}{T} = \overset{\leftrightarrow}{c} : \overset{\leftrightarrow}{S} \tag{2.30}$$

where $\overset{\leftrightarrow}{c}$ is the fourth-rank *elastic stiffness tensor*. Similarly, where stress is taken to be the independent variable, it follows that

$$S_{ij} = \left(\frac{\partial S_{ij}}{\partial T_{kl}}\right)_{T_{kl}=0} T_{kl} = s_{ijkl}T_{kl} \tag{2.31}$$

or

$$\overset{\leftrightarrow}{S} = \overset{\leftrightarrow}{s} : \overset{\leftrightarrow}{T} \tag{2.32}$$

where $\overset{\leftrightarrow}{s}$ is the fourth-rank *elastic compliance tensor*. Note that the double scalar product of two tensors in Eqs. (2.30) and (2.31) reduces the rank of the fourth-rank tensors by two. A fourth-rank tensor has at most $3^4 = 81$ constants. Due to symmetry in solids, the maximum number of constants necessary to completely specify $\overset{\leftrightarrow}{c}$ or $\overset{\leftrightarrow}{s}$ reduces to 21 for any elastic body. Equations (2.29) and (2.31) can also be written in abbreviated from[1] as

$$T_I = c_{IJ}S_J \tag{2.33}$$

$$S_I = s_{IJ}T_J \tag{2.34}$$

Using the transformation of indices given by Eq. (2.5) and symmetry conditions in solids[1] given by

$$c_{ijkl} = c_{jikl} = c_{ijlk} = c_{jilk}$$

$$c_{ijkl} = c_{klij} \tag{2.35}$$

it can be shown that

$$c_{IJ} = c_{ijkl} \tag{2.36}$$

and

$$s_{IJ} = s_{ijkl} \times \begin{cases} 1, & \text{if } I \text{ and } J = 1,2,3 \\ 2, & \text{if } I \text{ or } J = 4,5,6 \\ 4, & \text{if } I \text{ and } J = 4,5,6 \end{cases} \tag{2.37}$$

Equations (2.29) and (2.30) are generalizations of Hooke's law, Eq. (1.8) and, using Eqs. (2.6) and (2.18), can be written in matrix form as

$$[T] = [c][S] \tag{2.38}$$

$$[S] = [s][T] \tag{2.39}$$

where $[c]$ and $[s]$ are 6×6 matrices with elements defined by Eqs. (2.36) and (2.37). Using the last two equations it is easy to show that

$$[s][c] = [I] \tag{2.40}$$

where $[I]$ is the identity matrix,[†] or

$$[s] = [c]^{-1} \tag{2.41}$$

Note that both compliance and stiffness matrices are symmetric, Eq. (2.35), that is, $c_{LK} = c_{KL}$, or

$$
\begin{bmatrix}
c_{11} & c_{12} & c_{13} & c_{14} & c_{15} & c_{16} \\
c_{12} & c_{22} & c_{23} & c_{24} & c_{25} & c_{26} \\
c_{13} & c_{14} & c_{33} & c_{34} & c_{35} & c_{36} \\
c_{14} & c_{24} & c_{34} & c_{44} & c_{45} & c_{46} \\
c_{15} & c_{25} & c_{35} & c_{45} & c_{55} & c_{56} \\
c_{16} & c_{26} & c_{36} & c_{46} & c_{56} & c_{66}
\end{bmatrix}
\tag{2.42}
$$

thus reducing the 36 unknown constants in the 6×6 matrix to 21 only.

The symmetry conditions given by Eqs. (2.35) can be derived from the assumption that the body is in thermal equilibrium; that is, the process is *isothermal* (no change in temperature) and reversible. Thus a compressed (or stretched) material must be allowed enough time to return to the original temperature. This happens only at very low frequencies. The elastic constants of interest to us are those derived under reversible *adiabatic conditions*[‡] (increase of temperature during compression, decrease of temperature during rarefaction, both reversible processes, without loss of heat to the surroundings). The adiabatic elastic constants are applicable to high frequencies and as such are the only ones used in the book. Both sets of constants are of the same form and obey the same symmetry conditions. The frequency range considered determines the use of the appropriate set of constants. The derived constants are also called *second-order elastic constants*. *Third-order elastic constants* given by the third (nonlinear) term in Eq. (2.28) are important in the theory of heat.[3]

All the derived equations are written in terms of Eulerian (spatial) coordinates, the only system of coordinates used in this volume. Lagrangian (material) coordinates are often used in the theory of fluids in motion.

The second-order symmetrical tensors such as T_{kl} and S_{ij} can be represented by elliptical surfaces. By rotating such a surface, which is equivalent to linear transformation of coordinates, it is possible to diagonalize such tensors. In

[†] The identity matrix is also called the *unit tensor* δ_{jk} or the Kronecker δ symbol. Sometimes it is called the *substitution tensor* because of the property $\delta_{jk} u_k = u_j$; the Kronecker δ symbol results in $\delta_{jk} = 1$ for $j = k$, and $\delta_{jk} = 0$ for $j \neq k$.

[‡] More precisely, constant-entropy conditions. Entropy is a thermodynamic concept that can be equated in solids with the disorder in the lattice of the material.

other words, only the diagonal components (S_{11}, S_{22}, S_{33} in S_{ij}) are nonzero. In this case the new axes are called the *principal axes* and the diagonal components the *principal strains*. Note that the sum of diagonal elements, $S_{ii} = S_{11} + S_{22} + S_{33}$, called the *trace* of the tensor, is an invariant (has identical values in all coordinate systems). In the next section it is shown that $S_{ii} = \Delta$, the change of volume per unit volume (*dilatation*).

In Eq. (2.20) the only external forces assumed are the body forces. It can be shown, by using the generalized divergence theorem,[1] that (external) surface forces $\vec{T}_e$ acting on the surface of the specimen can be reduced to equivalent body forces $\vec{F}_{eq}^b = \nabla \cdot \ddot{T}_e$.

2.2 ELASTIC ISOTROPY CONDITIONS AND HOOKE'S LAW

By definition, physical constants of an isotropic solid must be independent of the choice of the reference coordinate system.

It is interesting to examine the definition of isotropy in real materials. Under sufficient magnification, it can be seen that all real isotropic materials are polycrystalline, that is, they consist of small grains or small crystals randomly oriented. Some of them, like glasses or polymers, consist of molecular chains (short-range order). They differ from crystals in that the crystals have well-defined (infinite) lattice structure (long-range order).

For wave propagation purposes, it is often more convenient to compare the wavelength of the propagating acoustic wave λ_a, to the mean grain diameter d. The acoustic particle, defined in Chapter 1, does not have to be larger than the size of grains. When an acoustic wave propagates in a polycrystalline material, part of its energy is scattered by the boundaries of the grains.[2] At low frequencies, when $\lambda_a \gg d$, the acoustic wave does not see the grain, since the acoustic particle contains many grains. When $\lambda_a \simeq d$, a very complex scattering process results. In the case $\lambda_a \ll d$, each boundary of the grains can be treated as a reflection problem, Chapter 1; that is, specular reflections result.

The requirement that, in an isotropic solid, the physical constants must be independent of the choice of the reference coordinate system imposes further constraints on the 21 constants of the stiffness (compliance) tensor c_{ijkl} (s_{klij}). In this case, it can be shown that each stiffness constant can be represented by two parameters.[5] The latter are termed *Lamé coefficients* and are denoted by λ and μ, where λ is called the *first Lamé constant* and μ is known as the *modulus of rigidity*. Thus

$$c_{ijkl} = \lambda \delta_{ij} \delta_{kl} + \mu \left(\delta_{ik} \delta_{jl} + \delta_{il} \delta_{jk} \right) \tag{2.43}$$

where δ_{ij} is the Kronecker δ symbol.[†]

[†]Note that $\delta_{ij}\delta_{kl}$, $\delta_{ik}\delta_{jl} + \delta_{il}\delta_{jk}$, and $\delta_{ik}\delta_{jl} - \delta_{il}\delta_{jk}$ are the only three independent isotropic tensors of rank four.

Using Eqs. (2.5), (2.36), and (2.43), it follows that

$$c_{11} = c_{22} = c_{33} = \lambda + 2\mu$$

$$c_{12} = c_{23} = c_{13} = \lambda$$

$$c_{44} = c_{55} = c_{66} = \mu = \tfrac{1}{2}(c_{11} - c_{12}) \qquad (2.44)$$

while all other constants are equal to zero. Thus for an isotropic solid, the stiffness matrix, Eq. (2.42), reduces to

$$[c_{KL}] = \begin{bmatrix} \lambda + 2\mu & \lambda & \lambda & 0 & 0 & 0 \\ \lambda & \lambda + 2\mu & \lambda & 0 & 0 & 0 \\ \lambda & \lambda & \lambda + 2\mu & 0 & 0 & 0 \\ 0 & 0 & 0 & \mu & 0 & 0 \\ 0 & 0 & 0 & 0 & \mu & 0 \\ 0 & 0 & 0 & 0 & 0 & \mu \end{bmatrix} \qquad (2.45)$$

Hooke's law, given by Eqs. (2.29) and (2.33), can now be written for normal components (compressive and extensional stresses) as

$$T_{ii} = \lambda(S_{xx} + S_{yy} + S_{zz}) + 2\mu S_{ii}, \qquad i = x, y, z \qquad (2.46)$$

or in abbreviated subscripts as

$$T_J = \lambda(S_1 + S_2 + S_3) + 2\mu S_J, \qquad J = 1, 2, 3 \qquad (2.47)$$

and for tangential components (shear stresses) as

$$T_{ij} = 2\mu S_{ij} \qquad i \neq j, \qquad i, j = x, y, z \qquad (2.48)$$

or

$$T_J = \mu S_J \qquad J = 4, 5, 6 \qquad (2.49)$$

The *dilatation* Δ is defined, using Eq. (2.47), as

$$\Delta \equiv S_1 + S_2 + S_3 \qquad (2.50)$$

and represents the change of volume per unit volume. It has been pointed out earlier that the two Lamé constants λ and μ completely define the elastic response of an isotropic body. In practice, however, four elastic constants are routinely used because they relate better to the physics of materials testing (note that λ has no physical meaning). These are Young's modulus E_Y, the bulk modulus (also called the coefficient of incompressibility) K, Poisson's ratio σ, and the rigidity (shear) modulus, which is identically equal to Lamé's constant μ.

Young's modulus E_Y (see also Section 1.2) is defined as the ratio of the applied longitudinal stress to the longitudinal strain (fractional extension) when a rod is subjected to a uniform stress over its end planes and its lateral surface is free to expand. For instance, if the z axis coincides with the rod axis, the only applied stress is $T_{zz} = T_3$, while the other five components are equal to zero. From Eq. (2.47) it follows that

$$0 = (\lambda + 2\mu)S_1 + \lambda(S_2 + S_3)$$

$$0 = (\lambda + 2\mu)S_2 + \lambda(S_1 + S_3)$$

$$T_3 = (\lambda + 2\mu)S_3 + \lambda(S_1 + S_2) \tag{2.51}$$

yielding

$$E_Y = \frac{T_3}{S_3} = \frac{\mu(3\lambda + 2\mu)}{\lambda + \mu} \tag{2.52}$$

In fact, when a longitudinal wave compresses the material, its lateral distortions will depend on the dimensions of the specimen. These lateral distortions may cause excitation of shear waves. In large solids, where the transverse dimensions are much larger than the acoustic wavelength (*free-field condition*), the longitudinal and shear waves may be considered separately. If the wavelength is comparable to the dimensions of the solid (*waveguide condition*), both shear and longitudinal waves must be considered simultaneously. In the limit of wavelength being much larger than the transverse dimensions (*thin-waveguide approximation*), shear waves may be neglected.

Equations (2.51) are also used to define *Poisson's ratio*, which is equal to the ratio of the lateral contraction to the longitudinal extension of the rod, that is,

$$\sigma = -\frac{S_1}{S_3} = -\frac{S_2}{S_3} = \frac{\lambda}{2(\lambda + \mu)} \tag{2.53}$$

The coefficient of incompressibility is defined as the ratio between the applied pressure and the fractional change in volume when the solid is subjected to uniform compressive stress. From Eq. (2.47), for $T_1 = T_2 = T_3 = -T$, $T_4 = T_5 = T_6 = 0$, it follows that

$$S_1 = S_2 = S_3 = -T(3\lambda + 2\mu) \tag{2.54}$$

and on using Eqs. (2.50) and (2.54) we obtain

$$K = -\frac{T}{\Delta} = \lambda + \frac{2\mu}{3} \tag{2.55}$$

The coefficient of incompressibility is routinely used in acoustics of liquids and gases.

The Lamé constant μ (*rigidity shear modulus*) is, in fact, the ratio of transverse stress to transverse strain. Thus it is the elastic constant determining the propagation of pure shear waves under free-field conditions. Its role is analogous to Young's modulus, which is the elastic constant determining the propagation properties of longitudinal waves under thin-waveguide conditions.

It is now possible to classify materials according to the values of the four constants E_Y, K, σ, and μ. Poisson's ratio σ is dimensionless, and the remaining three constants, like stiffnesses c_{IJ}, are measured in units of N/m^2. For instance, $\sigma = 0.35$ for aluminum, 0.3 for stainless steel, and 0.05 for beryllium. For plastics, $\sigma = 0.4$ (Lucite), for ceramics 0.2, and for soft rubber 0.5. Since nonviscous liquids, such as water, cannot support shear stresses, it follows that $\mu = 0$, resulting in $E_Y = 0$ and $\sigma = 0.5$. Thus for isotropic materials $0 < \sigma \leqslant 0.5$. Young's modulus is 21×10^{10} for steel, 12×10^{10} for copper, 7×10^{10} for aluminum, 6×10^{10} for glass, 4×10^9 for Lucite, 5×10^6 for rubber, and zero for water. The rigidity modulus varies from 8×10^{10} for steel to 7×10^5 for rubber and is zero for water. The coefficient of incompressibility K varies from 17×10^{10} for steel to 10^9 for soft rubber and is equal to 2.2×10^9 for fresh water.

The coefficient of compressibility, Eq. (1.56), is also used. It is defined as $\kappa = 1/K$.

In practice, many materials, due to methods of production, exhibit a certain amount of anisotropy, and various identities between constants must be used with care to avoid errors.

2.3 BULK ACOUSTIC WAVES IN ISOTROPIC MATERIALS

It was pointed out before that a crystalline material, when no external forces are applied, is in thermodynamic equilibrium with its surroundings. At the atomic level, the *specific heat* that the material possesses can be expressed in terms of the incoherent energy of phonons, also called *thermal sound waves* or *thermoelastic waves*. Phonons, like acoustic waves, are classified as either shear or longitudinal. The lowest frequency of phonons in a material depends on the greatest linear dimension l of the material (for instance, for $l = 1.5$ m and $v_a = 3$ km/s, the lowest phonon frequency is on the order of 1 kHz), while the highest frequency corresponds to the frequency of lattice vibrations (on the order of 10^{13} Hz). (At present, the highest frequency in an acoustic wave that can be excited is on the order of 10^{12} Hz; see Chap. 5.) Phonons can be detected using Brillouin scattering methods.[6]

Bulk longitudinal and shear acoustic waves are known to seismologists as the P (primary) and S (secondary) waves, respectively. This designation refers to the time of arrival, the longitudinal waves being faster than the shear waves. In what follows, some general characteristics of bulk waves are considered.

With no body forces, $\vec{F}^b = 0$, the general wave equation is obtained by combining Eqs. (2.8), (2.26) (2.29), and (2.35) and is given by

$$\rho_m \frac{\partial^2 u_i}{\partial t^2} = c_{ijkl} \frac{\partial^2 u_l}{\partial x_j \partial x_k} \qquad (2.56)$$

With the symmetry conditions of Eqs. (2.43) and (2.44) imposed on stiffnesses this can be written as

$$\rho_m \frac{\partial^2 u_i}{\partial t^2} = \left[\frac{c_{11} + c_{12}}{2} \right] \frac{\partial^2 u_l}{\partial x_i \partial x_l} + \left[\frac{c_{11} - c_{12}}{2} \right] \nabla^2 u_i \qquad (2.57)$$

where ∇^2 is the Laplacian given by

$$\nabla^2 = \frac{\partial^2}{\partial x^2} + \frac{\partial^2}{\partial y^2} + \frac{\partial^2}{\partial z^2} \qquad (2.58)$$

Since

$$\frac{\partial u_l}{\partial x_l} = \nabla \cdot \vec{u} \qquad (2.59)$$

and

$$\nabla^2 u_i = (\nabla^2 \vec{u})_i \qquad (2.60)$$

Eq. (2.57) can be written as

$$\rho_m \frac{\partial^2 \vec{u}}{\partial t^2} = \left[\frac{c_{11} + c_{12}}{2} \right] \nabla(\nabla \cdot \vec{u}) + \left[\frac{c_{11} - c_{12}}{2} \right] \nabla^2 \vec{u} \qquad (2.61)$$

The quantities

$$v_L = \left(\frac{c_{11}}{\rho_m} \right)^{1/2} \quad \text{and} \quad v_S = \left(\frac{c_{11} - c_{12}}{2\rho_m} \right)^{1/2} = \left(\frac{c_{44}}{\rho_m} \right)^{1/2} \qquad (2.62)$$

represent, as will be shown later, the phase velocities of the longitudinal and shear waves, respectively, and Eq. (2.61) is written as

$$\frac{\partial^2 \vec{u}}{\partial t^2} = v_S^2 \nabla^2 \vec{u} + \left(v_L^2 - v_S^2 \right) \nabla(\nabla \cdot \vec{u}) \qquad (2.63)$$

In general it is always possible to decompose the vector $\vec{u}$ into a divergenceless vector $\vec{u}_S$ and an irrotational vector $\vec{u}_L$. Therefore, it follows that

$$\vec{u} = \vec{u}_S + \vec{u}_L \qquad (2.64)$$

with

$$\nabla \cdot \vec{u}_S = 0 \qquad \text{or} \qquad \vec{u}_S = \nabla \times \vec{\psi} \tag{2.65}$$

and

$$\nabla \times \vec{u}_L = 0 \qquad \text{or} \qquad \vec{u}_L = \nabla \phi \tag{2.66}$$

where ψ and ϕ are termed the vector and scalar potentials. In many boundary value problems, such as those considered in Chapters 4, 9 and 10, the use of potential functions, in addition to making it possible to separate variables, as already demonstrated, allow the solutions to be simply related to other quantities such as stress (pressure) and particle velocity. On combining Eqs. (2.63) and (2.64) it follows that

$$\frac{\partial^2 \vec{u}_S}{\partial t^2} + \frac{\partial^2 \vec{u}_L}{\partial t^2} = v_S^2 \nabla^2 \vec{u}_S + v_S^2 \nabla^2 \vec{u}_L + \left(v_L^2 - v_S^2\right) \nabla(\nabla \cdot \vec{u}_L) \tag{2.67}$$

However, since

$$\nabla(\nabla \cdot \vec{u}_L) = \nabla^2 \vec{u}_L + \nabla \times \nabla \times \vec{u}_L = \nabla^2 \vec{u}_L \tag{2.68}$$

Eq. (2.67) can be written as

$$\left(\frac{\partial^2 \vec{u}_S}{\partial t^2} - v_S^2 \nabla^2 \vec{u}_S\right) + \left(\frac{\partial^2 \vec{u}_L}{\partial t^2} - v_L^2 \nabla^2 \vec{u}_L\right) = 0 \tag{2.69}$$

Since $\vec{u}_S$ is divergenceless and $\vec{u}_L$ is irrotational, the expressions in parentheses can each be set equal to zero, yielding

$$\frac{\partial^2 \vec{u}_S}{\partial t^2} - v_S^2 \nabla^2 \vec{u}_S = 0 \tag{2.70}$$

and

$$\frac{\partial^2 \vec{u}_L}{\partial t^2} - v_L^2 \nabla^2 \vec{u}_L = 0. \tag{2.71}$$

Thus in an infinite, isotropic solid, two acoustic waves can exist. These are a compressional (longitudinal) wave propagating with phase velocity v_L and a shear (transverse) wave propagating with phase velocity v_S. The ratio of the two velocities is

$$\frac{v_L}{v_S} = \left(\frac{2c_{11}}{c_{11} - c_{12}}\right)^{1/2} = \left(\frac{c_{11}}{c_{44}}\right)^{1/2} = \left(\frac{\lambda + 2\mu}{\mu}\right)^{1/2} = \left(\frac{1 - \sigma}{0.5 - \sigma}\right)^{1/2}$$

$$\tag{2.72}$$

and since $0 < \sigma < 0.5$, $v_L/v_S > \sqrt{2}$, in all isotropic materials.

Applying the divergence operation to Eq. (2.71), we find that

$$\frac{\partial^2(\Delta)}{\partial t^2} - v_L^2 \nabla^2(\Delta) = 0$$

where $\Delta = \nabla \cdot \vec{u}_L$ is given by Eq. (2.50). Therefore the longitudinal wave causes a change in the volume of the material. This wave is also called a *compressional*, *dilatational*, *primary* (P), and *irrotational* wave. Applying the curl operation to Eq. (2.70) and defining

$$\vec{\Omega} = \tfrac{1}{2} \nabla \times \vec{u}_S$$

where $\vec{\Omega}$ is the rotation vector, we find that

$$\frac{\partial^2 \vec{\Omega}}{\partial t^2} - v_S^2 \nabla^2 \vec{\Omega} = 0.$$

Therefore the shear wave propagation involves "rotation" of the volume of material without requiring a change in the volume. The shear waves are called *rotational*, *secondary* (S), and *solenoidal* waves.

Equations (2.70) and (2.71) can be written in terms of displacement potentials, defined by Eqs. (2.65) and (2.66), as

$$\frac{\partial^2 \nabla \phi}{\partial t^2} - v_L^2 \nabla^2(\nabla \phi) = 0 \tag{2.73}$$

and

$$\frac{\partial^2 \nabla \times \vec{\psi}}{\partial t^2} - v_S^2 \nabla^2(\nabla \times \vec{\psi}) = 0 \tag{2.74}$$

or

$$\nabla^2 \phi - \frac{1}{v_L^2} \frac{\partial^2 \phi}{\partial t^2} = 0 \tag{2.75}$$

and

$$\nabla^2 \vec{\psi} - \frac{1}{v_S^2} \frac{\partial^2 \vec{\psi}}{\partial t^2} = 0 \tag{2.76}$$

and Eq. (3.64) can now be written, in general, as

$$\vec{u} = \nabla \phi + \nabla \times \vec{\psi}. \tag{2.77}$$

In what follows, two simple examples of plane longitudinal and plane shear wave propagation are considered.

For the plane longitudinal wave propagating in the $+z$ direction, the scalar potential ϕ is a function of z and t only. Therefore, from Eq. (2.77) it follows that

$$\vec{u} = \hat{z} u_z = \hat{z} \frac{\partial \phi}{\partial z} \tag{2.78}$$

and from Eq. (2.8), that

$$S_{zz} = S_3 = \frac{\partial u_z}{\partial z} \tag{2.79}$$

The associated stress is obtained from Eqs. (2.29) and (2.47) as

$$T_{zz} = T_3 = c_{zzzz} S_{zz} = c_{33} S_3 = (\lambda + 2\mu) S_3 \tag{2.80}$$

The geometry and the fields for a plane longitudinal wave are shown in Fig. 2.4(a), and the grid diagram in Fig. 1.1(a). For harmonic perturbation, where all the fields change

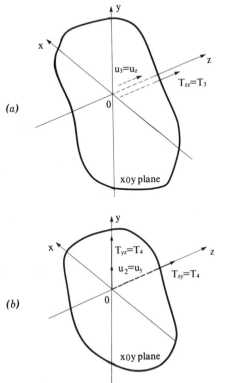

(a)

(b)

Figure 2.4. (a) A longitudinal wave propagating in the $+z$ direction with displacement u_z normal to $x0y$ plane. (b) A y-polarized shear wave propagating in the $+z$ direction with displacement u_y in the $x0y$ plane.

as $\exp j(\omega t - \beta z)$ with $\beta = \omega/v_L$, the wave impedance is given by

$$Z_L = -\frac{T_3}{v_3} = \frac{c_{33}\beta}{\omega} = \sqrt{\rho_m c_{33}} = \frac{(\lambda + 2\mu)}{v_L} = \frac{\rho_m \omega}{\beta} = \rho_m v_L \qquad (2.81)$$

and the complex Poynting vector by

$$\Pi = -\frac{v_3{}^* T_3}{2} = \frac{c_{33}\omega^2 |u_3|^2}{2v_L} \qquad (2.82)$$

A plane shear wave propagating in the $+z$ direction can be represented by a vector potential $\vec{\psi}$ dependent on z and t only. If the displacement is assumed to be in the y direction, as shown in Fig. 2.4(b), only the x component of $\vec{\psi}$ is nonzero, that is, $\vec{\psi} = \hat{x}\psi(t, z)$. Therefore,

$$\vec{u} = \hat{y}u_y = \hat{y}\frac{\partial \psi}{\partial z} \qquad (2.83)$$

$$S_4 = 2S_{yz} = \frac{\partial u_y}{\partial z} = \frac{\partial^2 \psi}{\partial z^2} = 2S_{zy} \qquad (2.84)$$

and

$$T_4 = T_{yz} = \mu S_4 = T_{zy} \qquad (2.85)$$

The shear wave impedance is given by

$$Z_S = -\frac{T_{yz}}{v_y} = \frac{\mu}{v_S} = \sqrt{\mu \rho_m} = \left[\frac{c_{11} - c_{12}}{2}\rho_m\right]^{1/2} = \frac{\rho_m \omega}{\beta} = \rho_m v_S \qquad (2.86)$$

The grid diagram of this wave is shown in Fig. 1.1(b), and the fields in Fig. 2.4(b). Another shear wave can be defined by choosing the displacement in the x direction rather than in the y direction. In that case,

$$\vec{u} = \hat{x}u_x \qquad (2.87)$$

$$S_5 = 2S_{xz} = 2S_{zx} \qquad (2.88)$$

$$T_5 = T_{xz} = \mu S_5 = T_{zx} \qquad (2.89)$$

and

$$Z_S = -\frac{T_{xz}}{v_x} = \sqrt{\mu \rho_m} = \frac{\rho_m \omega}{\beta} = \rho_m v_S \qquad (2.90)$$

Equations (2.78) to (2.90) demonstrate that for an arbitrary propagating direction in isotropic material, there are at most three propagating waves. To prove this, consider a coordinate system with its z axis in the direction of wave propagation, in which case Eq. (2.70) reduces to

$$\frac{\partial^2 u_x}{\partial t^2} - v_S^2 \frac{\partial^2 u_x}{\partial z^2} = 0 \tag{2.91}$$

$$\frac{\partial^2 u_y}{\partial t^2} - v_S^2 \frac{\partial^2 u_y}{\partial z^2} = 0 \tag{2.92}$$

and Eq. (2.71) to

$$\frac{\partial^2 u_z}{\partial t^2} - v_L^2 \frac{\partial^2 u_z}{\partial z^2} = 0 \tag{2.93}$$

Therefore two shear waves, one polarized in the x direction and the other in the y direction are propagating in the z direction with the same velocity v_S (*degenerate modes*). The third wave, propagating with velocity v_L in the z direction, is the longitudinal wave.

In addition to v_S and v_L, the two velocities just defined for a wave propagating in a material of infinite extent (free-field approximation; with the dimensions of the specimen much larger than the acoustic wavelength), extensional wave velocity v_e defined for wave propagation along the rod with cross-sectional dimensions much smaller than the wavelength (thin-waveguide approximation; see Sect. 1.3) can also be expressed in terms of Poisson's ratio σ. Using Eq. (2.52) and substituting E_Y for c_{11} in Eq. (2.62), it follows that

$$v_e = \left(\frac{E_Y}{\rho_m} \right)^{1/2} = v_L \left[\frac{(1 + \sigma)(1 - 2\sigma)}{1 - \sigma} \right]^{1/2} \tag{2.94}$$

Typical values of stiffnesses for some isotropic materials are given in Table 3. In materials that exhibit no experimental evidence of shear-wave propagation, $v_S = 0$, and from Eq. (2.62), $c_{44} \approx 0$.

2.4   EXPONENTIAL HORN

An important device used in practice for excitation of bulk waves and acoustic matching purposes is the exponential horn[7,8] with cross section shown in Fig. 2.5, where the z axis coincides with the rotational symmetry axis of the horn. This device is also used to increase the magnitude of the strain up to the fatigue limit of the material from which the horn is made. The radial cross-sectional dimensions are in practice less or comparable to the acoustic wavelength, and in the approximate plane-wave analysis which follows, the elastic constant used is Young's modulus.

TABLE 3 Stiffness Constants and Densities of Some Isotropic Materials[a]

Material	c_{11}, $\times 10^{10}$ N/m^2	c_{44}, $\times 10^{10}$ N/m^2	ρ_m, kg/m^3
Aluminum	11.1	2.5	2695
Beryllium	28.8	13.6	1850
Gold	20.7	2.85	19300
Lead	4.4	0.54	11400
Iron	27.7	8.2	7830
Nickel	32.4	8.0	8905
Silver	13.95	2.7	10490
Titanium	16.59	4.4	4500
Tungsten	58.1	13.4	19100
Glass T-40	6.3	2.26	3390
Glass, Pyrex	7.3	2.5	2320
Glass, crown	5.82	1.81	2243
Glass, flint	6.13	2.18	3879
Fused silica	7.85	3.12	2200
Lucite	0.848	0.143	1182
Polyethylene	0.340	0.026	900
Polystyrene	0.58	0.12	1056
Silicone rubber (GE, RTV 615)	0.107	~ 0	1010
Teflon	0.388	0.03	2160
Brass	15.9	3.65	8100

[a] The listed parameters for some materials should be considered average values. The parameters of these vary depending on the production method used.

For the longitudinal wave propagation along the z direction, where $A(z)$ is the cross-sectional area of the horn, Fig. 2.5, Eq. (1.3) can be written as

$$\frac{\partial F}{\partial z} = \rho_m A(z) \frac{\partial^2 u}{\partial t^2} \tag{2.95}$$

and Eq. (1.8) as

$$F = A(z) E_Y \frac{\partial u}{\partial z} \tag{2.96}$$

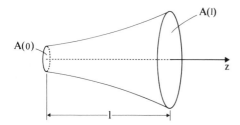

Figure 2.5. An exponential horn used for acoustic matching purposes.

Differentiating Eq. (2.96) and combining the result with Eq. (2.95), the wave equation is obtained as

$$\frac{\partial^2 u}{\partial z^2} + \frac{1}{A}\frac{\partial A}{\partial z}\frac{\partial u}{\partial z} - \frac{1}{v_e^2}\frac{\partial_u^2}{\partial t^2} = 0 \tag{2.97}$$

where $v_e = (E_Y/\rho_m)^{1/2}$ is the extensional wave velocity, Eq. (2.94). If the cross section is chosen to vary in the exponential fashion as

$$A(z) = A_0 e^{2az} \tag{2.98}$$

where a is a constant, and the excitation is assumed as $\exp(j\omega t)$, then it is easy to show that the solution of Eq. (2.97) is given by

$$u(t, z) = U^+ e^{-az}e^{j(\omega t - kz)} + U^- e^{-az}e^{j(\omega t + kz)} \tag{2.99}$$

where

$$k^2 + a^2 = \left(\frac{\omega}{v_e}\right)^2 \tag{2.100}$$

and where U^+ and U^- are the magnitudes of the displacement waves traveling in the $+z$ and $-z$ direction, respectively. For transducers such as horns and some other acoustic devices (see Chap. 5), the acoustic impedance is often defined[†] as

$$Z = -\frac{F}{v} \tag{2.101}$$

where F is the force. If the length l of the horn is chosen so that $kl = \pi$ and the load impedance at $z = l$ of the medium in intimate contact with the horn is Z_L, then the input impedance of the horn $Z(0)$ at $z = 0$ is given by

$$Z(0) = \frac{A(0)}{A(l)} Z_L = e^{-2al}Z_L \tag{2.102}$$

This relation is similar to the one obtained for an electrical transformer, $Z_0/Z_L = n^2$, where n^2 is the transformer turns ratio, and therefore the exponential transformer can be used for acoustic matching purposes. In Eq. (2.102) it was assumed that the impedance of the load Z_L is larger than the generator (driver) impedance $Z(0)$ and that the wave travels in the $+z$ direction. In most practical cases, however, the acoustic impedance of the

---

[†] The impedance defined by Eq. (1.15) can more precisely be called "specific acoustic impedance," since its dimensions are $-(F/\text{area})/v = -T/v$.

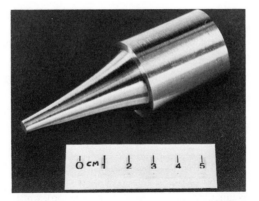

Figure 2.6. An exponential horn transducer used for nondestructive evaluation of materials. The diameters of the exponential section are 6 mm and 26 mm, and the length of the exponential section is 50 mm. The material is stainless steel and the frequency of operation is 50 kHz. The cylindrical section houses the generator, a piezoelectric plate of acoustic impedance equal to 32 $kg/m^2 s$.

driver is higher than that of the load. For these cases the driver is situated in the space $z \geq l$ and the load in the space $z \leq 0$ in Fig. 2.5, and the wave travels in the $-z$ direction. This is illustrated in Fig. 2.6, which shows an exponential horn transducer used for nondestructive testing of materials.

REFERENCES

1. B. A. Auld, *Acoustic Fields and Waves in Solids*, Vol. I, Wiley, New York, 1973, Chaps. 1, 2, and 3.
2. H. F. Pollard, *Sound Waves in Solids*, Methuen, New York, 1977, Chaps. 2 and 8.
3. H. J. Pain, *The Physics of Vibrations and Waves*, 2nd ed., Wiley, New York, 1976, Chap. 11.
4. E. Dieulesaint and D. Royer, *Elastic Waves in Solids*, Wiley, New York, 1980, Chap. 4.
5. F. I. Fedorov, *Theory of Elastic Waves in Crystals*, Plenum, New York, 1968, Chap. 1.
6. G. B. Benedek, J. B. Lastovka, K. Fritsch, and T. Greytak, *J. Opt. Soc. Am.*, **54**, 1284 (1964).
7. W. P. Mason and R. F. Wick, *J. Acoust. Soc. Am.*, **23**, 209 (1951).
8. F. J. Jackson, *J. Acoust. Soc. Am.*, **32**, 1387 (1960).

EXERCISES

1. Consider Fig. 2.2(c). The two forces shown would result in a net torque (couple), causing a rotation.

    (a) Sketch another pair of stresses $\pm T_{yx}$ resulting in forces that will cause the rotational equilibrium.

    (b) Show that the shear stresses in Fig. 2.2(c) are equivalent to an extension $T_{x'x'} = T_{xy}$ acting parallel to the diagonal $0B$ and a compression $T_{y'y'} = -T_{xy}$ acting parallel to the diagonal AC.

2. Of the four quantities K, σ, μ, and E_Y, only two are independent.
   (a) Show that $E_Y = 9\mu K/(3K + \mu)$.
   (b) Show that $-1 < \sigma < \frac{1}{2}$ if it is required that K, μ, and E_Y must be positive.

3. Express the stiffnesses of an isotropic solid in terms of E_Y and σ.

4. (a) Using Eqs. (2.40) and (2.45), find the compliances of an isotropic solid in terms of stiffnesses.
   (b) Express the compliances in terms of E_Y and σ.

5. (a) Show that

$$\mu = \frac{E_Y}{2(1 + \sigma)} \quad \text{and} \quad \lambda = \frac{E_Y \sigma}{(1 + \sigma)(1 - 2\sigma)}$$

   (b) Consider a rectangular plate of length l (x direction), width w (y direction), and thickness t (z direction). A uniform stress T_{xx} is applied at the ends, and a uniform stress T_{yy} on both sides, so that the width w remains unchanged. Using Hooke's law, Eq. (2.46), find the plate Poisson's ratio defined as $-S_{zz}/S_{xx}$ and the plate Young's modulus defined as T_{xx}/S_{xx}.
   (c) Express the computed quantities in terms of E_Y and σ.

6. Consider Eqs. (1.4) and (1.9). Setting $v = V(z)e^{j\omega t}$ and $T = T(z)e^{j\omega t}$, reduce these equations to

$$\frac{dT}{dz} - j\left(\frac{\omega}{v_a}\right)Z_a V = 0 \quad \text{and} \quad Z_a \frac{dV}{dz} - j\left(\frac{\omega}{v_a}\right)T = 0 \qquad \text{(E2.1)}$$

   where Z_a and v_a are the acoustic impedance and velocity, respectively. Equations (E2.1) can be solved in the conventional manner. Often the formulation in terms of normal mode amplitudes a_i is more convenient. The *normal modes* are linear combinations of V and T usually defined as

$$a_1 = \frac{1}{2}(T - Z_a V) \quad \text{and} \quad a_2 = \frac{1}{2}(T + Z_a V)$$

   (a) Show that the wave equation in terms of normal modes becomes

$$\frac{da_i}{dz} \pm j\left(\frac{\omega}{v_a}\right)a_i = 0$$

   with solutions $a_i = a_i^0 \exp(\pm j\omega z/v_a)$, $i = 1, 2$, where the minus sign in the last equation corresponds to the progressive wave (wave propagating in the $+z$ direction).
   (b) Using the definitions for the acoustic impedance and the power density, show that if only the progressive wave is present, $a_2 = 0$ and $P_a = |a_1|^2/2Z_a$.
   (c) Redefine the normal mode amplitudes as $b_i = a_i/\sqrt{2Z_a}$, and see whether this will affect the derived equations.

7. Using Eq. (2.99) and (2.96), find the expressions for the particle velocity and the force of the exponential horn. Apply these two relations at $z = 0$ and at $z = l = \pi/k$, and derive Eq. (2.102).

8. Consider the exponential transformer shown in Fig. 2.5. Find the cutoff frequency of the horn defined as the lowest frequency at which the horn will operate.

(a) Consider the horn as a resonant structure (with no forces or load applied), and find the resonant frequencies. (*Note*: The length of the horn must be an integer number of half-wavelengths.)

(b) From $F = 0$ at the ends of the horn, find a relation between U^+ and U^-. Use this relation to find the value of the strain along the horn.

(c) For $k = \pi/l$, $l = 3.14$ cm, and $a = 0.3$ cm^{-1}, plot the magnitude of the strain versus z, and find its maximum.

Chapter 3

Electrodynamic Excitation and Detection of Bulk Acoustic Waves and Dispersive Filters

Having become acquainted with the concepts presented in Chapters 1 and 2, it is now possible to consider mechanisms for conversion of electromagnetic into acoustic energy and vice versa. In one of the simplest (electrodynamic) methods for acoustic wave excitation, the Lorentz force mechanism, Eq. (1.32), is employed. Electrodynamic methods of acoustic wave transduction[1] have, in practice, an advantage over other methods in that the excitation of acoustic waves can be achieved without physical contact with the propagation medium (contactless configuration). The analysis of contactless configurations, however, tends to be complicated. In this chapter a much simpler configuration is considered. Nevertheless, many conclusions and formulas derived here will be applicable to practical contactless configurations considered in Chapter 4. The employed configuration is also used for demonstration of a rather general approach in modeling and finding the electrical input impedance and impulse response of acoustic devices. The basic concepts and design equations of the pulse-echo method used in radar and sonar are also developed and then demonstrated by applying the underlying principles to dispersive filters made with meander lines. The phenomenon of magnetostriction is also considered, and the equations developed are used to demonstrate bulk-wave excitation in ferromagnetic materials.

On the atomic scale the effect of the Lorentz force mechanism can be explained as follows.[2] The conducting state of a solid can be understood in terms of the *dielectric relaxation frequency* $\omega_R = \sigma_c/\varepsilon_L$, where σ_c is the DC conductivity and ε_L is the permittivity of the material. For metals, $\omega_R \simeq 10^{18}$ rad/s. If an external electric field of frequency ω is applied, since $\omega_R \gg \omega$ the field will be quickly screened out by redistribution of electrons within a distance of one Debye length. The Debye length for metals is on the order of 10^{-9} m. The redistribution of electrons will follow various cycles of the field of frequency ω and will cause, on average, a finite current density J_s [A/m^2]. If an external DC magnetic flux density B_0 is applied, the electron trajectories will become helical. This effect is taken into account by defining the *cyclotron frequency* $\omega_c = qB_0/m_0$, where q is the magnitude of the electronic charge and m_0 is the free electron mass (in most metals the effective mass of the electron m^* is equal to the free electron mass). For $B_0 = 1$ T, $\omega_c \simeq 10^{11}$ rad/s. In conductors the electrons also undergo collisions among themselves and with ions imbedded in the crystal lattice. The collisions can be characterized with collision frequency ν, which for metals is on the order of 10^{13} rad/s. Thus good conductors, $\omega_R \gg \omega$, are collision dominated, $\nu \gg \omega$, and for low DC magnetic flux densities, $B_0 \simeq 1$ T, $\nu \gg \omega_c$, so that the effect of B_0 on conductivity can be neglected.

The simultaneous application of an RF electric field, causing the current density $\vec{J}_s(\vec{r}, t)$, and the DC magnetic flux density $\vec{B}_0$ will generate the Lorentz force,

$$\vec{F}(\vec{r}, t) = \vec{J}_s(\vec{r}, t) \times \vec{B}_0$$

and thus impart a momentum to the electrons. However, due to high collision frequency between the electrons and the lattice ions, the momentum will be transferred to the lattice and acoustic waves will be excited. The same mechanism can be used for detection of acoustic waves. The particle velocity $\vec{v}(\vec{r}, t)$ of an acoustic wave propagating in a conductor immersed in the magnetic flux density $\vec{B}_0$ will cause the RF current density

$$\vec{J}(\vec{r}, t) = \sigma_c \left[\vec{v}(\vec{r}, t) \times \vec{B}_0 \right]$$

which can be directly detected or can be detected through the associated RF electric field.

In practice it is often required to design a device for excitation of acoustic waves in (nonpiezoelectric) dielectrics. In applying the Lorentz force mechanism to device design, the simplest approach would be to make an acoustic bond between the dielectric and a current-carrying conductor and to insert the resulting structure, also called the *transducer*, in a DC magnetic field.[3] The time-varying current in the conductor will produce a time-varying stress in the dielectric. This simple structure has two major drawbacks in that its conversion efficiency is low and its frequency selectivity is not well defined. These drawbacks can be remedied by using a meander line, Fig. 3.1(a). A sheet meander line, assumed to be made of a good conductor, is firmly bonded between several pieces of dielectric in which we wish to excite acoustic waves. In order to generate body forces within the material, the device is immersed in a DC magnetic flux density B_0 as shown in Fig. 3.1(b). The meander line is excited with a time-varying current density of magnitude J_s^0 [A/m^2] assumed constant over the cross section of conductor of length w, width l, and thickness ϵ such that $\epsilon \ll D$, where D is the spatial period. A time-varying current density J_s will generate a time-varying body force

$$\vec{F}^b(t, z) = \vec{J}_s \times \vec{B}_0 \tag{3.1}$$

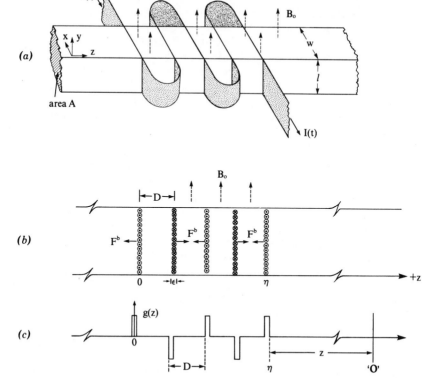

Figure 3.1. (a) Sheet meander line embedded in a nonpiezoelectric dielectric. (b) A $y - z$ cross section of (a) showing the direction of current $I(t)$, magnetic flux density B_0, and body forces F^b. (c) The source shape function.

as shown in the figure. The presence of the body force in the source region, $0 < z < \eta$, will cause rarefaction and compression of material. Outside the source region, due to symmetry of the structure, two waves of equal magnitude will be excited, one in region $z < 0$ propagating in the $-z$ direction, and another in region $z > \eta$ propagating in the $+z$ direction. In order to derive various physical quantities of interest, T is eliminated from the equations

$$\frac{\partial T}{\partial z} = \rho_m \frac{\partial v}{\partial t} - F^b \qquad (1.33)$$

$$\frac{1}{c} \frac{\partial T}{\partial t} = \frac{\partial v}{\partial z} \qquad (1.9)$$

and the wave equation is obtained as

$$\frac{\partial^2 v}{\partial z^2} - \frac{1}{v_a^2} \frac{\partial^2 v}{\partial t^2} = -\frac{1}{c} \frac{\partial F^b}{\partial t} \qquad (3.2)$$

The solution of the wave equation[4] for zero initial and boundary conditions is given by the convolution integral

$$v(t, z) = + \frac{v_a}{2c} \int \int \chi \left[\frac{t - \tau}{T} - \frac{|z - \xi|}{D} \right] \frac{\partial F^b \left[\frac{\tau}{T}, \frac{\xi}{D} \right]}{\partial \tau} d\tau \, d\xi \qquad (3.3)$$

where the integration is performed over the space-time volume occupied by the source, that is, $0 \leqslant \xi \leqslant \eta$, $-\infty < \tau \leqslant t$, and where $\chi(t)$ is the familiar step function, equal to unity for $t \geqslant 0$ and zero otherwise. The quantities D and T are the spatial period, as shown in Fig. 3.1(a), and the wave transit time over the spatial period $T = D/v_a$, respectively. The response $v(t, z)$ in position z at instant t arises from the source located in position ξ at instant τ. The source function in Eq. (3.3) can be written as

$$\frac{\partial F^b}{\partial \tau} = F_0 g \left(\frac{\xi}{D} \right) \frac{d}{d\tau} f \left(\frac{\tau}{T} \right) \qquad (3.4)$$

where $F_0 = B_0 J_s^0$, g is the source-shape function as shown in Fig. 3.1(c), and f is the source-time function.[†] When the impulse response is sought, $J_s = J_s^0 \delta(t)$ and

$$\frac{\partial F^b}{\partial \tau} = F_0 g \left(\frac{\xi}{D} \right) T \delta'(\tau) \qquad (3.5)$$

since $\delta(\tau/T) = T\delta(\tau)$.

Each of the conductor cross sections shown in Fig. 3.1(b) can be identified as a "single source" of acoustic waves. In that case it is more convenient to evaluate Eq. (3.3) for a single source and then sum up all the contributions of individual sources. For instance, for a source located at $\xi = 0$, the integral is evaluated over the limits $-D/2 \leqslant \xi \leqslant D/2$, $-\infty < \tau \leqslant t$. Combining Eqs. (3.3), (3.5), and the relation

$$\int \delta'(x) \mu(x + a) dx = -\mu'(a) \qquad (3.6)$$

it follows that

$$v(t, z) = -\frac{F_0 v_a}{2c} \int_{-D/2}^{+D/2} \delta \left[\frac{t}{T} - \frac{|z - \xi|}{D} \right] g \left(\frac{\xi}{D} \right) d\xi \qquad (3.7)$$

or

$$v(t, z) = -\frac{DF_0 v_a}{2c} \left[g \left(\frac{z - tv_a}{D} \right) + g \left(\frac{z + tv_a}{D} \right) \right] \qquad (3.8)$$

[†] Note that the same symbol is used to denote frequency, f. The text determines the meaning of f.

Thus a single source generates two oppositely propagating acoustic waves in the $\pm z$ direction. Since it is assumed that $\epsilon \ll D$ in Fig. 3.1(b), a good approximation to Eq. (3.8) is given by

$$v(t, z) \simeq -\frac{F_0 v_a D^2}{2c} \left[\delta(z - tv_a) + \delta(z + tv_a) \right] \qquad (3.9)$$

with associated stress for the progressive wave (traveling in the $+z$ direction) for $t < D/2v_a$, by

$$T(t, z) = -Z_a v(t, z) \qquad (1.15)$$

where $Z_a = \rho_m v_a$ is the acoustic impedance.

To find the impulse response due to M sources, consider the wave propagating in the $+z$ direction only, assuming noninteracting sources and neglecting diffraction and losses. The particle velocity at some plane O, normal to the z axis, in the region $z > \eta$ as shown in Fig. 3.1(c), is given by

$$v(t, z) = v_0 \sum_{m=1}^{M} (-1)^m \delta\left(t - \frac{z_m}{v_a} \right) \qquad (3.10)$$

where $v_0 = -F_0 D^2 v_a / 2c$, and where $z_m = z + Dm - D$, $m = 1, 2, \ldots, M$ determines the location of sources. The Fourier transform of Eq. (3.10) is given by

$$\tilde{v}(\omega, z) = v_0 \exp(-j\omega z/v_a) \sum_{m=1}^{M} (-1)^m \exp\left(\frac{-j\omega Dm}{v_a} \right) \qquad (3.11)$$

where the factor z/v_a represents, approximately, the time delay for wave propagation between the closest source and the surface O. In Eq. (3.11), when $\omega = \omega_0 \equiv v_a \pi / D$ the angular frequency ω_0 is called the *synchronous* or *resonant angular frequency of the transducer*. At that frequency the contributions of all M sources add coherently at plane O, resulting in the maximum possible signal. At some other frequencies that are different from ω_0, destructive interference between the contributions from various sources sets in resulting in a response of smaller magnitude.

The sum in Eq. (2.11) can be written as

$$\sum_{m=1}^{M} \exp\left[-jm\pi\left(\frac{f - f_0}{f_0} \right) \right] \approx M \frac{\sin X_M}{X_M} \exp\left[j\left(\frac{M\pi}{2} \right) - j\left(\frac{M\pi f}{2f_0} \right) \right] \qquad (3.12)$$

where $f = \omega/2\pi$ and $f_0 = \omega_0/2\pi$ are the excitation and resonant frequencies, respectively, $\sin[\pi(f - f_0)/2f_0] \simeq \pi(f - f_0)/2f_0$ is used, and $X_M = M\pi(f - f_0)/2f_0$. The transfer function and the impulse response of a transducer, consisting of M conductor segments at distance z from the transducer, are thus given by

$$\tilde{H}(\omega, z) = \frac{\tilde{v}(\omega, z)}{v_a} = j^M \frac{J_s^0 B_0 D^2}{2c} M \frac{\sin X_M}{X_M} \exp\left(-\frac{jM\pi f}{2f_0}\right) \exp\left(-\frac{j\omega z}{v_a}\right)$$

$$(3.13)$$

and

$$h(t, z) = \frac{v(t, z)}{v_a} = \frac{J_s^0 B_0 D^2}{2c} \sum_{m=1}^{M} (-1)^m \delta\left(\frac{t - z_m}{v_a}\right) \qquad (3.14)$$

From Eq. (3.13) it is seen that the presence of the meander line has caused a signal M times higher in magnitude than the simple transducer discussed at the beginning of the chapter, and that the function $(\sin X_m)/X_M$ determines the frequency selectivity of the meander-line transducer.

3.1 MAGNETOMECHANICAL COUPLING FACTOR

The power of the wave generated in the source region, $0 < z < \eta$, as it propagates in the $+z$ (or $-z$) direction, is not conserved but, at the resonant frequency, increases with distance at the expense of the stored magnetic energy, which is in turn supplied by the generator. By applying the acoustic Poynting's theorem in the transducer region over a volume consisting of an acoustic beam cross section of area A and length D, under the assumption of no losses, zero electric energy, and equal strain and kinetic energies, it can be shown that

$$\Delta P_a = -\frac{\Delta W_m}{\Delta t} \qquad (3.15)$$

where ΔP_a is the increase of the acoustic power in both waves propagating in the $\pm z$ direction, after each wave has traveled the length D in a time $\Delta t \, (= T)$, and where $\Delta W_m/\Delta t$ is the rate of decrease of the peak stored magnetic energy. The magnitude of the latter is equal to the power supplied by the generator. The depletion of stored energy at the resonant frequency $f_0 = 1/2T$ as the wave passes the length D can be characterized by

$$\frac{\Delta W_m'}{W_m'} = -k_m^2 \qquad (3.16)$$

where k_m^2 is defined as the magnetomechanical coupling coefficient. For both waves,

$$\frac{\Delta W_m}{W_m} = -2k_m^2 \qquad (3.17)$$

and on combining Eqs. (3.15) and (3.17) it follows that

$$\Delta P_a = \frac{2k_m^2 W_m}{T} = 4f_0 k_m^2 W_m = 4f_0 k_m^2 \left(\frac{wL_D I_0^2}{2} \right) \tag{3.18}$$

where I_0 is the peak current $(= AJ_s^0)$, L_D is the inductance per single D-long section per unit aperture width w, and $W_m = wL_D I_0^2/2$ is the peak magnetic energy stored in L_D. The RF resistance R of a D-long section, due to the presence of both waves, is found from

$$R_0 = \frac{2\Delta P_a}{I_0^2} = 4f_0 k_m^2 L_D w \tag{3.19}$$

and if the point of view is taken that in the source region the amplitude of the wave increases linearly, rather than exponentially,[5] with distance in the direction of propagation, the power of the wave will be increasing quadratically with distance (that is, as the square of the number of sections). Therefore for both waves, the resistance $R_0^{(M)}$ of M, D-long sections will be

$$R_0^{(M)} = 4f_0 k_m^2 L_D w M^2. \tag{3.20}$$

To determine the constant k_m^2, consider the power at resonance, $f = f_0$, of the transducer excited with peak current $I_0(t) = \delta(t)$ A. Using Eqs. (1.47), (1.15), and (3.13), and considering the power of a single wave, it follows that

$$P_a = \Pi A = -\frac{v^* T}{2} A = \frac{Z_a |v|^2 A}{2} = \frac{Z_a A |\tilde{H}(\omega_0, z)|^2 v_a^2}{2}$$

$$= \frac{Z_a A M^2}{2} \left(\frac{J_s^0 B_0 D^2}{2c} \right)^2 v_a^2 \tag{3.21}$$

where $Z_a = \rho_m v_a$. However, from Eq. (3.19) it is seen that

$$P_a = \frac{R_0^{(M)} |\tilde{I}_0(\omega_0)|^2}{4} = \frac{R_0^{(M)}}{4} \tag{3.22}$$

since the transducer is excited by current $\tilde{I}_0(\omega) = 1$. Using Eqs. (3.21) and (3.22), it is found that

$$k_m^2 = \frac{B_0^2 v_a D^4}{32 \rho_m f_0 L_D w^2 l} \tag{3.23}$$

Since the magnetomechanical coupling factor is the measure of the energy transfer from magnetic fields to acoustic fields via the Lorentz-force equation, several conclusions can be drawn from Eq. (3.23). The efficiency of acoustic wave generation is directly proportional to the square of DC magnetic field B_0. Since

TABLE 4 Figure of Merit for Electromagnetic Generation of Acoustic Waves in Various Polycrystalline Materials

Material	F/F_{Al} [a]
Beryllium	41
Aluminum	1
Magnesium	0.94
Titanium	0.45
Iron[b]	0.23
Zircaloy (Zr–2.5 Nb)	0.086
Tungsten	0.065
Brass	0.05
Silver	0.011
Gold	0.005
Lead	0.0005

[a] $F = v_a^5/\rho_m$, normalized to aluminum.
[b] Magnetostrictive effect not taken into account.

$D = v_a/2f_0$ at resonant frequency, the coupling factor is proportional to v_a^5/ρ_m and inversely proportional to f_0^5. Therefore, materials having a high v_a^5/ρ_m factor will be more efficient in the transduction of electromagnetic to acoustic energy and vice versa. This factor, identified as a material's "figure of merit," is listed in Table 4 for various materials. Of the materials listed, beryllium with its high velocity performs the best, and lead with its high mass density and low velocity the worst.

3.2 MEANDER-TYPE TRANSDUCERS

The transducer shown in Fig. 3.1(a) belongs to a group of devices commonly known as meander-type transducers.[1] Planar meander-type transducers are considered in Chapter 4. In this section, impedance characteristics of sheet meander-type transducers are derived.

Using Eqs. (3.13), (3.21), and (3.22) off-resonance, it follows that the acoustic radiation resistance of the meander-type transducer consisting of M, D-long sections is given by

$$R(\omega) = R_0^{(M)} \frac{\sin^2 X_M}{X_M^2} \qquad (3.24)$$

where $R_0^{(M)}$ is obtained from Eqs. (3.20) and (3.23) and is given by

$$R_0^{(M)} = \frac{v_a B_0^2 M^2 D^4}{8\rho_m wl}. \qquad (3.25)$$

Using the Hilbert transform,[6] relating the real and imaginary parts of the impedance function $Z(\omega) = R(\omega) + jX(\omega)$ as

$$X(\omega) = -\frac{1}{\pi} \int_{-\infty}^{+\infty} \frac{R(\omega)}{\omega - \xi} d\xi \qquad (3.26)$$

the acoustic radiation reactance is found to be

$$X(\omega) = R_0^{(M)} \frac{\sin(2X_M) - 2X_M}{2X_M^2} \qquad (3.27)$$

The static inductance and resistance and the skin-effect inductance and resistance are not included in Eqs. (3.24) and (3.27). These quantities must be considered separately. The input equivalent circuit is shown in Fig. 3.2(a), where L_E is total electrical inductance and R_E the total electrical resistance due to static and skin effects. Both of these quantities, due to the skin effect, are frequency dependent. The behavior of the acoustic radiation resistance and reactance is shown in Fig. 3.2(b).

The particle displacement is obtained by integration of Eq. (3.10), where the integrand is normalized with respect to T and is given by

$$u(t, z) = Tv_0 \sum_{m=1}^{M} (-1)^m \chi\left(t - \frac{z_m}{v_a}\right)$$

$$\simeq \frac{2Dv_0}{\pi v_a} [\chi(z - tv_a) - \chi(z - tv_a - MD)] \cos[k_0(z - tv_a)] \qquad (3.28)$$

where $k_0 = \pi/D$ is the transducer wave number, and where the DC term and the higher order Fourier harmonics at $3f_0, 5f_0, \ldots$ have been neglected. Therefore, in this approximation, which agrees well with experimental evidence, the impulse response in terms of particle displacement consists of an acoustic wave train of frequency $f_0 = k_0 v_a/2\pi$, as shown in Fig. 3.3(a).

Having derived the impulse response of a single transducer, it is now possible to consider a device consisting of two meander-line transducers, since in practice it is necessary to convert the acoustic wave train into an electric signal. This is shown in Fig. 3.3(a), where the output transducer consists of a single conductor. Wax layers are added at both endfaces of the substrate to suppress unwanted reflections. The open-circuit voltage e_O at the output will be given by

$$e_O(t) = -\frac{d\phi}{dt} = -B_0 w \frac{du}{dt} = -B_0 wv \qquad (3.29)$$

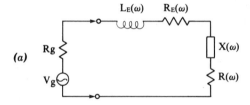

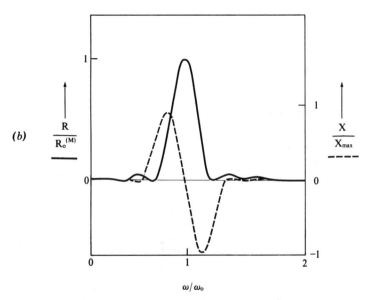

Figure 3.2. (*a*) Equivalent circuit for the input impedance of the meander-type transducer shown in Fig. 3.1(*a*). (*b*) Frequency dependence of the normalized acoustic radiation resistance and reactance.

where ϕ is the magnetic flux[†] and v is obtained from Eq. (3.28) as

$$v(t, d) \approx 2v_0\left[\chi(d - tv_a) - \chi(d - tv_a - MD)\right]\sin\left[k_0 d - \omega_0 t\right] \quad (3.30)$$

Therefore $e_O(t)$ will also be a finite-duration sinusoidal signal of constant amplitude as shown by the inset in Fig. 3.3(*a*). In the frequency domain, the open-circuit output voltage $\tilde{e}_O(\omega)$ is obtained from Eq. (3.13) and the Fourier

[†]Note that ϕ is also used to denote electrical phase and potential functions. The text determines the meaning of ϕ.

transform of Eq. (3.29). This can be written as

$$\tilde{e}_O(\omega) = -B_0 w \tilde{\upsilon}(\omega, d)$$

$$= j^M \frac{J_s^0 B_0^2 wD\upsilon_a}{2c} M \frac{\sin X_M}{X_M} \exp\left(-\frac{jM\pi f}{2f_0}\right) \exp\left(-\frac{j\omega d}{\upsilon_a}\right) \quad (3.31)$$

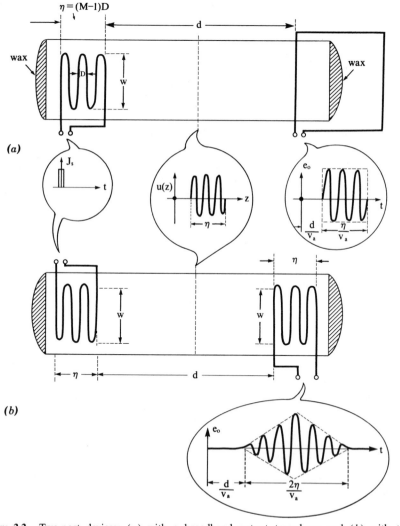

Figure 3.3. Two-port devices: (*a*) with a broadband output transducer and (*b*) with two identical transducers.

The bandwidth of a single transducer consisting of M sources can be derived from Eq. (3.13) by normalizing the transfer function with respect to its value at resonant frequency. Therefore,

$$\left| \frac{\tilde{H}(\omega, z)}{\tilde{H}(\omega_0, z)} \right| = \frac{\sin X_M}{X_M} \tag{3.32}$$

and equating this with $1/\sqrt{2}$ results in a 3-dB relative bandwidth of

$$\frac{\Delta f}{f_0} \simeq \frac{1.8}{M}. \tag{3.33}$$

Thus for a single-segment transducer, such as the one shown in Fig. 3.3(a), the relative bandwidth is 180%, and for $M = 2$ it is 90%. The output transducer shown in Fig. 3.3(a) will therefore reproduce the incoming impulse response of the input transducer without distortions provided the latter has $M > 1$. The total delay provided by the device at $f = f_0$ is found from Eq. (3.31) to be $[d + (\eta/2)]/v_a$. The space on the substrate occupied by the response, Eq. (3.30), equals the length of the input transducer, resulting in a signal of duration η/v_a at the output of the device.

For two transducers, consisting of number of segments M_i and M_o, resonant frequencies f_0^i and f_0^o, and lengths η_i and η_o, respectively, located at distance d apart, the open-circuit output voltage in the frequency domain is given by

$$\tilde{e}_0(\omega) = j^{M_i} \left(\frac{J_s^0 B_0^2 w}{2c} \right) M_i M_o \frac{\sin X_{M_i}}{X_{M_i}} \frac{\sin X_{M_o}}{X_{M_o}} \exp \left\{ -\frac{j\omega [d + (\eta_i + \eta_o)/2]}{v_a} \right\} \tag{3.34}$$

The output, given by Eq. (3.34), is equal to the product of the input and output transducer transfer functions. In the time domain, the output impulse is therefore a convolution of input and output transducer impulse responses. When two identical transducers are used as shown in Fig. 3.3(b), the envelope of the output impulse response is of triangular form, due to convolution of two square envelopes, one corresponding to the acoustic wave train given by Eq. (3.28) and the other to the impulse response of the output transducer itself. The envelope maximum occurs at $(d + \eta)/v_a$, as shown on the inset in Fig. 3.3(b). From Eq. (3.34) it is seen that the normalized transfer function of two identical transducers is equal to $(\sin^2 X_M)/X_M^2$, resulting in a 3-dB relative bandwidth of

$$\frac{\Delta F}{f_0} \simeq \frac{1.3}{M} \tag{3.35}$$

So far only the wave propagation outside the transducer region, $0 < z < \eta$ in Fig. 3.1(b), has been considered. In the transducer region, when the input transducer is operating at the resonant frequency ω_0, the wave propagating in the $+z$ (or $-z$) direction experiences a continuous growth in magnitude due to the action of the generator. It can be shown[5] that the growth in magnitude is exponential with distance, that is, of the form $\exp(\gamma z)$, $\gamma > 0$ for waves propagating in the $+z$ direction. The coupling coefficient, defined by Eq. (3.16), is closely related to the growth of the wave and can be expressed in terms of the stored acoustic energy in the wave. The increase of the average stored energy or energy density in the acoustic wave is equal to the magnitude of the decrease of the magnetic energy, Eq. (3.16), and is given by

$$k_m^2 = \frac{\delta U}{U} = \frac{\delta u_v}{u_v} \tag{3.36}$$

where u_v is the average acoustic energy density given by Eq. (1.49). It follows that the coupling coefficient is the relative difference of the energy densities at the entrance and exit of a D-long section,

$$k_m^2 = \frac{u_v(D) - u_v(0)}{u_v(0)} = \exp(2\gamma D) - 1 \simeq 2\gamma D \tag{3.37}$$

resulting in the particle velocity of the wave traveling in the $+z$ direction of

$$v(t, z) \approx v' \exp[j(\omega_0 t - \beta_a z)] \exp\left[\frac{k_m^2 z}{2D}\right] \tag{3.38}$$

where v' is a constant. A schematic of the wave magnitudes inside and outside the source (transducer) region are shown in Fig. 3.4.

Meander-type transducers can also be used for excitation of shear waves. For instance, if the direction of the applied DC magnetic flux density in Fig.

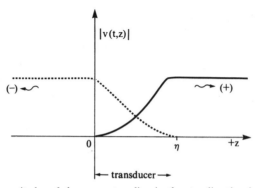

Figure 3.4. The magnitudes of the waves traveling in the $\pm z$ direction inside and outside the transducer region in a lossless material.

3.1(a) is chosen along the z axis, the Lorentz forces given by Eq. (3.1) will be directed in the $\pm y$ direction, causing excitation of y-polarized shear waves. All the formulas derived for longitudinal waves can be used, with appropriate changes in the values of the various constants, for shear waves as well.

3.3 TRANSDUCER APODIZATION

During the study of the impulse response of two-port devices, shown in Fig. 3.3, it has been established that the envelope of the transducer pattern bears a one-to-one correspondence with the envelope of its impulse response. It is therefore possible to alter the time-domain (or frequency-domain) characteris-

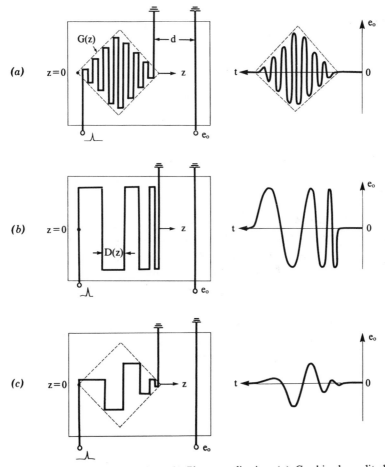

Figure 3.5. (a) Amplitude apodization. (b) Phase apodization. (c) Combined amplitude and phase apodization.

tics of the response by altering the envelope of the segments in the transducer pattern. This procedure is commonly known as *amplitude apodization* or *weighting.*[7] An amplitude-apodized[†] transducer is shown in Fig. 3.5(a). It follows from Eqs. (3.29) and (3.30) that the impulse response for that device can be written as

$$e_O(t) \approx -\frac{2B_0wv_0}{\pi}G(d - tv_a)\sin[k_0 d - \omega_0 t] \qquad (3.39)$$

In this case the spacing D between conductor segments is constant. Another type of apodization, called phase (or spectral) apodization, is shown in Fig. 3.5(b), where the G function (a square function) is unaltered, while the spacing between the segments, D, is a function of z,

$$D = D(z) \qquad (3.40)$$

In that case the resonant frequency is different for each pair of adjacent segments,

$$f_0(z) = \frac{v_a}{2D(z)} \qquad (3.41)$$

For instance, if $D(z) = D_0(1 \pm az)$, where a is a constant, Eq. (3.41) can be written approximately as

$$f_0(z) \approx \frac{v_a}{2D_0}(1 \mp az) = f_0(1 \mp \mu t) = f(t) \qquad (3.42)$$

where $z = v_a t$ and $\mu = av_a$. A signal with a frequency that varies linearly with time, Eq. (3.42), is termed the *linear chirp* or the *linear FM signal*, and the corresponding transducer, the linear chirp or dispersive transducer. The phase of the signal is obtained from $d\phi/dt = \omega(t)$ and is given by

$$\phi = \omega_0 \left(t \mp \tfrac{1}{2}\mu t^2 \right) \qquad (3.43)$$

In this case the impulse response of the device, shown in Fig. 3.5(b), is obtained from Eqs. (3.29) and (3.30) as

$$e_O(t) \approx -2B_0wv_0[\chi(d - tv_a) - \chi(d - tv_a - MD)]$$

$$\times \sin\left[k_0 d - \omega_0 \left(t \mp \frac{\mu t^2}{2} \right) \right] \qquad (3.44)$$

[†]*Apodized* means "without feet." Apodization, within the context of the book, relates to getting rid of sidelobes on a response by adjusting phase or amplitudes of the sources.

corresponding to the up-chirp[†] signal for a plus sign and to the down-chirp signal for a minus sign.

A transducer apodized in both amplitude and phase is shown in Fig. 3.5(c). In general, if the amplitude apodization function $G(z)$ and the phase apodization function $D(z)$ are assumed continuous, then the frequency response of the device, for instance, shown in Fig. 3.5(c), is simply obtained as

$$\tilde{e}_O(\omega) = 2B_0 w v_0 \int_0^\eta G(-z)\sin\left[\pi\int_0^\eta \frac{dz}{D(z)}\right]\exp\left(-\frac{j\omega z}{v_a}\right)\frac{dz}{v_a} \quad (3.45)$$

where Eq. (3.41), $z = v_a t$, and Eq. (3.44) with $d = 0$ were used.

3.4 DISPERSIVE TRANSDUCERS

In many applications such as radar, sonar, and NDT (nondestructive testing), the pulse-echo method[7] is used to detect objects (in NDT imperfections of materials). The pulse-echo method consists, in its simplest form, of a transmitter emitting short pulses, which are reflected from an object and then received by a receiver. Obviously, if the propagation velocity of the wave is known, by measuring the transit time of the pulse to the object and back, it is possible to determine the range (distance) of the object. In practice, the echo received is a combination of the reflections from the object and its neighborhood and noise, which, for instance, may be an inherent property of the propagation medium. Under these circumstances, for good range resolution it is necessary to use as short a pulse as possible, and for the best signal-to-noise ratio as high a peak power as possible. However, due to high peak power requirements this type of design results in very difficult demands on the transmitter components. In 1950 it was shown[8] that instead of using a short-duration signal of high peak power, it is possible to work with a long-duration signal of low peak power and to obtain an output signal in the form of a narrow pulse, provided that a matched filter is used at the receiver input.

Many radars and sonars use waveforms of the type shown in Fig. 3.5(b). An important practical problem is the optimum detection (in the presence of noise) of such a waveform upon its reflection from the target. If the signal shown in Fig. 3.5(b) is passed through a filter whose time delay characteristic is such that it delays in an adequate way the high frequencies at the beginning of the waveform more than the low frequencies which follow, a compression of the signal, that is, an accumulation of energy on a reduced time interval, will occur, Fig. 3.6(a). In other words, if the duration of the signal is T, the duration of the compressed pulse will be $\tau < T$. If the filter is lossless, the energy of the signal Q will be

[†]The reader might whistle from low to high frequency rapidly to appreciate the characterization "up-chirp."

conserved. Therefore, the ratio of peak powers at the output and the input of the filter (gain) will be $(Q/\tau)/(Q/T) = T/\tau > 1$, as shown in Fig. 3.6($a$). The procedure just described is a particular case of a very general method known as *matched filtering*, which can be applied to any waveform of finite duration.

It is well known[7] that a matched filter for a signal $s(t)$ is obtained if the impulse response of the filter is given by

$$h(t) = As(b - t) \qquad (3.46)$$

where A and b are constants, as shown in Fig. 3.6(b). The constant A is the magnitude factor, and b is the time delay of the signal through the filter due to the finite velocity of wave propagation. In the frequency domain, if the Fourier

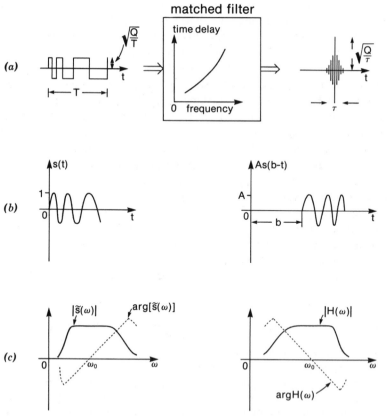

Figure 3.6.   (a) Illustration of signal compression. See text for more details. (b) The input signal $s(t)$ and the impulse response $As(b - t)$ of the corresponding matched filter. (c) The magnitude and argument of the input signal $\tilde{s}(\omega)$ versus frequency, and the corresponding quantities of the matched filter with transfer function $\tilde{H}(\omega)$ and with the time delay $b = 0$.

transform of the signal is $\tilde{s}(\omega)$, then the transfer function of the matched filter is given by

$$\tilde{H}(\omega) = Ae^{-j\omega b}\tilde{s}*(\omega) \tag{3.47}$$

where $\tilde{s}*$ is the complex conjugate of $\tilde{s}$. This is illustrated in Fig. 3.6(c) with $b = 0$ for the sake of clarity. Thus the output of the matched receiver is obtained as

$$e_0(t) = \int_{-\infty}^{+\infty} s(\tau)h(t - \tau)d\tau$$

$$= A\int_{-\infty}^{+\infty} s(\tau)s(b + \tau - t)d\tau \tag{3.48}$$

and the frequency response as

$$\tilde{e}_0(\omega) = Ae^{-j\omega b}\tilde{s}(\tau)\tilde{s}*(\tau) \tag{3.49}$$

Equation (3.48) with $A = 1$ and $b = 0$ is the expression for the *autocorrelation function*,

$$e_0(t) = \int_{-\infty}^{+\infty} s(\tau)s(\tau - t)d\tau \tag{3.50}$$

which is an even function of t with its maximum at $t = 0$, and whose duration is equal to twice the duration of the signal $s(t)$. The value of the maximum is

$$e_0(0) = \int_{-\infty}^{+\infty} s^2(\tau)d\tau \tag{3.51}$$

which is seen to be proportional to the total energy in the signal. Thus, a finite-time autocorrelation function maximum of a certain value may be obtained with a short-duration signal of high peak power or with a long-duration signal of low peak power. One way of generating the latter is with *chirp-type transducers*.

Consider the waveform emitted upon impulse excitation by the input transducer shown in Fig. 3.7(a). Using Eqs. (3.30) and (3.44) the expression for the waveform (called the *expanded pulse*) is written at $d = 0$ as

$$v_i(t) = 2v_0\text{rect}\left[\frac{t}{T}\right]\exp\left[2\pi j\left(f_0t + \frac{pt^2}{2}\right)\right] \tag{3.52}$$

where the origin of time is chosen to coincide with the center of the pulse, as shown in Fig. 3.7(b), $T \equiv MD/v_a$, $p \equiv \mu f_0$, and the function rect is defined as

$$\text{rect}(x) = \begin{cases} 1, & |x| \leq \frac{1}{2} \\ 0, & |x| \geq \frac{1}{2} \end{cases} \tag{3.53}$$

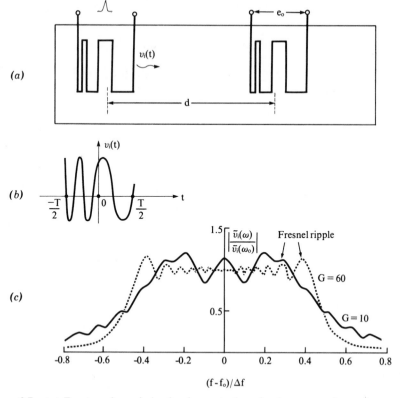

Figure 3.7. (*a*) Two-transducer device for demonstration of pulse compression and expansion. (*b*) Input transducer waveform versus time at $d = 0$. (*c*) The normalized frequency response of (*b*) for two values of gain G. Δf is the bandwidth and f_0 is the midband frequency.

During the duration T of the pulse, also called the *dispersion*, the instantaneous frequency

$$f(t) = \frac{1}{2\pi} \frac{d\phi}{dt} = f_0 + pt \tag{3.54}$$

changes from $f_0 - \frac{1}{2}pT$ to $f_0 + \frac{1}{2}pT$, and the difference between these two values is defined as the *frequency sweep*

$$\Delta f = pT \tag{3.55}$$

which, as it will be shown later, is approximately equal to the *bandwidth of the signal*. Another parameter of interest, the *processing gain* or *time-bandwidth product*, is defined as

$$G = T\Delta f \tag{3.56}$$

The frequency response $v_i(t)$ is obtained by use of the Fourier transform as

$$\tilde{v}_i(\omega) = \int_{-T/2}^{+T/2} v_i(t) e^{-j\omega t}\, dt$$

$$= 2v_0 \left(\frac{T}{2\Delta f}\right)^{1/2} \exp\left[-\frac{j\pi(f - f_0)^2}{p}\right] \left[\tilde{Z}(u_2) - \tilde{Z}(u_1)\right] \quad (3.57)$$

where $\tilde{Z}(u)$ is the complex Fresnel integral given by

$$\tilde{Z}(u) = C(u) + jS(u) = \int_0^u e^{j\pi\alpha^2/2}\, d\alpha \qquad (3.58)$$

and

$$u_2 = \left(\frac{G}{2}\right)^{1/2}\left[\frac{2(f - f_0)}{\Delta f} - 1\right] \quad \text{and} \quad u_1 = \left(\frac{G}{2}\right)^{1/2}\left[\frac{2(f - f_0)}{\Delta f} + 1\right]$$

$$(3.59)$$

The normalized amplitude $|\tilde{v}_i(\omega)/\tilde{v}_i(\omega_0)|$ is shown in Fig. 3.7(c) for various values of the gain G. Numerical evaluation of the normalized amplitude shows that for $G \simeq 10$ and $G \simeq 100$, 95 and 98%, respectively, of the spectral energy is contained in the frequency band Δf. In that case a good approximation to Eq. (3.57) valid for $G \geqslant 10$ is given by

$$\tilde{v}_i(\omega) \simeq 2v_0\sqrt{\frac{T}{2D}}\ \text{rect}\left[\frac{f - f_0}{\Delta f}\right]\exp\left[-\frac{j\pi(f - f_0)^2}{p}\right] \qquad (3.60)$$

It is not intended to design the output transducer of the device, shown in Fig. 3.7(a), as the matched filter to the signal $\tilde{v}_i(\omega)$. From Eq. (3.47) it follows that the transfer function of the output transducer must be of the form

$$\tilde{H}(\omega) = Ae^{-j\omega b}\tilde{v}_1^*(\omega)$$

$$= A2v_0\exp\left[-j\frac{\omega d}{v_a}\right]\sqrt{\frac{T}{2D}}\ \text{rect}\left[\frac{f - f_0}{\Delta f}\right]\exp\left[\frac{j\pi(f - f_0)^2}{p}\right] \quad (3.61)$$

where $b \simeq d/v_a$, d being the center-to-center distance between the input and output transducers. The frequency response of the induced voltage at the terminals of the output transducer is obtained from Eqs. (3.29) and (3.61) as

$$\tilde{e}_0(\omega) = -B_0 w\tilde{H}(\omega)\tilde{v}_i(\omega)$$

$$= -AB_0 w(2v_0)^2\exp\left[-j\frac{\omega d}{v_a}\right]\left(\frac{T}{2\Delta f}\right)\text{rect}^2\left[\frac{f - f_0}{\Delta f}\right] \qquad (3.62)$$

Using the approximation $\text{rect}^2 \approx \text{rect}$, its time response is obtained by the inverse Fourier transform as

$$e_o(t) = -B_0 w(2v_0)^2 \frac{AT}{2} \exp\left[j\omega_0\left(t - \frac{d}{v_a}\right)\right]\text{sinc}\left[\left(t - \frac{d}{v_a}\right)\Delta f\right] \quad (3.63)$$

where the sinc function is defined as

$$\text{sinc } x \equiv \frac{\sin \pi x}{\pi x} \quad (3.64)$$

If the propagating medium is assumed lossless and all the energy of the input signal is absorbed by the output transducer, it follows that the normalized signals of Eqs. (3.52) and (3.63) must have the same energy, that is,

$$\int_{-\infty}^{+\infty}\left[\frac{v_i(t)}{2v_0}\right]^2 dt = \int_{-\infty}^{+\infty}\left[\frac{e_o(t)}{2v_0 B_0 w}\right]^2 dt \quad (3.65)$$

resulting in the value of the constant A,

$$A = \frac{\pi}{v_0}\left(\frac{\Delta f}{T}\right)^{1/2} \quad (3.66)$$

and the expression for the compressed pulse,

$$e_o(t) = -B_0 w\left(\frac{2v_0}{\pi}\right)\sqrt{G}\exp\left\{ j\omega_0\left[t - \frac{d}{v_a}\right]\right\}\text{sinc}\left[\left(t - \frac{d}{v_a}\right)\Delta f\right] \quad (3.67)$$

From Eq. (3.67) it is seen that the peak wave energy, proportional to the signal magnitude squared, has been increased (compressed) by the factor $G = T\Delta f$, defined in Eq. (3.56) as the time-bandwidth product (or gain), Δf being the bandwidth and T the duration of the expanded pulse given by Eq. (3.52). The duration of the compressed pulse, computed from the sinc function in Eq. (3.67) as the time interval around $t = d/v_a$, where the amplitude of the signal falls by 3 dB, is given by

$$\tau = \frac{1}{\Delta f} \quad (3.68)$$

and the gain parameter can be written as

$$G = \frac{Q/\tau}{Q/T} = \frac{T}{\tau} \quad (3.69)$$

where Q is the energy of the input pulse. Equation (3.69) indicates that the gain-bandwidth product is approximately equal to the ratio of the peak power at the output to the peak power at the input which, since the energy of the pulse is constant, is equal to the ratio of the duration of the expanded (input) pulse and the duration of the compressed (output) pulse. The sinc function in Eq. (3.67) has its first time sidelobes 13 dB down from its peak value and a null width of $2/\Delta f$ as shown in Fig. 3.8. In order to reduce the time sidelobes, in addition to phase apodization, amplitude apodization is usually employed at the receiving transducer. For instance, Hamming-type[7, 9] amplitude apodization is capable of 42-dB theoretical time-sidelobe suppression with null-width broadening by a factor of 2 and duration of compressed pulse broadening by a factor of approximately 1.3. Figure 3.9 shows an FM-apodized planar meander-line transducer and the corresponding matched filter response.

Another device made of dispersive transducers and exhibiting dispersive properties is shown in Fig. 3.10(a). In this case the impulse responses of the transducers are equal. The device overall transfer function is obtained from Eqs. (3.29) and (3.57) as

$$\tilde{e}_O(\omega) = -B_0 w (2v_0)^2 A\left(\frac{T}{2\Delta f}\right)\exp[-j\omega d/v_a]$$

$$\times [\tilde{Z}(u_2) - \tilde{Z}(u_1)]^2 \exp\left[-\frac{j2\pi(f-f_0)^2}{p}\right] \qquad (3.70)$$

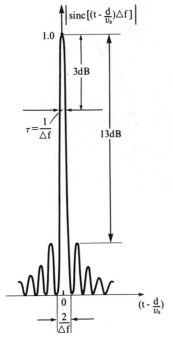

Figure 3.8. The properties of sinc function in Eq. (3.67) giving the waveform of the compressed pulse. The compression gain is $G = T/\tau$, where T is the duration of the expanded pulse and τ is the width of the compressed pulse.

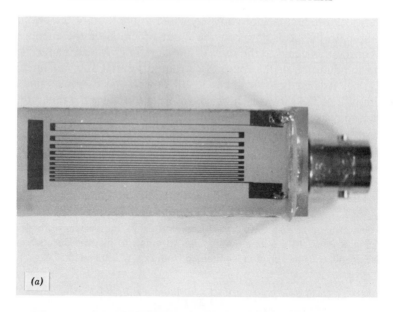

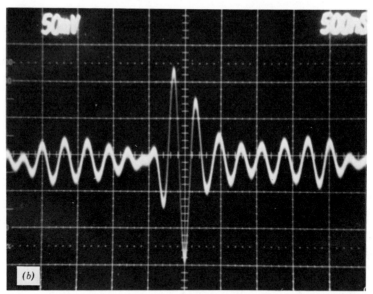

Figure 3.9. (*a*) An FM apodized planar meander-line transducer. (*b*) The corresponding matched filter response.

where A is an unknown constant. In this dispersive device, each frequency component f can be thought of as being generated at a position along the transducer where the spacing is half the acoustic wavelength, $\lambda/2 = D = v_a/2f$. The high frequencies are generated and detected where the meander-line spacing is small, near the interior of the device, and the low frequencies are generated and detected near the outer ends, where the spacing is large. Thus low-frequency waves travel a larger distance and have more

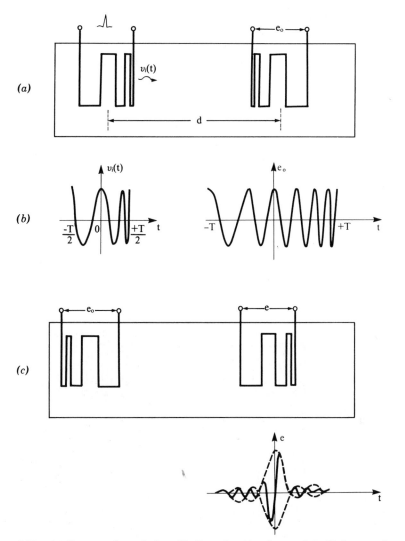

Figure 3.10. (a) Two-transducer device with dispersion (signal expander). (b) Input and output transducer waveform versus time. (c) Complementary device to (a) used to compress the signal e_O from (a). The voltage e is the compressed signal.

delay than the high-frequency waves, this being the reason that the total structure is dispersive. In this case, the voltage $e_O(t)$ is also a chirp-type signal but is now dispersed over duration $2T$. Note that in this symmetric device, half of the operation is carried out in each meander line. Each meander line responds to the full frequency range of the chirped pulse but provides only one-half of the total time dispersion. The complementary (compression) device is shown in Fig. 3.10(c). When the expanded pulse e_O, from the device shown in Fig. 3.10(a) is applied to the input of the device shown in Fig. 3.10(c), the output e_O of the device results in a signal of the type given by Eq. (3.67), that is, the compressed pulse.

An additional insight into the operation of unapodized transducers can be gained by evaluating the response of the device shown in Fig. 3.11(a) using criteria established in this section for dispersive transducers. The signal, shown in Fig. 3.11(b), is generated by impulse excitation at the input of the device and is obtained from Eq. (3.30) at $d = 0$ as

$$v_i(t) \approx 2v_0 \text{rect}\left[\frac{t}{T}\right] e^{j\omega_0 t} \tag{3.71}$$

where $T \equiv MD/v_a$, $\omega_0 = v_a/D$, and MD is the physical length of the transducers. The corresponding frequency response is given by

$$\tilde{v}_i(\omega) = 2v_0 T \text{sinc}\left[(f - f_0)T\right] \tag{3.72}$$

and it follows that the frequency response of the voltage at the output of the device is given by

$$\tilde{e}_0(\omega) = -B_0 w (2v_0)^2 T^2 A \exp\left[-j\omega \frac{d}{v_a}\right] \text{sinc}^2\left[(f - f_0)T\right] \tag{3.73}$$

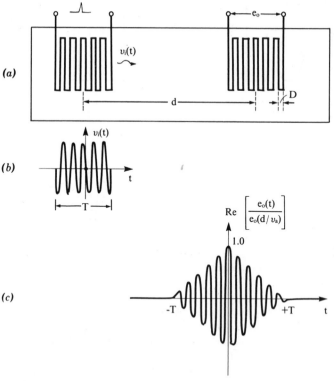

Figure 3.11. (a) Two-port device with identical input and output transducers. (b) The time behavior of particle velocity generated at $d = 0$ upon impulse excitation. (c) The normalized output voltage versus time; $T = l/v_a$, where $l = MD$ is the physical length of the transducer.

where A is a constant to be determined. The time-domain behavior of the output voltage is obtained using the inverse Fourier transform as

$$e_O(t) = -B_0 w (2v_0)^2 TA \exp\left[j\omega_0\left(t - \frac{d}{v_a}\right)\right]\text{tri}\left[\frac{t - d/v_a}{T}\right] \qquad (3.74)$$

where the function tri is defined as

$$\text{tri}(x) = \begin{cases} 1 - |x|, & |x| < 1 \\ 0, & |x| > 1 \end{cases} \qquad (3.75)$$

Using Eqs. (3.65), (3.71), and (3.74), the constant A is determined as $A = \sqrt{3}/(2v_0 T\sqrt{2})$, and Eq. (3.74) is written as

$$e_O(t) = -Bw(2v_0)\sqrt{\frac{3}{2}}\exp\left[j\omega_0\left(t - \frac{d}{v_a}\right)\right]\text{tri}\left[\frac{t - d/v_a}{T}\right] \qquad (3.76)$$

which is shown in Fig. 3.11(c). It is seen that compression is negligible in this case, corresponding to a peak energy increase of only 1.5. Figure 3.12 shows an unapodized planar meander-line transducer and the corresponding response upon detection with an identical transducer.

In building practical devices, the segment-positioning law must be found. The phase of the chirp signal, Eq. (3.52), is given by

$$\phi(t) = 2\pi\left(f_0 t + \frac{\Delta f}{2T}t^2\right) \qquad (3.77)$$

from which the segment positions are specified so as to place the segments at times for which the desired chirp waveform is real, that is, at times for which the sample phase is either 0 or π, as shown in Fig. 3.13. Therefore the sample time t_m is determined by solving the equation $\sin\phi(t) = 0$, or

$$\phi(t_m) = m\pi + C \qquad (3.78)$$

where m is the sample ($=$ segment) number and C is an unknown constant. Combining Eqs. (3.77) and (3.78), the constant C is determined to put the time t_m at the origin of the time scale, that is, $t_m = 0$. If the total number of segments is M, this can be written as

$$2\left[f_0 t_m + \frac{\Delta f}{2T}t_m^2\right] = m - \frac{M}{2} \qquad (3.79)$$

where by inspection of Fig. 3.13 it can be concluded that

$$M = 2f_0 T + 1 \qquad (3.80)$$

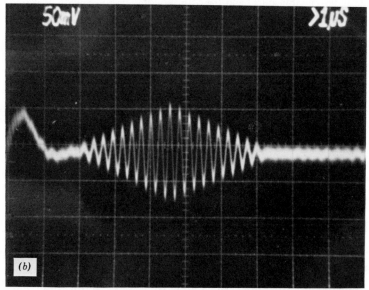

Figure 3.12. (*a*) An unapodized meander-liner transducer. (*b*) The corresponding response upon detection with an identical transducer.

Solving Eq. (3.79) for t_m and using $z_m = v_a t_m$, the segment positions are found as

$$z_m = v_a f_0 \frac{T}{\Delta f} \left\{ -1 + \left[1 - \frac{1}{f_0^2} \frac{\Delta f}{T} \left(\frac{M}{2} - m \right) \right]^{1/2} \right\} \qquad (3.81)$$

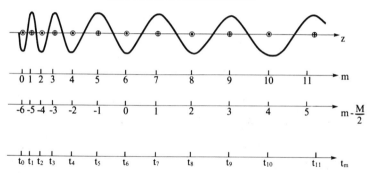

Figure 3.13. Segment positioning for sampling an arbitrary signal at time t_m. M is the total number of samples (segments in meander line). See text for more details.

Besides the linear-FM waveform, Eq. (3.77), many others are also used in various applications. For instance, the hyperbolic-FM (chype) waveform is given by[10]

$$\phi(t) = \frac{\omega_0^2}{\mu} \ln\left(1 + \frac{\mu t}{\omega_0}\right), \qquad |t| < \frac{T}{2} \tag{3.82}$$

where the logarithmic phase variations result in hyperbolic frequency modulation, and where $\mu = 2\pi\Delta f/T$. Another waveform that is sometimes used to avoid Fresnel ripple, shown in Fig. 3.7(c), is[11]

$$f(t) = \frac{1}{2\pi}\frac{d\phi(t)}{dt} = \frac{\Delta f}{2}\tan\left(\frac{\pi t}{T}\right), \qquad |t| < \frac{T}{2} \tag{3.83}$$

where Δf is the 3-dB bandwidth of the resulting FM pulse.

Linear FM filters, Eq. (3.77), should be used whenever $G \geqslant 100$, since in that case the associated Fresnel ripple is small. The ripple can severely affect the response of the filter when the latter is used as correlator. In the time domain, the ripple produces some distortions of the correlation peak and near sidelobes and also causes the far sidelobes to be relatively high. For $G \approx 10$ the ripple becomes so severe that even Hamming-type amplitude apodization cannot remedy the situation. Therefore for low G's it is desirable to choose phase apodization functions with no spectral ripple and low correlation sidelobes. One of these functions is given by Eq. (3.83). Linear FM filters are also sensitive to high Doppler shifts in frequency that arise in the process of reflection from fast-moving objects. In that case, the hyperbolic-FM waveform, Eq. (3.82), is usually employed. Another apodization function[12] known as the *S-law* provides the mainlobe-to-sidelobe ratio of about 20 dB for relatively low gains ($\approx$ 16).

Both amplitude and phase apodization obey the sampling theorem. For instance, for the case of the transducer shown in Fig. 3.5(a), the amplitude sampling is done according to

$$g(z) \sum_{m=0}^{M} \delta(z - mD) = \sum_{m=0}^{M} g(mD)\delta(z - mD) \qquad (3.84)$$

which is referred to as *uniform sampling*, since the samples are at a distance equal to an integer multiple of D. In the case of nonuniform sampling, such as encountered in linear-FM filters, the amplitude function obtained as the real part of Eq. (3.52) is sampled according to

$$\cos \phi(t) \sum_{m} \delta[\phi(t) - m\pi] \qquad (3.85)$$

where $\phi(t)$ is given by Eq. (3.77). Denoting by $\phi'(t) = d\phi(t)/dt$ and using

$$\delta[\phi(t) - m\pi] = \frac{\delta(t - t_m)}{|\phi'(t_m)|} \qquad (3.86)$$

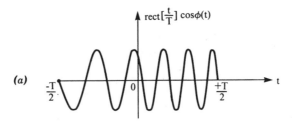

(a)

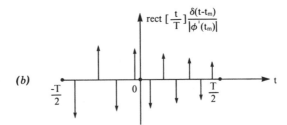

(b)

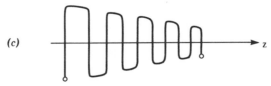

(c)

Figure 3.14. (*a*) Linear-FM waveform. (*b*) Nonlinear sampling of (*a*). (*c*) Meander-line transducer corresponding to (*b*).

where t_m are solutions of Eq. (3.78), and combining with Eq. (3.85) it follows that

$$\cos \phi(t) \sum_m \delta[\phi(t) - m\pi] = \sum_m \frac{\cos \phi(t_m) \delta(t - t_m)}{|\phi'(t_m)|} \qquad (3.87)$$

It is seen from Eq. (3.87) that the major contribution to the spectrum occurs at times with a minimal rate of phase change (oscillation), Fig. 3.14. Because $\phi(t)$ is the instantaneous phase of the RF signal, which is monotonically increasing, $\phi'(t)$ is always positive and nonzero. Therefore, in order to obtain a chirp signal with envelope of constant magnitude as given by Eq. (3.71) and shown in Fig. 3.14(a), Eq. (3.87) requires that the amplitude apodization of the device must also be implemented according to the dictate of the function $g(z_m) = 1/|\phi'(z_m)|$, $z_m = v_a t_m$, as shown in Fig. 3.14(b) with a corresponding meander-type transducer shown in Fig. 3.14(c). Since the relation given by Eq. (3.87) applies in general, it is applicable to other nonuniform sampling functions, such as those given by Eqs. (3.82) and (3.83). In practical devices the effect described will be affected by losses which increase with frequency and are predominant. Therefore the required amplitude apodization will be an increasing function of z, that is, opposite to the one shown in Fig. 3.14(c).

3.5   MAGNETOSTRICTIVE EXCITATION OF BULK ACOUSTIC WAVES

Bulk acoustic wave excitation was considered at the beginning of this section, where the Lorentz-force mechanism, Eq. (3.1), was used for generating body forces in the meander lines embedded in dielectric materials. Some of the conclusions derived are also applicable to Lorentz-force excitation of bulk acoustic waves in nonferromagnetic metals. In particular, as seen from Eq. (3.9), the particle (RF) velocity magnitude for low values of the magnetic flux density B_0 is linearly dependent on the value of B_0. However, this is not the case in ferromagnetic materials. In ferromagnetic materials, such as iron or nickel, the two dominant mechanisms available for acoustic wave excitation are the Lorentz-force mechanism, already familiar to us, and the *magnetostrictive mechanism*. A characteristic of the magnetostrictive mechanism is that, for low values of the B_0, the excited wave amplitude is a complex function of B_0 and, as shown in Fig. 3.15, usually exhibits a prominent peak. The latter is of considerable technical importance, since relatively large values of the wave magnitude can be obtained for relatively low values of the magnetic flux density B_0, thus requiring smaller biasing magnets.

Experimentally, magnetostriction manifests itself through a change in dimensions of a specimen when the external magnetic field is varied. For instance, the iron-nickel compound Invar 36 experiences a relative change of

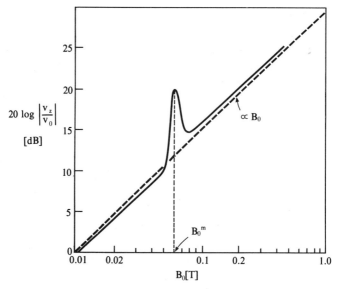

Figure 3.15. A schematic of the relative particle velocity of the excited acoustic wave versus magnetic flux density B_0 in magneto-strictive materials.

length $\Delta l/l \simeq 16 \times 10^{-6}$, if the DC magnetic field strength of 2×10^4 A/m is applied. Magnetostriction is a nonlinear effect, and the derivation of the constitutive relations, the counterparts of $T = cS$ and $B = \mu_{em}H$, is rather[†] complicated.[13] In all cases considered here it is assumed that the specimen is biased with DC field H_0 such that $H_0 \gg |H|$, where $H = |H|\exp(j\omega t)$. Under these circumstances, and under the assumption that all field quantities, u, T, S, H, H_0, and B, have components in the z direction only, it follows that

$$T = c^H S + mH_z \qquad \text{and} \qquad B_z = nS + \mu^S H_z \qquad (3.88)$$

where c^H is the elastic constant measured for $H_z = 0$, μ^S is the permeability measured for $S = 0$, m is the *inverse magnetostrictive stress constant*, and n is the *direct magnetostrictive stress constant*. All the field quantities in Eq. (3.88) are time dependent, that is, they are all proportional to $\exp(j\omega t)$. The magnetostrictive constants m and n are functions of the frequency, the magnitude of H_z, the magnetic field bias H_0, and the magnetic history of the material. The wave equation is obtained by eliminating H_z from Eq. (3.88) and substituting the resulting expressions into

$$\frac{\partial T}{\partial z} = \rho_m \frac{\partial^2 u}{\partial t^2} - F^b \qquad (3.89)$$

[†] In this chapter the permeability is denoted by μ_{em}, μ^S, or μ_0 in order to avoid confusion with the rigidity modulus μ.

where $u = u_z$. It follows that

$$T = c^H\left[1 - \frac{nm}{\mu^S c^H}\right]S + \frac{m}{\mu^S}B_z \qquad (3.90)$$

Using Eq. (3.90), the elastic constant at constant magnetic flux density ($B_z = 0$) can be defined as

$$c^B = c^H\left[1 - k_n^2\right] \qquad (3.91)$$

where $k_n^2 \equiv nm/\mu^S c^H$ is the *magnetostrictive coupling factor*. Physically a nonzero magnetostrictive coupling factor means that it is possible to excite acoustic waves in the material by applying an outside RF magnetic field. Using $S = \partial u/\partial z$, Eq. (1.33), and Eq. (3.90), the wave equation is obtained as

$$\frac{\partial^2 u}{\partial z^2} - \frac{1}{v^{B2}}\frac{\partial^2 u}{\partial t^2} = -\frac{m}{c^B\mu^S}\frac{\partial B_z}{\partial z} - \frac{F^b}{c^B} \qquad (3.92)$$

where $v^B \equiv v_a^B$ is the acoustic wave velocity at constant magnetic flux density,

$$v^B = v_a\left[1 - k_n^2\right]^{1/2} = \left[\frac{c^H}{\rho_m}\left(1 - k_n^2\right)\right]^{1/2} \qquad (3.93)$$

which in the presence of magnetostriction is smaller than the wave velocity of an acoustically equivalent material with no magnetostriction. In materials used for magnetostrictive devices, a typical value is $k_n \approx 0.45$, resulting in a reduction of velocity by about 10%.

The wave equation (3.92) can be written as

$$\frac{\partial^2 u}{\partial z^2} - \frac{1}{v^{B2}}\frac{\partial^2 u}{\partial t^2} = -\left[\frac{(m/c^H\mu^S)}{1 - k_n^2}\frac{\partial B_z}{\partial z} + \frac{F^b}{c^H\left(1 - k_n^2\right)}\right] \qquad (3.94)$$

and should be compared with Eq. (3.2), which can also be expressed as

$$\frac{\partial^2 u}{\partial z^2} - \frac{1}{v_a^2}\frac{\partial^2 u}{\partial t^2} = -\frac{F^b}{c} \qquad (3.95)$$

Thus, it is seen that the presence of the magnetostriction causes another source term to appear in Eq. (3.94). Since m, μ^S, k_n, and $F^b = J_s B_0$ are all functions of the biasing DC magnetic field H_0 (or the magnetic flux density B_0), for certain values of $B_0 = B_0^m$ the coefficient in front of $\partial B_z/\partial z$ has a local maximum (or several maxima) corresponding to the prominent peak shown in Fig. 3.15. Beyond B_0^m the value of k_n does not change very much, and the second source term, $F_b = J_s B_0$, becomes dominant. Another important point in

Eq. (3.94) is that the gradient of the magnetic flux density is the origin of sound in magnetostrictive materials.

Equation (3.94) can be solved by the methods used in the beginning of the chapter for meander-line transducers, and the relevant design quantities can be determined by following identical mathematical procedures. A practical configuration[14] for a magnetostrictive device is shown in Fig. 3.16, where input and output transducers in the form of coils are wound around a cylindrical magnetostrictive material. A time-varying current will generate an axial magnetic field and azimuthal (circumferential) electric field due to time-varying flux. The electric field will generate azimuthal currents in the rod in such a direction as to cause a magnetic field that opposes the applied magnetic field. These currents, known as *eddy currents*, will contribute to the appearance of the Lorentz-force term in Eq. (3.94). In addition to that force, a magnetostrictive force derived from Eq. (3.92) as

$$F^m = \frac{m}{\mu^S} \frac{\partial B_z}{\partial z} \tag{3.96}$$

will also cause generation of acoustic waves. By modeling the magnetic field inside the material as shown in Fig. 3.16(b), this force can be sketched as shown in the figure. For impulse-type excitation, where $B(t, z) = B(z)\delta(t)$, Eq. (3.94) is then solved for the displacement due to a single turn of the coil. It follows that each turn generates two displacement waves given by

$$u_z(t, z) = \frac{v^B(m/c^H\mu^S)}{2(1 - k_n^2)} \times \begin{cases} B_z\left[\dfrac{tv^B - z}{l}\right], & z > l \\[4mm] B_a\left[\dfrac{tv^B + z}{l}\right], & z < 0 \end{cases} \tag{3.97}$$

as shown in Fig. 3.16(c).

3.6   GENERATION OF BULK WAVES AND IMPEDANCE MATCHING

In this section, certain aspects of bulk wave generation in conducting materials using the Lorentz-force and magnetostrictive mechanisms[1] will be explored. Figure 3.17 shows two configurations for bulk wave excitation in materials of semiinfinite extent. Both configurations are contactless and use either a periodic meander line[15] and a permanent magnet or a periodically polarized magnet[16] and a uniformly wound coil. When the uniformly wound coil or the periodic meander line, which are in close proximity to the conducting material, are excited with an RF current, eddy currents will be induced in the surface of the material. Since a static magnetic field is present, Lorentz forces (and in

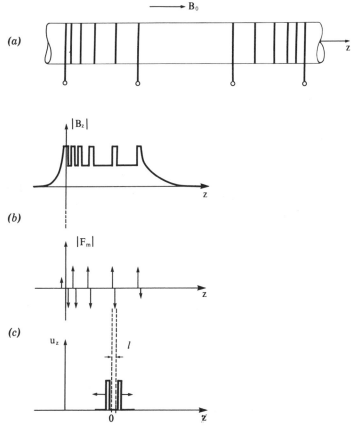

Figure 3.16. (a) Magnetostrictive dispersive line. (b) Schematics used for wave generation study. (c) Displacement waves generated by a single turn.

ferromagnetic materials the magnetostrictive force) will be exerted. These forces act as mechanical stresses on the surface of the material, causing the generation of acoustic waves at the frequency of the RF current. The structures shown in Fig. 3.17 are known as *electromagnetic-acoustic transducers* (EMATs) and are used extensively in nondestructive evaluation of materials. They can also be used as detectors of acoustic waves, since the motion of a conducting surface immersed in a static magnetic field generates an eddy current on the surface. The latter, being inductively coupled to the coil, will generate an RF current in the coil.

In general, each EMAT generates the three bulk waves and, in addition, a surface wave propagating along the surface. By optimizing the configuration of an EMAT it is possible to cause only one of the four types of waves to be strongly excited and the other three to be substantially suppressed. In the contractless configuration it is not possible to do the acoustic matching in the sense discussed

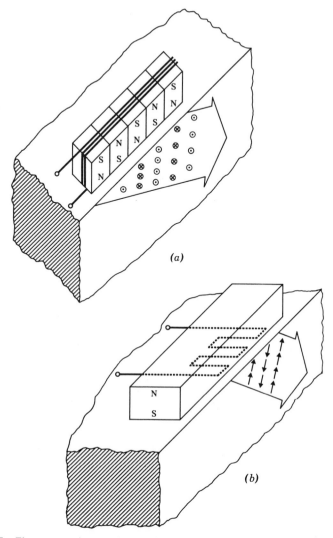

Figure 3.17. Electromagnetic-acoustic transducers (EMATs), (*a*) for the generation of shear horizontal waves and (*b*) for the generation of shear vertical waves.

in Chapter 1, and therefore EMAT operation is strongly dependent on the characteristics of the conducting substrate used. In practice,[17] the efficiency of the operation for various conducting substrates closely obeys the listing shown in Table 4 (see Sect. 3.1). The static magnetic flux density employed is usually 0.5 to 1 T.

The input impedance of an EMAT consists of a series connection of an inductance L and a resistance R, which are on the order of 1 μH and 1 Ω, respectively, Fig. 3.18. The proper impedance matching of this load to the generator resistance is of great importance for successful operation. The most widely used matching networks are shown in Fig. 3.18. A series or shunt capacitance C_2 is used to resonate the inductance L at the resonant frequency, and the resulting resistance is transformed with transformer T to match the resistance of the generator. For narrow-band matching (relative bandwidth ~ 20%), circuits (*a*) and (*b*) in Fig. 3.18 are usually employed. In the latter, the inductance L_2 and capacitance C_3 are chosen to "wrap around" the zero, the locus of the input reflection coefficient in the frequency band of interest as shown on the Smith chart in Fig. 3.18(*b*). For broadband matching the network shown in Fig. 3.18(*c*) is usually employed. Other methods leading to the realization of either Butterworth, Chebyshev, or elliptic broadband equalizers[18] can also be used.

In essence, *ferromagnetism* can be explained in terms of ordered magnetic moments (spins) of unpaired electrons associated with the magnetic ions of the

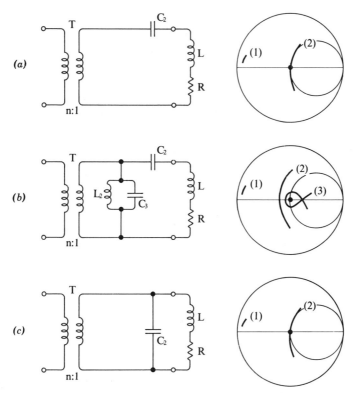

Figure 3.18. Matching networks used for matching the impedance (R, L) of EMATs. The Smith charts indicate (1) the impedance of EMAT with no matching, (2) after matching with C_2 and transformer T and, (3) with C_2, T, and the shunt network L_2, C_3.

lattice in metals such as Ni, Co, Fe, and rare earths such as gadolinium (Gd). For the sake of simplicity, it will be assumed that magnetic moments, represented by polar vectors, are localized, that is, the unpaired electrons are attached to the lattice of the material (Heisenberg model). Below a certain temperature (their Curie point), ferromagnetic elements and some of their compounds exhibit mutual attraction. In ferromagnetics the magnetic moments tend spontaneously to align parallel to each other in the same direction, this being the state of lowest energy. From a macroscopic point of view, a ferromagnetic material is considered continuous and is characterized by the nonzero magnetization (magnetic moment density) $\vec{M}$ in the constitutive relation $\vec{B} = \mu_0(\vec{H} + \vec{M})$. Single-crystal ferromagnetic materials immersed in an external magnetic field, below the saturation point, experience strain. This strain is generated by rotation of the magnetic moments in the lattice toward the external field direction. The imparted strain may change a single dimension or the volume of the crystal; this is called magnetostriction. Therefore magnetostriction is a body force. An RF external field $|H|\exp j\omega t$ will produce RF strains in the material, thus exciting acoustic waves. However, due to instrinsic nonlinearity[19] of the process, the resulting acoustic waves will be of frequency 2ω. In order to make the mechanism useful for linear devices, a bias field H_0 is employed, $|H_0| \gg |H|$. In polycrystalline ferromagnetics, consisting of ferromagnetic grains and other impurities, the magnetostriction is understood in terms of the rotation of magnetic moments of individual grains in the direction of the applied field. In polycrystalline (conducting) ferromagnetics, for the inductive coupling shown in Fig. 3.16, the excitation of acoustic waves occurs under the action of both the Lorentz force and the magnetostrictive force. Both forces are dependent on the RF field penetration depth (skin depth).

In *antiferromagnetics* the magnetic moments tend spontaneously to line up antiparallel to each other, this being the reason that the net macroscopic magnetization M is zero. However, under action of an external stress, some antiferromagnets, such as MnF_2, become weak ferromagnets ($M \neq 0$). This effect, called *piezomagnetism*, is a linear process. Piezomagnetism, because of its weak manifestation, has no practical importance.

REFERENCES

1. H. M. Frost, "Electromagnetic-Ultrasound Transducers: Principles, Practice, and Applications," in *Physical Acoustics* (W. P. Mason and R. N. Thurston, Eds.), Vol. XIV, Academic, New York, 1979, Chap. 3.

2. E. Burstein, H. Talaat, J. Schoenwald, J. J. Quin, and E. G. H. Lean, "Direct EM Generation and Detection of Surface Elastic Waves on Conducting Solids. I. Theoretical Considerations," *1973 Ultrason. Symp. Proc.*, pp. 564–568, 1973.

3. A. Barone and A. Giacomini, *Acustica*, **4**, 182 (1954).

4. For an elementary approach, see, for instance, C. A. Coulson and A. Jeffrey, *Waves*, 2nd ed., Longmans, New York, 1977, Chap. 1; for a more rigorous approach, see P. M. Morse and H. Feshbach, *Methods of Theoretical Physics*, Vol. 2, McGraw-Hill, New York, 1953.

5. The transducer can be regarded as an exponential coupler; see, for instance, W. H. Louisell, *Coupled Mode and Parametric Electronics*, Wiley, New York, 1960.

6. See, for instance, H. J. Blinchikoff and A. I. Zverev, *Filtering in the Time and the Frequency Domains*, Wiley, New York, 1976, Chap. 2.

7. C. E. Cook and M. Bernfeld, *Radar Signals*, Academic, New York, 1967, Chaps. 1 and 7.

8. J. R. Klauder, A. C. Price, S. Darlington, and W. J. Albersheim, "The Theory and Design of Chirp Radars," *BSTJ*, **39** (4), 745–808 (1960).

9. J. J. G. McCue, "A Note on the Hamming Weighting of Linear-FM Pulses," *Proc. IEEE (Corr.)*, **67**(11), 1575–1577 (1979).

10. C. Atzeni and L. Masotti, "Acoustic Surface Wave Processor for Hyperbolic F.M. Signals," *Elect. Lett.*, **7**(23), 693–694 (1971).

11. J. C. Worley, "Implementation of Nonlinear FM Pulse Compression Filters Using Surface Wave Delay Lines," *Proc. IEEE (Corr.)*, **59**(11), 1618–1619 (1971).

12. F. Hauser, G. R. Dubois, H. Licht, and V. M. Ristic, "A New Nonlinear FM Coded EMAT for Nondestructive Testing of Materials," *1981 Ultrason. Symp. Proc.*, 989–991 (1981).

13. R. B. Thompson, "A Model for the Electromagnetic Generation of Ultrasonic Guided Waves in Ferromagnetic Metal Polycrystals," *IEEE Trans. Sonics Ultrason.*, **SU-25**(1), 7–15 (1978).

14. S. Davidson, "Wire and Strip Delay Lines," *Ultrasonics*, July–September, 136–146 (1965).

15. R. B. Thompson, "Noncontact Transducers," *1977 Ultrason. Symp. Proc.*, 74–82 (1977).

16. C. F. Vasile and R. B. Thompson, "Periodic Magnet Non-Contact Electromagnetic Acoustic Wave Transducer—Theory and Applications," *1977 Ultrason. Symp. Proc.*, 84–88 (1977).

17. V. M. Ristic, F. Hauser, G. Dubois, and H. Licht, "Nonlinear FM Coded SAW EMAT's for Nondestructive Testing of Materials," *1980 Ultrason. Symp. Proc.*, **2**, 898–901 (1980).

18. W. K. Chen, *"Theory and Design of Broadband Matching Networks,"* Pergamon, New York, 1976, Chap. 4.

19. L. E. Kinsler and A. R. Frey, *Fundamentals of Acoustics*, 2nd ed., Wiley, New York, 1962, Chap. 12.

EXERCISES

1. Find higher-order Fourier harmonics of $u(t, z)$ using Eq. (3.28). See whether the existence of the higher harmonics will restrict the bandwidth of a transducer with $M = 1$. Comment on the validity of the theoretical approach used to compute the bandwidth (a) for $M = 1$ and (b) for $M = 10$.

2. Find the transfer function and the impulse response due to M sources, using Eq. (3.8) for the particle velocity. Comment how the width of the function g will affect the computed quantities.

3. The impulse response of an amplitude-apodized device is given by Eq. (3.39). (a) Find an expression for the impulse response of a phase-apodized device.
   (b) A device of length η is apodized in both phase and amplitude. Consider the impulse response at $d = 0$ and find a justification for the expression given by Eq. (3.45).

4. Find the segment positions for a chirp (linear-FM) device of bandwidth $\Delta f = 4$ MHz and duration $T = 3$ μs operating at the center frequency of $f_0 = 3$ MHz in the material with acoustic velocity $v_a = 3$ km/s.

5. Change the direction of the B_0 field in Fig. 3.1(a) in order to excite y-polarized shear waves. Repeat the derivation given by Eqs. (3.2) to (3.14) for this case.

6. Figure 3.1(a) shows a structure for excitation of acoustic waves in nonpiezoelectric dielectric materials. In practice, the excitation of acoustic waves in metals is also of interest (Table 4). How would you modify the structure shown in Fig. 3.1(a) in order to excite the acoustic waves in metals?

7. For $R = 3.5 \ \Omega$ and $L = 0.7 \ \mu H$, find C_2 in Fig. 3.18(a) and (c) so that the resonance occurs at $f_0 = 3$ MHz.

   (a) Find the transformer turns ratio in both cases if the load is to be matched at 50 Ω at the resonant frequency.

   (b) Find the input reflection coefficient r of both the circuit and the power transfer function defined as $G(\omega) = 1 - |r|^2$.

   (c) Plot $G(\omega)$ and find the 3-dB bandwidth of both circuits defined by $G(\omega \pm) = \frac{1}{2}$.

8. Consider the circuit in Fig. 3.18(b). Use $R = 3.5 \ \Omega$, $L = 0.7 \ \mu H$, and adjust C_2 to resonate with L at $f_0 = 3$ MHz.

   (a) Disregard for the moment L_2 and C_3, and adjust the transformer turns ratio so that the load is matched to 40 Ω at the resonant frequency.

   (b) Find the reflection coefficient r of the circuit, taking the generator resistance of 50 Ω.

   (c) By requiring that the imaginary part of the reflection coefficient is zero, obtain a cubic equation in ω^2.

   (d) Using the well-known relations valid for cubic equations, impose the conditions that all three roots are real. Choose one root to be $\omega_1^2 \approx \omega_0^2$, $\omega_0 = 2\pi f_0$, and the second to be $\omega_2^2 \approx (0.8\omega_0)^2$. Determine the values of L_2 and C_3 and the third root.

   (e) Plot the power transfer function $G(\omega) = 1 - |r|^2$ versus frequency, and determine the 3-dB bandwidth.

Chapter 4

Surface Acoustic Waves and Electrodynamic Transducers

In 1885, J. Strutt (Lord Rayleigh) made the theoretical prediction[1] that acoustic waves can propagate over a plane boundary between a semiinfinite solid and a vacuum or a sufficiently rarefied medium (for instance, air), as shown in Fig. 4.1(a), where the amplitude of the waves decays rapidly in the $-y$ direction. Since then these waves have been found to exist in many other circumstances[2-6] and are now called *Rayleigh waves*, Fig. 4.1(b). Rayleigh waves are of considerable technical importance. Their applications range from detection of surface cracks in nondestructive testing,[2] to very sophisticated signal processing in modern communications,[4,5] to spectral analysis of light signals in integrated optics.[7,8]

Rayleigh waves or *surface acoustic waves* (SAW) consist of combined longitudinal motion, Fig. 4.2(a), and shear motion, Fig. 4.2(b), which are coupled by the boundary surface, Fig. 4.2(c). In terms of the geometry shown in Fig. 4.1(a), where a plane SAW is assumed to propagate in the $+z$ direction with its amplitude decaying in the $-y$ direction, the field quantities will be independent of the x coordinate. The potentials defined by Eq. (2.77) are dependent on y and z coordinates only, $\phi = \phi(y, z)\exp(j\omega t)$, and the vector potential $\vec{\psi}$, which has only an x component, can be expressed as $\vec{\psi} = \hat{x}\psi = \hat{x}\psi(y, z)\exp(j\omega t)$.

In that case, Eqs. (2.75) and (2.76) can be written as[†]

$$\frac{\partial^2 \phi}{\partial y^2} + \frac{\partial^2 \phi}{\partial z^2} + k_L^2 \phi = 0 \quad \text{and} \quad \frac{\partial^2 \psi}{\partial y^2} + \frac{\partial^2 \psi}{\partial z^2} + k_S^2 \psi = 0 \qquad (4.1)$$

where k_L and k_S are the wave numbers of the longitudinal and transverse

---

[†] The existence of a Rayleigh wave is derived here by means of the so-called direct method. Other methods can also be used. See, for instance, L. M. Brekhovskikh, *Waves in Layered Media*, Academic Press, New York, 1960, Chap. 4.

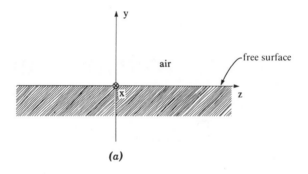

(a)

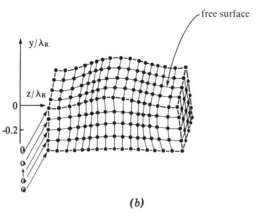

(b)

Figure 4.1. (a) Cross section through semiinfinite solid on which SAW propagates. (b) Grid (deformation) diagram of a material supporting SAW propagation in the z direction. The polarization of the wave is retrograde down to $y/\lambda_R \simeq -0.2$ and then becomes reversed.

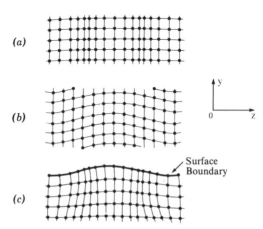

Figure 4.2. Comparison of (a) longitudinal (bulk) wave, (b) Y-polarized shear (bulk) wave, and (c) Rayleigh (SAW) wave, all propagating in the z direction. The polarization of the SAW wave is in the $y0z$ plane.

modes, respectively, given by

$$k_L = \frac{\omega}{v_L} \quad \text{and} \quad k_S = \frac{\omega}{v_S} \tag{4.2}$$

The components of the displacement are obtained from Eq. (2.77) and are given by

$$u_y = \frac{\partial \phi}{\partial y} + \frac{\partial \psi}{\partial z} \quad \text{and} \quad u_z = \frac{\partial \phi}{\partial z} - \frac{\partial \psi}{\partial y} \tag{4.3}$$

Using Eqs. (2.8), (2.17), and (4.3), Hooke's law, Eqs. (2.46) to (2.49), can now be written as

$$T_{xx} = T_1 = \lambda \left(\frac{\partial^2 \phi}{\partial y^2} + \frac{\partial^2 \phi}{\partial z^2} \right) \tag{4.4}$$

$$T_{yy} = T_2 = \lambda \left(\frac{\partial^2 \phi}{\partial y^2} + \frac{\partial^2 \phi}{\partial z^2} \right) + 2\mu \left(\frac{\partial^2 \phi}{\partial y^2} + \frac{\partial^2 \psi}{\partial y \, \partial z} \right) \tag{4.5}$$

$$T_{zz} = T_3 = \lambda \left(\frac{\partial^2 \phi}{\partial y^2} + \frac{\partial^2 \phi}{\partial z^2} \right) + 2\mu \left(\frac{\partial^2 \phi}{\partial z^2} - \frac{\partial^2 \psi}{\partial y \, \partial z} \right) \tag{4.6}$$

and

$$T_{yz} = T_4 = \lambda \left(2 \frac{\partial^2 \phi}{\partial y \, \partial z} + \frac{\partial^2 \psi}{\partial z^2} - \frac{\partial^2 \psi}{\partial y^2} \right) \tag{4.7}$$

The stress T_{xx}, not containing the shear wave potential ψ, is completely decoupled from SAW fields and need not to be considered further. For harmonic solution of a wave propagating in the $+z$ direction, the potentials can be specified as

$$\phi = F(y)\exp(-j\beta z) \quad \text{and} \quad \psi = G(y)\exp(-j\beta z) \tag{4.8}$$

where β is the propagation constant (as yet, unknown) of SAW waves. Combining Eqs. (4.1) and (4.8), it follows that

$$\frac{d^2 F}{dy^2} - \left(\beta^2 - k_L^2 \right) F(y) = 0 \quad \text{and} \quad \frac{d^2 G}{dy^2} - \left(\beta^2 - k_S^2 \right) G(y) = 0 \tag{4.9}$$

Denoting by

$$\gamma_L^2 = \beta^2 - k_L^2 \quad \text{and} \quad \gamma_S^2 = \beta - k_S^2 \tag{4.10}$$

the solutions are obtained as

$$\phi = A_1 e^{j(\omega t - \beta z)} e^{\gamma_L y} \qquad (4.11)$$

and

$$\psi = B_1 e^{j(\omega t - \beta z)} e^{\gamma_S y} \qquad (4.12)$$

for $y < 0$ under assumptions of $\gamma_L^2 > 0$ and $\gamma_S^2 > 0$. It is seen that both potentials decay with depth; that is, the SAW energy is concentrated at the solid–air interface. For the boundary at $y = 0$, free from external stresses, the normal components of stresses must be zero, namely,

$$T_{yy} = 0 \qquad (4.13)$$

and

$$T_{yz} = 0 \qquad (4.14)$$

On combining Eqs. (4.7), (4.11), (4.12), and (4.14), we obtain

$$\phi = A_1 e^{j(\omega t - \beta z)} e^{\gamma_L y} \qquad (4.15)$$

and

$$\psi = -jA_1 \frac{2\beta\gamma_L}{\beta^2 + \gamma_S^2} e^{j(\omega t - \beta z)} e^{\gamma_S y} \qquad (4.16)$$

The dispersion (characteristic) equation is obtained by combining the other boundary condition, Eq. (4.13), with Eqs. (4.5), (4.11), and (4.12) and is given by

$$\left(\beta^2 + \gamma_S^2 \right)^2 - 4\beta^2 \gamma_L \gamma_S = 0 \qquad (4.17)$$

Denoting the SAW velocity by $v_R = \omega/\beta$, and using Eqs. (4.10) and (4.17), it follows that

$$\left(\frac{v_R}{v_S} \right)^6 - 8\left(\frac{v_R}{v_S} \right)^4 + 8\left[3 - 2\left(\frac{v_S}{v_L} \right)^2 \right]\left(\frac{v_R}{v_S} \right)^2 - 16\left[1 - \left(\frac{v_S}{v_L} \right)^2 \right] = 0 \quad (4.18)$$

Since the ratio v_S/v_L can be expressed in terms of σ, Eq. (2.72), with $0 < \sigma < 0.5$, analysis shows that only one root of Eq. (4.18) is relevant. This (real) root is given approximately by[2]

$$\frac{v_R}{v_S} = \frac{k_S}{\beta} \approx \frac{0.87 + 1.12\sigma}{1 + \sigma} \tag{4.19}$$

or

$$v_R \approx \left(\frac{c_{11} - c_{12}}{2\rho_m}\right)^{1/2} \frac{0.87c_{11} + 2c_{12}}{c_{11} + 2c_{12}} \tag{4.20}$$

In the more general case of an anisotropic substrate, the value of v_R is usually found by numerical techniques.[9] Using Eq. (4.19) it can be shown that when σ varies from 0 to 0.5, the SAW velocity varies from $0.87v_S$ to $0.96v_S$. Since it was already established that shear wave velocity is always less than longitudinal wave velocity, it follows that the SAW velocity is always less than the longitudinal wave velocity. Note that this was implicit in Eq. (4.10), which led to the proper choice of solutions for potentials.

The plane containing the displacements u_y and u_z is called the *sagittal plane*. Sometimes on anisotropic materials, in addition to displacement components u_y and u_z, there may be a displacement component u_x perpendicular to the sagittal plane. These waves are termed *generalized Rayleigh SAW waves*.[5] Also on anisotropic substrates the pure SAW modes (containing components in the sagittal plane only) are usually found in the planes of crystal symmetry. Many other types of surface waves can exist at the boundary between two materials or in a layered solid bonded to material of semiinfinite extent.[6] For certain crystal cuts or at the interface between two materials, it may happen that $\gamma_S^2 < 0$ in Eq. (4.10), that is, that the SAW velocity is larger than the shear wave velocity.[10] Those waves are termed *pseudo-surface waves* or *leaky waves*,[5] because as they propagate they leak their energy continuously toward the interior of the solid.

Using Eq. (4.3), (4.15), and (4.16), the displacements are found as

$$u_y = A_1 \gamma_L \left[e^{\gamma_L y} - \frac{2\beta^2}{\beta^2 + \gamma_S^2} e^{\gamma_S y} \right] e^{j(\omega t - \beta z)} \tag{4.21}$$

and

$$u_z = -jA_1 \beta \left[e^{\gamma_L y} - \frac{2\gamma_L \gamma_S}{\beta^2 + \gamma_S^2} e^{\gamma_S y} \right] e^{j(\omega t - \beta z)} \tag{4.22}$$

From Eq. (4.17) it is seen that

$$\frac{2\gamma_L \gamma_S}{\beta^2 + \gamma_S^2} = \frac{\beta^2 + \gamma_S^2}{2\beta^2} = a \tag{4.23}$$

which is for convenience denoted by a, and on introducing a new constant

$A = -j\beta A_1$ it follows that

$$u_y = jA\left[1 - \left(\frac{v_R}{v_L}\right)^2\right]^{1/2}\left[e^{\gamma_L y} - \frac{1}{a}e^{\gamma_S y}\right]e^{j(\omega t - \beta z)}$$

and

$$u_z = A\left[e^{\gamma_L y} - ae^{\gamma_S y}\right]e^{j(\omega t - \beta z)} \qquad (4.24)$$

The physically meaningful quantities are

$$\text{Re}[u_y] = U_y(y)\sin(\omega t - \beta z) \quad \text{and} \quad \text{Re}[u_z] = U_z(y)\cos(\omega t - \beta z)$$

$$(4.25)$$

where

$$U_y(y) = -A\left[1 - \left(\frac{v_R}{v_L}\right)^2\right]^{1/2}\left[e^{\gamma_L y} - \frac{e^{\gamma_S y}}{a}\right]$$

and

$$U_z(y) = A\left[e^{\gamma_L y} - ae^{\gamma_S y}\right] \qquad (4.26)$$

From Eqs. (4.25) and (4.26) it is readily shown that, up to a depth

$$y_0 = \frac{\ln a}{\gamma_S - \gamma_L} \approx -(0.15 \text{ to } 0.25)\lambda_R$$

where λ_R is the SAW wavelength, the particle motion in the sagittal plane is in an ellipse with rotation in the retrograde direction. At the depth y_0, as shown in Fig. 4.3(a), $U_z = 0$ and polarization is linear, while for $y < y_0$ the polarization is again elliptical, but rotation is in the reverse direction. This is illustrated in Fig. 4.1(b).

In studies of SAW generation mechanisms, such as Lorentz force, the displacement ratio at $y = 0$ is often of interest. Denoting the ratio by b, it is easy to show from Eqs. (4.21) and (4.22) that

$$b = \frac{U_z(0)}{U_y(0)} = \frac{2D - 1 - 2[(D - C)(D - 1)]^{1/2}}{[1 - C/D]^{1/2}} \qquad (4.27)$$

where

$$C = \frac{k_L^2}{k_S^2} = \frac{0.5 - \sigma}{1 - \sigma}$$

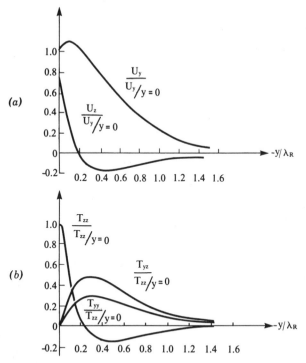

Figure 4.3. Typical fields in a SAW on an isotropic material. (*a*) Normalized displacements. (*b*) Corresponding normalized stresses.

and

$$D = \frac{\beta^2}{k_S^2} = \left[\frac{1 + \sigma}{0.87 + 1.12\sigma} \right]^2$$

The displacement ratio varies from 0.76 for beryllium ($\sigma = 0.05$) to 0.56 for polyethylene ($\sigma = 0.46$). For most metals, $\sigma = 0.25$ to 0.34 with corresponding values of displacement ratios of 0.7 to 0.6. Using Eqs. (4.5) to (4.7), the expressions for stresses in the Rayleigh wave are easily evaluated. Their general behavior and the corresponding displacements are shown in Fig. 4.3. Typical SAW velocities of certain materials are given in Table 5. The magnitude of displacement is small.[6] For instance, in a device operating at 100 MHz with 10 mW average power in a beam 1 cm wide on a substrate with SAW velocity $v_R = 3$ km/s, the wavelength is $\lambda_R = 30$ μm, with the peak vertical displacements on the order of 10^{-10} m.

For materials with a very "smooth" surface and whose attenuation constants for bulk waves vary linearly with frequency (for instance, Lucite), it was shown[11, 12] that the attenuation constant of the surface waves α_R can be related to the

TABLE 5   SAW (Rayleigh) Velocities for Some Isotropic Materials[a]

Material	v_R, m/s
Beryllium	7830
Ceramics (silicon nitride)	5800
Chromium	3678
Fused silica	3411
Glass (soda-lime)	3200
Nickel	2990
Aluminum	2905
Steel 4340	2890
Stainless steel 316	2780
Magnesium AZ31B	2730
Copper	2235
Zircaloy 2 (Zr–2Nb)	2190
Zircaloy 2.5 (Zr–2.5Nb)	2100
Silver	1658
Lead	710

[a] The listed velocities should be considered as average. Metals are assumed to be polycrystalline.

attenuation constants of the bulk waves, α_S and α_L, by the relation $\alpha_R = A_S \alpha_S + A_L \alpha_L$, where A_S and A_L are constants dependent on Poisson's ratio. The attenuation of Rayleigh waves propagating at an interface between a material exhibiting viscous losses, Eq. (1.25), and vacuum is given by $\alpha_R = Af^2$, where A is a constant. If instead of vacuum the adjacent medium is a gas (for instance, air), then $\alpha_R = Af^2 + Bf$, where the linear frequency term is due to gas loading. It has been found[13] that

$$B\left[\frac{\text{Np}}{\text{Hz} \cdot \text{m}}\right] = \frac{1}{\rho_m v_R^2}\left[p\left(\frac{\gamma M}{RT}\right)^{1/2}\right]$$

where ρ_m is the mass density of the substrate. The bracketed term contains parameters of the gas loading the substrate. These are the gas pressure p, the molecular weight M, the ratio of the specific heat at constant pressure to the specific heat at constant volume γ, the absolute temperature of the gas T, and the universal gas constant R. For instance, for Rayleigh waves propagating on the plane defined by the crystal axis Y, along the direction defined by the crystal axis Z in a $LiNbO_3$ crystal,

$$\alpha\left[\frac{\text{Np}}{\text{m}}\right] = 29f^2 + 6.27f$$

where f is in GHz.

4.1 ENERGY CONSERVATION THEOREM FOR ACOUSTIC MATERIALS

The one-dimensional concept of the energy conservation theorem for longitudinal waves has been introduced in Sect. 1.5. In what follows, the concept is generalized to plane acoustic waves propagating in acoustically isotropic materials.[14]

For waves with arbitrary time variations, Eq. (1.40) can now be written as

$$\iint_A \vec{\Pi} \cdot d\vec{A} + \frac{\partial U}{\partial t} + P_d = P_s \tag{4.28}$$

where the acoustic Poynting vector is given by

$$
\begin{aligned}
\vec{\Pi} &= \hat{x}\Pi_x + \hat{y}\Pi_y + \hat{z}\Pi_z \\
&= -\vec{v} \cdot \vec{\vec{T}} \\
&= -\hat{x}\left(v_x T_{xx} + v_y T_{yx} + v_z T_{zx} \right) - \hat{y}\left(v_x T_{xy} + v_y T_{yy} + v_z T_{zy} \right) \\
&\quad -\hat{z}\left(v_x T_{xz} + v_y T_{yz} + v_z T_{zz} \right)
\end{aligned}
\tag{4.29}
$$

describing the power density flow along the three coordinate directions. The integrand in Eq. (4.28) given by

$$\vec{\Pi} \cdot d\vec{A} \tag{4.30}$$

is the power flow through the surface $d\vec{A} = \vec{a}\, dA$, where $\vec{a}$ is the unit vector along the normal to the surface dA. The total stored elastic and kinetic energy U is given by

$$U = \iiint_V (u_s + u_v)\, dV \tag{4.31}$$

where the elastic energy density is given by[14]

$$u_s = \frac{\vec{\vec{S}} : \overset{\leftrightarrow}{c} : \vec{\vec{S}}}{2} = \tfrac{1}{2} S_I c_{IJ} S_J = \tfrac{1}{2} S_{ij} c_{ijkl} S_{kl} \tag{4.32}$$

and the kinetic energy density by

$$u_v = \tfrac{1}{2} \rho_m v^2 \tag{4.33}$$

Furthermore, the total viscous power loss is given by

$$P_d = \iiint_V \frac{\partial \vec{\vec{S}}}{\partial t} : \overset{\leftrightarrow}{\eta} : \frac{\partial \vec{\vec{S}}}{\partial t}\, dV \tag{4.34}$$

and the power supplied to the volume V by the sources is

$$P_s = \iiint_V \vec{F}^b \cdot \vec{v} \, dV \tag{4.35}$$

The integrand in Eq. (4.34) can be developed in either full tensor or abbreviated notation in the manner shown in Eq. (4.32). The double scalar (dot) product of two tensors reduces the rank of the product by two, as shown in Eq. (4.32). The instantaneous radiated acoustic power is now given by

$$P_a = \iint_A \vec{\Pi} \cdot d\vec{A} \tag{4.36}$$

where, in general, the integral need not be evaluated over a closed surface.

The complex Poynting theorem,[14] Eq. (1.43), applicable to time-harmonic excitation of the form exp $j\omega t$ is now generalized to

$$\iint_A \vec{\Pi} \cdot d\vec{A} - j\omega[U_s - U_v] + \langle P_d \rangle = P_s \tag{4.37}$$

where

$$\vec{\Pi} = -\frac{\vec{v}^* \cdot \overleftrightarrow{T}}{2} \tag{4.38}$$

is the complex Poynting vector, which can be developed in component form as shown by Eq. (4.29),

$$U_s = \iiint_V \frac{\overleftrightarrow{S} : \overleftrightarrow{c} : \overleftrightarrow{S}^*}{2} \, dV \tag{4.39}$$

is the peak elastic energy in the system,

$$U_v = \iiint_V \rho_m \frac{\vec{v} \cdot \vec{v}^*}{2} \, dV \tag{4.40}$$

is the peak kinetic energy stored,

$$\langle P_d \rangle = \omega^2 \iiint_V \frac{\overleftrightarrow{S} : \overleftrightarrow{\eta} : \overleftrightarrow{S}^*}{2} \, dV \tag{4.41}$$

is the time-averaged power loss due to viscous damping, and

$$P_s = \iiint_V \frac{\vec{v}^* \cdot \vec{F}^b}{2} \, dV \tag{4.42}$$

is the complex power supplied by sources. The average acoustic power carried by a wave through surface A is given by

$$\langle P \rangle = P_a = \mathrm{Re}\left\{ \iint\limits_{A} \vec{\Pi} \cdot d\vec{A} \right\} \tag{4.43}$$

where $\vec{\Pi}$ is given by Eq. (4.38). As in Sect. 1.5, it can be shown that peak elastic and kinetic energies are equal, $U_s = U_v$, in a plane wave propagating in a lossless medium, $\langle P_d \rangle = 0$. Under these circumstances for propagation far from the sources, $P_s = 0$ in Eq. (4.37), the energy of the wave is conserved, that is,

$$\iint\limits_{A} \vec{\Pi} \cdot d\vec{A} = 0 \tag{4.44}$$

4.2   THE ACOUSTIC IMPEDANCE CONCEPT

The concept of acoustic impedance[14] can be linked directly to the Poynting vector. The translational equation of motion, Eq. (2.20), without body forces is given by

$$\nabla \cdot \overleftrightarrow{T} = \rho_m \frac{\partial^2 \vec{u}}{\partial t^2} = \rho_m \frac{\partial \vec{v}}{\partial t} \tag{4.45}$$

For a plane wave propagating in the $\hat{\beta}$ direction, with propagation constant β, the wave fields will vary as

$$\exp j(\omega t - \vec{\beta} \cdot \vec{r}) \tag{4.46}$$

where

$$\vec{\beta} = \beta\hat{\beta} = \beta_x \hat{x} + \beta_y \hat{y} + \beta_z \hat{z} \tag{4.47}$$

and

$$\vec{r} = x\hat{x} + y\hat{y} + z\hat{z} \tag{4.48}$$

Under these circumstances the gradient operator becomes $\nabla = -j\vec{\beta}$, and Eq. (4.45) can be written as

$$-\vec{\beta} \cdot \overleftrightarrow{T} = \rho_m \omega \vec{v} \tag{4.49}$$

Since $\vec{\beta} \cdot \overleftrightarrow{T}$ is proportional to the force density acting on a surface element with the normal in the $\hat{\beta}$ direction, it follows that

$$\vec{T}_\beta = \hat{\beta} \cdot \overset{\leftrightarrow}{T} = -\frac{\rho_m \omega}{\beta} \vec{v} \qquad (4.50)$$

which can be written as

$$-\vec{T}_\beta = Z_a \vec{v} \qquad (4.51)$$

Using Eq. (4.38) for the complex Poynting vector, the power flow density in the $\hat{\beta}$ direction can be written as

$$\vec{\Pi} \cdot \hat{\beta} = -\left(\frac{\vec{v}^* \cdot \overset{\leftrightarrow}{T}}{2}\right) \cdot \hat{\beta} = -\frac{\vec{v}^* \cdot \vec{T}_\beta}{2} \qquad (4.52)$$

or, using Eq. (4.51), as

$$\vec{\Pi} \cdot \hat{\beta} = \frac{Z_a |v|^2}{2} = \langle \Pi_\beta \rangle \qquad (4.53)$$

giving the average power density in the $\hat{\beta}$ direction. The last equation is of considerable practical importance, since it enables us to calculate the RF (particle) velocity from average power density.

For instance, a power density of 1 W/cm² in tungsten ($Z_a = 105 \times 10^6$ in SI units) will produce a magnitude of particle velocity of 1.4 cm/s for longitudinal waves. In paraffin ($Z_a = 1.22 \times 10^6$ in SI units), the corresponding value is 13 cm/s. Since in a given material the shear wave velocity is always smaller than the longitudinal wave velocity, the shear wave acoustic impedance is also smaller than the longitudinal wave impedance. Thus for equal power densities in a shear wave and a longitudinal wave, the particle velocity for the shear wave will be larger. Note that RF particle acceleration, equal to $\omega^2 u$, is very large at high frequencies. For a device operating at 100 MHz with 10 mW average power in a beam 1 cm wide on a substrate with $v_R = 3$ km/s, the peak vertical displacement is on the order of 10^{-10} m and the particle acceleration on the order of 10^7 m/s².

In Sect. 2.3 it was shown that the concept of acoustic impedance introduced in Section 1.1 for longitudinal waves is also applicable to shear waves. For a y-polarized z-propagating shear wave, given by Eqs. (2.83) to (2.90), the transmission-line equation (1.22) can now be written as

$$\frac{\partial(-T_{yz})}{\partial z} = -j\omega \rho_m v_y \qquad (4.54)$$

$$\frac{\partial v_y}{\partial z} = -j\frac{\omega}{c_{44}}(-T_{yz}) \qquad (4.55)$$

with

$$Z_a^S = -\frac{T_{yz}}{v_y} = \frac{\rho_m \omega}{\beta} = \rho_m v_S = \sqrt{\rho_m c_{44}} = \frac{\mu}{v_S} \qquad (4.56)$$

The impedance-matching concepts, introduced in Sect. 1.3 for longitudinal waves, and the normal mode expansion (Exercise 6, Chap. 2) are also applicable to shear waves.

4.3   POYNTING VECTOR AND IMPEDANCE OF RAYLEIGH WAVES

For SAW studied at the beginning of this chapter, with the geometry shown in Fig. 4.1(a), the complex Poynting vector is given by

$$\vec{\Pi} = \hat{y}\Pi_y + \hat{z}\Pi_z \qquad (4.57)$$

with

$$\Pi_y = -\tfrac{1}{2}\left(v_y^* T_{yy} + v_z^* T_{zy}\right) \qquad (4.58)$$

and

$$\Pi_z = -\tfrac{1}{2}\left(v_y^* T_{yz} + v_z^* T_{zz}\right) \qquad (4.59)$$

where the values of the velocity and stress components can be computed from Eqs. (4.24) and (4.5) to (4.7), respectively. It can be shown that the power flow density Π_y is a pure imaginary number, that is, there is no real power flow in the $-y$ direction in a lossless material. The z component of the power flow density can be written as

$$\Pi_z(y) = \frac{\beta(\lambda + 2\mu)|v_z|^2}{2\omega} + \frac{\beta\mu|v_y|^2}{2\omega} + \frac{j}{2\omega}\left[\lambda v_z^*\frac{\partial v_y}{\partial y} + \mu v_y^*\frac{\partial v_z}{\partial y}\right]$$

$$(4.60)$$

or

$$\Pi_z \equiv \Pi_L + \Pi_S + \Pi_{LS} \qquad (4.61)$$

The first two terms are the longitudinal and shear powers, respectively, whereas the last term is the "cross-coupling" term containing the products of the longitudinal and shear RF velocities. Since in an acoustic medium of infinite extent the power of any type of acoustic wave can be expressed in

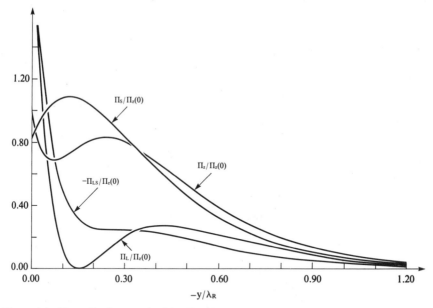

Figure 4.4. Normalized power densities $\Pi_L/\Pi_z(0)$, $\Pi_S/\Pi_z(0)$, and $\Pi_{LS}/\Pi_z(0)$ of longitudinal, shear, and cross-coupling terms in Eq. (4.61) for lead ($\sigma = 0.43$).

terms of two sums, corresponding to longitudinal and shear powers, the appearance of the cross-coupling term in Eq. (4.60) must be ascribed to the boundary condition requirements for the existence of the Rayleigh wave.

Figures 4.4 and 4.5 show the distribution of the power density Π_z as functions of the normalized depth y/λ_R for lead ($\sigma \simeq 0.43$) and beryllium ($\sigma \simeq .048$), respectively. The power densities due to longitudinal Π_L, shear Π_S, and cross-coupling Π_{LS} are also shown.

In practice, the best measure of the wave amplitude is the acoustic Poynting vector, and it is advantageous to express the SAW power flow density in the form given by Eq. (4.53). Furthermore, in many SAW excitation problems it is convenient to express various quantities in terms of normal modes.[†] In what follows the normal modes of a SAW are derived. The first two terms in Eq. (4.60) are easily amenable to that form; however, the cross-coupling term must be decomposed in some way. There is no unique procedure for achieving this. One of the possible decompositions is given by

$$\Pi_z(y) \equiv \Pi_{Lt}(y) + \Pi_{St}(y) \tag{4.62}$$

$$\Pi_{Lt}(y) = (\lambda + 2\mu)\left\{ \frac{\beta}{2\omega}|v_z|^2 + \frac{j}{2\omega}v_z^* \frac{\partial v_y}{\partial y} \right\}$$

$$\Pi_{St}(y) = \mu\left\{ \frac{\beta}{2\omega}|v_y|^2 + \frac{j}{2\omega}\left[v_y^* \frac{\partial v_z}{\partial y} - 2v_z^* \frac{\partial v_y}{\partial y} \right] \right\} \tag{4.63}$$

[†]See Exercise 6, Chapter 2.

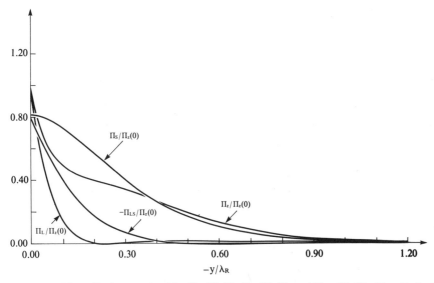

Figure 4.5. Normalized power densities $\Pi_L/\Pi_z(0)$, $\Pi_S/\Pi_z(0)$, and $\Pi_{LS}/\Pi_z(0)$ of longitudinal, shear, and cross-coupling terms in Eq. (4.61) for beryllium ($\sigma = 0.048$).

where Π_{Lt} and Π_{St} are the total power densities due to quasi-longitudinal and quasi-shear waves, respectively. It is seen that Π_{St} is directly proportional to μ, the rigidity modulus that determines the propagation of shear waves only, Eq. (4.56), while Π_{Lt} is directly proportional to $(\lambda + 2\mu)$, which determines the propagation of longitudinal waves, Eq. (2.81). The waves so defined are not pure modes, since they contain both components of the particle velocity. Figures 4.6 and 4.7 show the power densities in quasi-longitudinal and quasi-shear waves for lead and beryllium.

The total acoustic power P_a of a SAW beam of width w measured in the x direction is obtained by integration of Eq. (4.60) over the region $0 \leqslant x \leqslant w$, $-\infty < y \leqslant 0$, and is given by

$$P_a = \int_0^w \int_{-\infty}^0 \Pi_z(y) \, dx \, dy \tag{4.64}$$

or

$$P_a = \lambda_R w \left[\frac{\beta(\lambda + 2\mu)}{2\omega} \bar{v}_z^2 + \frac{\beta\mu}{2\omega} \bar{v}_y^2 \right] \tag{4.65}$$

where $\bar{v}_z$ and $\bar{v}_y$ are the RMS values of the quasi-longitudinal and quasi-shear wave velocities, respectively, defined as

$$\bar{v}_z^2 = \int_{-\infty}^0 \left[|v_z|^2 + \frac{j}{\beta} v_z^* \frac{\partial v_y}{\partial y} \right]^2 \left(\frac{dy}{\lambda_R} \right) \tag{4.66}$$

$$\bar{v}_y^2 = \int_{-\infty}^0 \left[|v_y|^2 + \frac{j}{\beta} \left(v_y^* \frac{\partial v_z}{\partial y} - 2 v_z^* \frac{\partial v_y}{\partial y} \right) \right]^2 \left(\frac{dy}{\lambda_R} \right) \tag{4.67}$$

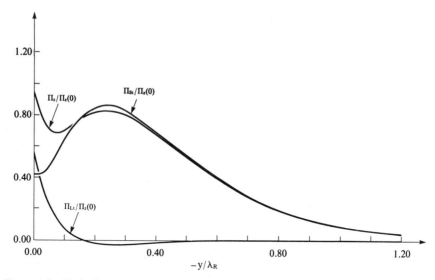

Figure 4.6. Normalized power densities $\Pi_{Lt}/\Pi_z(0)$ and $\Pi_{St}/\Pi_z(0)$ for Rayleigh waves in lead, as given by Eq. (4.62). The ratios of the normalized quasi-longitudinal and quasi-shear powers are $P_{Lt}/P_a = 0.03623$ and $P_{St}/P_a = 0.96377$, Eq. (4.68). The ratio of the RMS velocities, Eq. (4.66), is $\bar{v}_z/\bar{v}_y = 0.0673$.

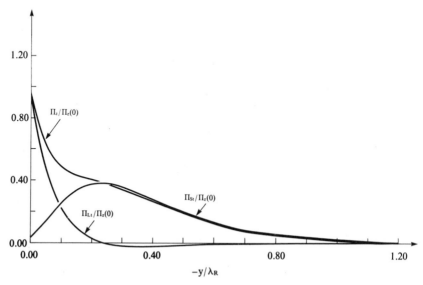

Figure 4.7. Normalized power densities $\Pi_{Lt}/\Pi_z(0)$ and $\Pi_{St}/\Pi_z(0)$ for Rayleigh waves on beryllium, as given by Eq. (4.62). The ratios of the normalized quasi-longitudinal and quasi-shear powers are $P_{Lt}/P_a = 0.234$ and $P_{St}/P_a = 0.766$, Eq. (4.68). The ratio of RMS velocities, Eq. (4.66), is $\bar{v}_z/\bar{v}_y = 0.38094$.

Equation (4.65) can now be written as

$$P_a = \overline{\Pi}_z A_S = P_{St} + P_{Lt} \qquad (4.68)$$

where $A_S = \lambda_R w$ is the equivalent cross-sectional area of the acoustic beam of power flow density

$$\overline{\Pi}_z = \frac{\beta(\lambda + 2\mu)}{2\omega}\bar{v}_z^2 + \frac{\beta\mu}{2\omega}\bar{v}_y^2 \qquad (4.69)$$

and P_{St} and P_{Lt} are the quasi-shear and quasi-longitudinal power, respectively. It is now possible to define two impedances, Eq. (4.53), related to the components of the Rayleigh wave as

$$Z_z^{LR} = \frac{\beta(\lambda + 2\mu)}{\omega} = \frac{\lambda + 2\mu}{v_R} \qquad (4.70)$$

associated with the quasi-longitudinal component, and

$$Z_a^{SR} = \frac{\beta\mu}{\omega} = \frac{\mu}{v_R} \qquad (4.71)$$

associated with the quasi-shear component. Therefore, the Rayleigh wave can be thought to consist of two homogeneous waves confined to a depth equal to λ_R from the boundary surface. Using the procedure given in Exercise 6 of Chap. 2, the normal modes of a SAW can be defined (see Exercise 5 at the end of this chapter).

The ratio of the quasi-longitudinal power P_{Lt} to the quasi-shear power P_{St} is found from Eq. (4.69) as

$$\frac{P_{Lt}}{P_{St}} = \left[\frac{1-\sigma}{0.5-\sigma}\right]\left[\frac{\bar{v}_z}{\bar{v}_z}\right]^2 \qquad (4.72)$$

where $\bar{v}_z/\bar{v}_y \simeq 0.33 - 0.58\sigma$.

4.4 EXCITATION OF SAW ON NONPIEZOELECTRIC SUBSTRATES

In practice, the two simplest mechanisms used for SAW excitation on nonpiezoelectric substrates[†] are the Lorentz-force mechanism and the magnetostrictive mechanism, already described in Chap. 3. Two types of substrates will be distinguished, a nonpiezoelectric dielectric and a nonpiezoelectric conductor.

Surface acoustic wave excitation on nonpiezoelectric dielectrics is achieved using the Lorentz force mechanism. Normally a conductive thin film pattern[15],

---

[†] The coupling prism is also used for SAW excitation and detection, see Chap. 9.

such as the meander line, is deposited on the substrate, which is then immersed in a DC magnetic field as shown in Fig. 4.8(a). For the acoustic wave propagation in the z direction, the DC magnetic bias must be in the sagittal plane; however, angle θ can be chosen to have any value from 0 to $\pm 90°$. The DC magnetic flux density is given by

$$\vec{B}_0 = \hat{y}B_0 \sin \theta + \hat{z}B_0 \cos \theta = \hat{y}B_y + \hat{z}B_z \qquad (4.73)$$

and the Lorentz body force, Eq. (3.1), acting on a single segment of meander line of width w is given by

$$\vec{F}^b = \vec{J}_s \times \vec{B}_0 = J_x \hat{x} \times \left(\hat{y}B_y + \hat{z}B_z \right) \qquad (4.74)$$

In order to assess which magnetic bias, $B_y = B_0$ ($\theta = 90°$) or $B_z = B_0$ ($\theta = 0$), yields the higher generation efficiency of SAW waves, the power supplied by the body forces, Eq. (1.47), is first computed as

$$P_S = \iiint_V \frac{\vec{F}^b \cdot \vec{v}^*}{2} \, dV \simeq \frac{\epsilon w \lambda_R}{2} J_x \left(B_y \bar{v}_z - B_z \bar{v}_y \right) \qquad (4.75)$$

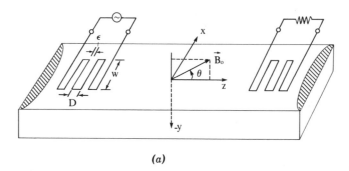

(a)

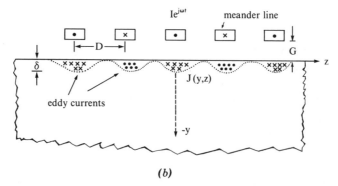

(b)

Figure 4.8. SAW excitation (a) on nonpiezoelectric dielectrics using contact configuration and (b) on nonpiezoelectric conductors using contactless configuration.

where ϵ and w are the width and length of each of the meander-line conductors, respectively, as shown in Fig. 4.8(a). The quantities $\bar{v}_y$ and $\bar{v}_z$ are the RMS quasi-shear and quasi-longitudinal velocities as given by Eqs. (4.66) and (4.67). In Eq. (4.75) the terms $J_x B_y \bar{v}_z$ and $J_x B_z \bar{v}_y$ can be thought of as being the quasi-longitudinal and the quasi-shear power density, respectively. It follows from Eq. (4.72) that in SAW the quasi-shear power P_{St} is always greater than the quasi-longitudinal power P_{Lt}. This suggests that the quasi-shear power supplied by the body forces will provide a larger amount of the total power necessary for SAW excitation. It follows that the efficiency is maximum when the static magnetic field B_0 is parallel to the direction of propagation ($\theta = 0$) and minimum when B_0 is oriented along the perpendicular to the surface ($\theta = 90°$). Similar results are also observed experimentally.

Using the concept of the SAW characterized by RMS velocities $\bar{v}_y$ and $\bar{v}_z$, Eqs. (4.66) and (4.67), most of the formulas derived in Chap. 3 for sheet-type meander lines, with appropriate modifications such as the acoustic beam cross section and relevant stiffness constants, can be used here as well. In particular, Eqs. (3.23) to (3.27), (3.32), (3.34), and (3.38), and material given in Sect. 3.3 and 3.4 are all applicable.

When the substrate is a nonpiezoelectric conductor, the meander line need not be bonded to the surface. In this contactless configuration, shown in Fig. 4.8(b), the meander line is at a distance G from the substrate. Due to the proximity of the meander-line conductors, eddy currents are induced in the substrate by inductive coupling. In order to use the theory developed in Chap. 3 for bulk waves, it is necessary to define another parameter to take into account the separation between the meander line and the substrate.[16] This parameter, called the coupling parameter C, is introduced in Eq. (3.15), which is written for this case[17] as

$$\Delta P_a = -C^2 \frac{\Delta W_m}{\Delta t} \tag{4.76}$$

where $C = C(G)$ is the gap-coupling parameter, dependent on the geometry of the device only. The value of C can be determined approximately as follows.

The induced current density in the substrate, see Fig. 4.8(b), can be approximated by considering a single segment of a meander line, assumed very long compared with the acoustic wavelength. Using the method of images, the magnetic field at the surface of the conductor is found for harmonic current excitation as

$$H_T = \frac{IGe^{j\omega t}}{\pi(z^2 + G^2)} \tag{4.77}$$

The current density is then given by

$$J = -\sigma_c H_T Z_e \exp[y(1 - j)\delta], \qquad y < 0 \tag{4.78}$$

where δ is the skin depth,[†] σ_c is the conductivity, and Z_c is the RF impedance of the conductor given by

$$Z_c = \left(\frac{\omega \mu_{em}}{\sigma_c}\right)^{1/2}\left(\frac{1+j1}{\sqrt{2}}\right), \qquad \delta = \left(\frac{2}{\omega \mu_{em}\sigma_c}\right)^{1/2} \tag{4.79}$$

where μ_{em} is the permeability of the conductor. Using Eqs. (4.77) and (4.78), it follows that

$$J = -\frac{I(1+j1)}{\pi\delta}\frac{G}{z^2 + G^2}\exp(j\omega t)\exp[y(1-j)/\delta], \qquad y < 0$$

$$= J_0(y)g_s(z)f(t) \tag{4.80}$$

where

$$g_s(z) = \frac{G}{z^2 + G^2} \tag{4.81}$$

is the source shape function for a single segment (current element) above an infinite conductor, see for instance Eq. (3.5), and

$$f(t) = e^{j\omega t}. \tag{4.82}$$

It is seen from Eq. (4.81) and Fig. 4.8(b) that for large gaps the surface current distribution spreads. When a meander line has several segments, the adjacent wires carry currents in opposite directions and cancellation occurs due to overlapping of the two adjacent current distributions, resulting in lower coupling to the SAW. In order to evaluate this effect and arrive at a value of the coupling factor, consider the total induced current at $z = 0$ in Fig. 4.8(b). Since all the segments carry the same current, only the g functions are relevant, the current density being proportional to

$$K(z) = \cdots + \frac{G}{G^2 + (z - 2D)^2} - \frac{G}{G^2 + (z - D)^2} + \frac{G}{G^2 + z^2}$$

$$- \frac{G}{G^2 + (z + D)^2} + \frac{G}{G^2 + (z + 2D)^2} - \cdots$$

$$= \sum_{m=-M}^{+M}\frac{G(-1)^m}{G^2 + (z - mD)^2} \tag{4.83}$$

[†]Note that δ is also used to denote the Dirac delta function, and to denote the increment of a quantity.

where it is assumed that the meander line consists of $2M$ sections of length D. In practice, Eq. (4.83) is seldom evaluated. Instead, an approximation is made, assuming an infinitely long meander line, $M \to \infty$, and for operation near resonant frequency only the first term of the Fourier series for K is used. Therefore, the first Fourier term is obtained as

$$K_1(G) = \frac{2}{D} \int_{-D/2}^{D/2} K(z)\cos\left(\frac{\pi}{D}z\right) dz$$

$$= \frac{2G}{D} \sum_{m=-\infty}^{\infty} (-1)^m \int_{-D/2}^{D/2} \frac{\cos[(\pi/D)z]}{G^2 + (z-mD)^2} dz$$

$$= \frac{2G\pi}{D^2} \sum_{m=-\infty}^{\infty} (-1)^m \int_{-\pi/2-m}^{\pi/2-m} \frac{\cos(\eta + m\pi)}{(\pi G/D)^2 + \eta^2} d\eta \qquad (4.84)$$

where

$$\eta = \frac{\pi}{D}z - m\pi. \qquad (4.85)$$

Equation (4.84) reduces to

$$K_1(G) = \frac{2G\pi}{D^2} \int_{-\infty}^{\infty} \frac{\cos \eta}{[(\pi/D)G]^2 + \eta^2} d\eta$$

$$= \frac{2\pi}{D} e^{-\pi G/D} \qquad (4.86)$$

In order to include the case when the meander line is in contact with the substrate, $K_1(G)$ is normalized to its value for $G = 0$, resulting in the gap-coupling parameter[16]

$$C = \frac{K_1(G)}{K_1(0)} = e^{-\pi G/D} \qquad (4.87)$$

The total current induced in the substrate under a segment of an infinite meander line, is obtained using Eq. (4.80) as

$$I_i = \int_{-D/2}^{D/2} \int_{-\infty}^{0} J(y, z) \, dy \, dz$$

$$= \bar{I}_0 \left[\int_{-D/2}^{+D/2} g_m(z) \frac{dz}{D} \right] e^{j\omega t} \qquad (4.88)$$

where $\bar{I}_0$ is the spatially averaged current given by

$$\bar{I}_0 = \int_{-\infty}^{0} J_0(y)\,dy \tag{4.89}$$

with

$$J_0(y) = -\frac{I(1+j)}{\pi\delta}e^{y(1-j)/\delta} \tag{4.90}$$

and where

$$g_s(z) \simeq \frac{2\pi}{D}e^{-\pi G/d}\cos(\pi z/D) \tag{4.91}$$

is the source shape function of a single segment of an infinite meander line above an infinite conductor. Equation (4.88) can now be written as

$$I_i = 4\bar{I}_0 e^{-\pi G/D}e^{j\omega t} \tag{4.92}$$

Again, using the concept of SAW normal modes (Sect. 4.3), most of the formulas derived in Chapter 3 can be used here as well, provided $k_m C$ is substituted for k_m. For instance, Eq. (3.20) is now written as[17]

$$R_0^{(M)} = 4f_0 k_m^2 C^2 L_D M^2 w \tag{4.93}$$

where l in Eq. (3.23) is now approximately equal to the skin depth δ. In addition, the C factor must be taken into account whenever the acoustic (output) variables in a transducer are computed from the electric (input) variables and vice versa. For instance, for the two transducer configurations shown in Fig. 4.9, the open-circuit output voltage resulting from the excitation of the input transducer with the spatially averaged current density $\bar{J}_0 = I_i/\lambda_R \epsilon$ is given (using the $\bar{v}_z$ component of the Rayleigh wave) by

$$\tilde{e}_o(\omega) \approx \left[\frac{\bar{J}_0 B_0^2 D_1 w}{2c_{11}}\right] v_R C_i C_o M_i M_o \frac{\sin x_{M_o}}{x_{M_o}}\frac{\sin X_{M_i}}{X_{M_i}}$$

$$\times \exp\left[-\frac{jM_i\pi f}{2f_0^i}\right]\exp\left[-\frac{jM_o\pi f}{2f_0^0}\right]$$

$$\times \exp\left[-\frac{j\omega(d+z_i/2+z_o/2)}{v_R}\right] \tag{4.94}$$

where, for instance, $B_0 = B_{0\parallel}$ as in Fig. 4.9, z_i and z_o are the lengths of the input and the output transducers, respectively, d is the distance between

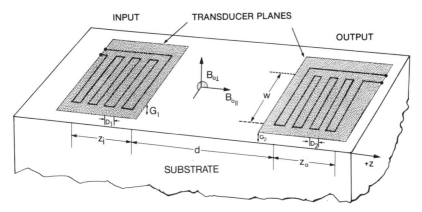

Figure 4.9. Two-transducer contactless configuration.

transducers with number of segments $M_o = z_o/D_2$ and $M_i = z_i/D_i$. Furthermore,

$$X_{M_o} = M_o \pi \frac{\left(f - f_0^o\right)}{2f_0^o}, \quad \text{and} \quad X_{M_i} = M_i \pi \frac{\left(f - f_0^i\right)}{2f_0^i},$$

with the resonant frequencies of the transducers given by $f_0^o = v_R/2D_2$, and $f_0^i = v_R/2D_1$, where v_R is the Rayleigh velocity. The gap-coupling factors are $C_i = \exp(-\pi G_1/D_1)$, which is due to the reduction of the induced current at the input, see Eq. (4.92), and $C_o = \exp(-\pi G_2/D_2)$, which is attributed to the reduction of the induced emf in each segment, Eq. (3.29), because the adjacent eddy currents propagate on the surface in opposite directions.

The gap-coupling parameter for a single transducer is usually expressed[16] as

$$C[\text{dB}] = 20 \log e^{-\pi G/D} \simeq -54.6 \frac{G}{\lambda_R} \tag{4.95}$$

where $\lambda_R = \lambda_0 = 2D$ is the resonant (Rayleigh) wavelength of the meander line. For $\lambda_R \approx G$, it is seen that gap-coupling losses amount to approximately 55 dB per transducer. For instance, on aluminum, with SAW velocity of 2905 m/s at 5 MHz, $\lambda_R = 0.58$ mm, requiring a gap of approximately 0.1 mm if the gap-coupling losses per transducer are to be kept to about 10 dB.

The magnetostrictive mechanism, described in Sect. 3.5 for bulk waves, can also be used in the contactless configuration for the generation of SAW on magnetostrictive substrates.[18, 19] This is demonstrated by the experimental results[18] shown in Fig. 4.10, where the efficiency of SAW generation is plotted versus magnetic flux density for various materials. In practice, such curves are obtained with a configuration similar to that shown in Fig. 4.9. It is seen from

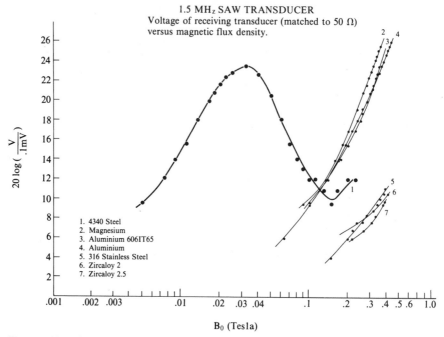

Figure 4.10. The efficiency of SAW generation versus magnetic flux density for various materials.

Fig. 4.10 that the magnetostrictive effect is very strong for 4340 steel substrate for low magnetic bias, $B_0 \approx 0.03$ T, and is comparable in strength with aluminum substrate for $B_0 \approx 0.3$ T.

Two SAW electromagnetic-acoustic transducers (EMATs) developed for NDT applications are shown in Figs. 3.9 and 3.12.

REFERENCES

1. J. W. Strutt, Lord Rayleigh, *The Theory of Sound*, Dover, New York, 1945.
2. I. A. Viktorov, *Rayleigh and Lamb Waves*, Plenum, New York, 1967.
3. B. A. Auld, *Acoustic Fields and Waves in Solids*, Vol. 2, Wiley, New York, 1973, Chap. 10.
4. H. Matthews, Ed., *Surface Wave Filters*, Wiley, New York, 1977.
5. R. M. White, "Surface Elastic Waves," *Proc. IEEE*, **58**, 1238–1276 (1970).
6. R. M. White, "Surface Elastic Waves," in *Methods of Experimental Physics* (P. D. Edmonds, Ed.), Vol. 19, Academic, New York, 1981, Chap. 10.
7. E. G. H. Lean, J. M. White, and C. D. Wilkinson, "Thin-Film Acousto-Optic Devices," *Proc. IEEE*, **64**, 779–788 (1976).
8. C. S. Tsai, M. A. Alhaider, L. T. Nguyen, and B. Kimm, "Wide-Band Guided-Wave Acousto-Optic Bragg Diffraction and Devices Using Multiple Tilted Surface Acoustic Waves," *Proc. IEEE*, **64**, 318–328 (1976).

9. G. W. Farnell, "Types and Properties of Surface Waves," in *Acoustic Surface Waves* (A. A. Oliner, Ed.), Springer-Verlag, New York, 1978.

10. N. G. Brower, D. E. Himberger, and W. G. Mayer, "Restrictions on the Existence of Leaky Rayleigh Waves," *IEEE Trans. Sonics Ultrason.*, **SU-26**(4), 306–308 (1979).

11. F. Press and J. Healy, "Absorption of Rayleigh Waves in Low-Loss Media," *J. Appl. Phys.*, **28**, 1323–1325 (1957).

12. R. M. White, "Surface Elastic-Wave Propagation and Amplification," *IEEE Trans. Electron Dev.*, **ED-14**, 181–189 (1967).

13. A. J. Slobonik, Jr., "Materials and Their Influence on Performance," in *Acoustic Surface Waves* (A. A. Oliner, Ed.), Springer-Verlag, New York, 1978, Chap. 6.

14. B. A. Auld, *Acoustic Fields and Waves in Solids*, Vol. I, Wiley, New York, 1973, Chaps. 5 and 7.

15. H. Talaat and E. Burstein, "Phase-Matched Electromagnetic Generation and Detection of Surface Elastic Waves in Nonconducting Solids," *J. Appl. Phys.*, **45**(10), 4360–4362 (1972).

16. R. B. Thompson, "A Model for the Electromagnetic Generation and Detection of Rayleigh and Lamb Waves," *IEEE Trans. Sonics Ultrason.*, **SU-20**(4), 340–346 (1973).

17. V. M. Ristic, "Magnetomechanical Coupling Coefficient for Periodic EMAT's," *IEEE Trans. Sonics Ultrason.*, **SU-29**, 229–231 (1982).

18. R. W. Voltmer, R. M. White, and C. W. Turner, "Magnetostrictive Generation of Surface Elastic Waves," *Appl. Phys. Lett.*, **15**(5), 153–154 (1969).

19. V. M. Ristic, F. Hauser, G. Dubois, and H. Licht, "Nonlinear FM Coded SAW EMAT's for Nondestructive Testing of Materials," *1980 Ultrason. Symp. Proc*, **2**, 898–901 (1980).

EXERCISES

1. An x-polarized shear wave is propagating in the $+z$ direction in a material of infinite extend.
   (a) Find the Poynting vector of the wave.
   (b) Repeat the calculation for a longitudinal wave propagating in the x direction.

2. A composite wave, consisting of an x-polarized shear wave, a y-polarized shear wave, and a longitudinal wave, is propagating in the z direction in a material of infinite extent.
   (a) Find the Poynting vector of the composite wave using the principle of superposition.
   (b) Explain the results obtained for (a) in terms of Eq. (4.29).

3. A Rayleigh wave of frequency ω is propagating in the z direction.
   (a) Express S_{yy}, S_{zz}, and S_{yz} in terms of v_y and v_z.
   (b) Find the corresponding expressions for stresses.
   (c) Using Eqs. (4.29) and (4.38), find the Poynting vector of the Rayleigh wave.

4. Using Eqs. (4.54) to (4.56), derive expressions for the normal modes of bulk shear wave. (See Exercise 6, Chap. 2.)

5. Using Eqs. (4.68) to (4.71), find the wave equations for the normal modes of the Rayleigh wave. (See Exercise 6, Chap. 2.)

6. Show that the solution for the Rayleigh wave of the form

$$u_z = Ae^{by}e^{j(\omega t - \beta z)} \quad \text{and} \quad u_y = jBe^{by}e^{j(\omega t - \beta z)}, \quad y < 0$$

where A and B are two unknown constants and b and β are real, cannot satisfy the boundary conditions, Eqs. (4.13) and (4.14).

7. (a) Design a meander line for operation on aluminum with a bandwidth of 30% and resonant frequency of 3 MHz.
    (b) Determine the number of segments M and the spacing D.
    (c) If the gap coupling losses must be less than 20 dB, determine the gap spacing.
    (d) Repeat (a), (b), and (c) for lead and beryllium.

8. Using Eqs. (3.23) and (4.93), find the radiation resistance of a meander line with the following characteristics: $w = 2$ cm, $D/G = 3.14$, $M = 10$, resonant frequency 1 MHz for operation on aluminum with $B_0 = 1$ T. The conductivity of aluminum is $\sigma_c = 3.7 \times 10^7$ S/m. *Note:* In Eq. (3.23), use v_R instead of v_a.

9. Compare the Rayleigh wavelengths with the corresponding skin depths at frequencies of 1, 5, and 50 MHz in the following materials: copper ($\sigma_c = 58$ MS/m), aluminum ($\sigma_c = 37$ MS/m), beryllium ($\sigma_c = 22$ MS/m), lead ($\sigma_c = 4.76$ MS/m), and Zircaloy 2 ($\sigma_c = 2.43$ MS/m). In all cases, assume $\mu_{em} = \mu_0 = 4\pi \times 10^{-7}$ H/m. What conclusion can you draw from these data?

10. The attenuation constant of Rayleigh waves propagating on YZ LiNbO$_3$ crystal is given by

$$\alpha \left[\frac{\text{Np}}{\text{m}} \right] = 29f^2 + 6.27f$$

where f is in GHz. Express the attenuation constant in dB/μs given the Rayleigh wave velocity $v_R = 3400$ m/s.

11. Show that Hooke's law, Eqs. (2.46) and (2.48), can be expressed for an isotropic solid as

$$T_{ik} = 2\mu S_{ik} + \lambda \, \Delta \delta_{ik}$$

where $\Delta = S_{ii} = S_{11} + S_{22} + S_{33}$ is the dilatation. Show that elastic (strain) energy of an isotropic body can be found as

$$U = \frac{T_{ik} S_{ik}}{2} = \mu S_{ik}^2 + \frac{\lambda \Delta^2}{2}$$

where

$$S_{ik}^2 = S_{11}^2 + S_{12}^2 + S_{13}^2 + S_{21}^2 + S_{22}^2 + S_{23}^2 + S_{31}^2 + S_{32}^2 + S_{33}^2$$

Chapter 5

Piezoelectric Bulk Wave Transducers

Piezoelectricity[1] is a linear property of certain materials that manifests itself by an onset of electrical polarization (generation of bound charges of opposite polarity) when the material is subjected to mechanical stress, and by the onset of deformation in the material subjected to an electric field. In the first case it is termed the *direct piezoelectric effect*, and in the second it is termed the *inverse piezoelectric effect*. The modeling of materials in terms of periodic structures[2] in order to explain various phenomena such as piezoelectricity is an important method for gaining a better understanding of the underlying physics. By considering the one-dimensional model of a piezoelectric and a nonpiezoelectric solid, it can be seen[3] that the piezoelectricity arises because of an interaction between Coulomb forces and the elastic restoring forces in a unit cell subject to either an electric field or a mechanical force. By calculating the resulting strain and the electric polarization, the form of the constitutive relation can be readily established.

Piezoelectricity is intimately related to crystallographic properties of materials. All known materials can be categorized into 32 crystallographic classes. Using symmetry arguments, it can be shown that the piezoelectric effect cannot exist in materials possessing central symmetry. Indeed, there are 21 classes without central symmetry, and the piezoelectric effect is found in 20 of them. In most materials, piezoelectricity can be explained by asymmetry of ionic polarization. However, pure elements selenium (Se) and tellurium (Te), exhibit piezoelectricity as well. In this case the electric polarization induced by strain is attributed to change in the electronic distribution.

It is well known that the electric field due to an electric dipole, shown in Fig. 5.1(a), on the axis of the dipole for $r \gg l$ is given by $E = p/2\pi\varepsilon_0 r^3$, where p is the electric dipole moment, $p = ql$. Consider a hypothetical solid,[4] shown in Fig. 5.1(b), where charged molecules are held by elastic forces represented by springs. This solid has no center of symmetry; that is, a line connecting the negatively charged molecule B and the center O will intersect a positively charged molecule G. For simplicity it is assumed that the elementary, two-dimensional cell consists of three noninteracting dipoles AB, HC, and

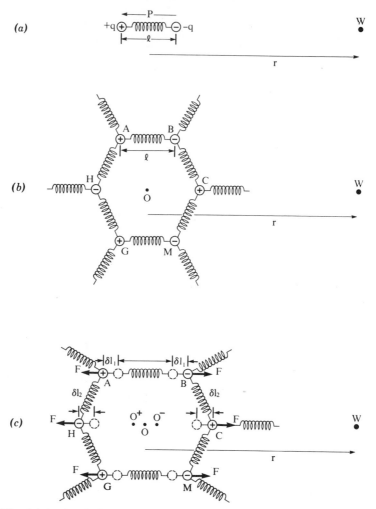

Figure 5.1. (*a*) An electric dipole of dipole moment $p = lq$; for $r \gg l$ the resulting electric field at point W is $E = p/2\pi\epsilon_0 r^3$. (*b*) Model of a hypothetical solid consisting of charged molecules of identical charge magnitudes. The elastic forces are represented by springs. The resulting electric field at W, due to all six molecules, is equal to zero. (*c*) The same solid under action of a symmetric force F. The resulting electric field at W is nonzero. The electric field is due to displacements of the center of positive charge from O to O^+ and the center of negative charge from O to O^-.

GM. In equilibrium the electric field of the cell at point W is given, for $r \gg l$, by

$$E \simeq \frac{ql - q2l + ql}{2\pi\varepsilon_0 r^3} = 0$$

Thus, the total electric dipole moment of the cell is zero. If a symmetric force F is applied to the solid, Fig. 5.1(c), the solid will be deformed. The displacements δl_1 for point B and δl_2 for point C will be different due to different spatial distribution of elastic forces with respect to the external force F. For simplicity it is assumed that $\delta l_1 = \delta l_2 = \delta l$. The electric field of the deformed cell at point W will now be

$$E = \frac{2q(2\delta l_1 - \delta l_2)}{2\pi\varepsilon_0 r^3} = \frac{2q\delta l}{2\pi\varepsilon_0 r^3}$$

where the dipole moment p can be written as

$$p = 2q\delta l = 2qlS$$

where $S = \delta l/l$ is the longitudinal strain. Assuming that the body contains N molecules per unit volume, it follows that the electric polarization P, that is, the dipole moment per unit volume, is given by

$$P = Nq = 2qlNS = eS$$

where e is by definition *the piezoelectric stress constant* $e = 2qlN$. Taking $N \simeq 5 \times 10^{-27}\text{m}^{-3}$, $q = 1.6 \times 10^{-19}$ C, and $l = 5 \times 10^{-10}$m, it is found that $e = 0.8$ C/m^2. This is comparable, for instance, to the piezoelectric stress constant of ZnO, equal to 0.5 C/m^2. The presence of the polarization P is taken into account by modifying the relation $D = \varepsilon E$ to

$$D = \varepsilon E + eS$$

Note that during the application of the force F, the centers of positive and negative charges, initially both in point O, Fig. 5.1(c), have been displaced to points O^+ and O^- respectively, thus producing a displacement current.

Similar considerations for the case when an external electric field E is applied show that an additional stress term, $-eE$, appears. This is taken into account by modifying Hooke's law, $T = cS$, to

$$T = cS - eE$$

If the charges of the positive and the negative molecules were different in magnitude, the model would have exhibited polarization even with no forces applied. This is the case of the *spontaneous polarization* that occurs in pyroelectrics.

Further elucidation of the model is achieved by assuming that each molecule, instead of being charged, has a dipole moment; that is, the material exhibits *ionic polarization*. In the case of selenium and tellurium only, materials having only one type of atoms, polarized atoms are placed in the model rather than polarized molecules. In this case it is said that the material exhibits *electronic polarization*.

Certain materials possess a natural (spontaneous) tendency to polarize. These are termed *pyroelectrics*. This permanent type of polarization is usually compensated by redistribution of free charges in the body or at the surface of the materials. Experimentally,[5] pyroelectricity is demonstrated by connecting an electroded specimen to an electric load (a resistor) and observing the charge flow as a function of temperature.[†] If the polarization of a pyroelectric can be varied by an applied electric field, the material is said to be *ferroelectric*. Out of 20 crystallographic classes exhibiting piezoelectricity, 10 are pyroelectric as well. All ferroelectrics are piezoelectric and pyroelectric. To summarize, a ferroelectric material is a crystal possessing an internal dipole moment that tends to align itself in an allowed direction closest to an applied electric field. The allowed directions (easy axes of polarization) can be deduced from symmetry considerations of the crystal cell.

As the temperature of the material is raised and a certain critical temperature is reached, the material changes its crystallographic class, entering a class with a higher order of symmetry. This process is termed *phase transition*. The newly acquired class may not be either ferroelectric, pyroelectric, or piezoelectric. As the temperature changes, various materials may exhibit several phase transitions. For instance, ferroelectric barium titanate ($BaTiO_3$) has four transitions in the temperature range from -70 to $+120°C$. In one of these phases, the material is piezoelectric without being ferroelectric. When a temperature is reached at which it undergoes a phase transition into a class possessing central symmetry, piezoelectricity ceases. This temperature is termed the *Curie point* (or Curie temperature). For instance, the Curie temperature of barium titanate is 120°C. Synthetic piezoelectric crystal lithium niobate ($LiNbO_3$) has a Curie point of approximately 1200°C. For quartz (SiO_2), the corresponding temperature is 573°C.

So far, only linear effects (piezoelectricity, pyroelectricity, and ferroelectricity) have been described. Materials exhibit nonlinear effects such as *electrostriction* as well. Electrostriction is characterized by quadratic dependence of stress on the electric field; that is, reversal of the electric field direction does not change the polarity of the strain. In all cases, in the temperature range where both piezoelectricity and electrostriction exist, electrostriction is a much weaker effect of no practical importance. Electrostriction is a property of all dielectrics, requiring application of very high fields (1 MV/cm) for the phenomenon to become detectable.

The counterpart of piezoelectricity is a piezomagnetism (a linear effect) found in antiferromagnetic materials, Sect. 3.6. Piezomagnetism, being a weak effect, is of little practical importance. However, magnetostriction, a nonlinear effect and the counterpart of electrostriction, is of considerable importance, as discussed in Chap. 3.

Since the discovery of piezoelectricity in natural quartz crystals in 1880, many synthetic piezoelectric crystals have been grown from water solutions,[6] including EDT, ethylenediamine tartrate ($C_6H_{14}N_2O_6$); ADP, ammonium dihydrogen phosphate ($NH_4H_2PO_4$); and Rochelle salt ($NaKC_4H_4O_6 \cdot 4H_2O$) in the period 1936 to 1956, and others from high-temperature melts such as lithium

---

[†] For instance, the pyroelectric constant of ZnO in the temperature range between 230 and 410 K is 9.5×10^{-10} C/cm$^2 \cdot °$ C.

niobate and lithium tantalate, since 1966. Presently most piezoelectric crystals, except for quartz, which became available in synthetic form in 1958, are grown from melts. In 1956, sintered ferroelectric ceramics became available. The piezoelectric (ferroelectric) ceramics[7] PZT is a combination of lead (Pb), zirconium (Zr), and titanium (Ti) oxides or salts and consists, after sintering, of ferroelectric domains, that is, small volumes in which the polarization is uniformly oriented. The chemical formula of the compound is $PbTi_{1-x}Zr_xO_3$, with $x \approx 0.5$. Before use, ceramics must be poled. The poling process consists of elevating the temperature close to the Curie point and then applying a very strong DC electric field (20 kV/cm) while slowly decreasing the temperature. The poling process aligns the great majority of individual domains.

Another important class of piezoelectric materials are those that can be deposited in a thin film on various substrates using different deposition techniques. The most prominent among these are zinc oxide (ZnO), cadmium sulfide (CdS), and aluminum nitride (AlN).

A relatively new development in the field of piezoelectric materials is the manufacture of ferroelectrics made out of thin plastics,[8, 9] such as polyvinylidene fluoride (PVF_2), which became available in 1970.

5.1 PIEZOELECTRIC CONSTITUTIVE RELATIONS

The piezoelectric constitutive relations relate electrical (E, D) and mechanical (S, T) variables. It has been pointed out at the beginning of the chapter that the form of the constitutive relations can be found from the one-dimensional model of the piezoelectric solid. Since there are four different choices for independent (dependent) variables, four such relations can be written. For instance, if it is assumed that

$$T = T(S, E) \quad \text{and} \quad D = D(S, E) \tag{5.1}$$

then for small fields, using Taylor's expansion (or from the one-dimensional model), we have

$$T = \left(\frac{\partial T}{\partial S}\right)_E S + \left(\frac{\partial T}{\partial E}\right)_S E \tag{5.2a}$$

$$D = \left(\frac{\partial D}{\partial S}\right)_E S + \left(\frac{\partial D}{\partial E}\right)_S E \tag{5.2b}$$

where all other effects, such as magnetic or thermal, have been neglected and where the nonlinear terms have been neglected as well.

The constants derived, in order to be true constants, must be evaluated under precise experimental conditions, that is, they are measured either at constant (zero) electric field as

$$c^E = \left(\frac{\partial T}{\partial S}\right)_E \quad e^E = \left(\frac{\partial D}{\partial S}\right)_E \tag{5.3}$$

or at constant (zero) strain as

$$e^S = \left(\frac{\partial T}{\partial E}\right)_S \qquad \varepsilon^E = \left(\frac{\partial D}{\partial E}\right)_S \tag{5.4}$$

Consider the case when an electric field is applied to a piezoelectric. The second terms in Eqs. (5.2a) and (5.2b) give the stress and electric displacement in the material. If the material is not confined mechanically, the strain will develop in response to the stress. This strain contributes to the first term in Eq. (5.2b) and therefore alters the relationship between D and E. Thus, measurement of the electrical properties of the piezoelectric is dependent on the mechanical constraints. A similar situation arises when strain is applied. In this case the measurement of mechanical properties of the material depends upon the imposed electrical constraints. Both cases demonstrate the essence of *piezoelectric coupling*. However, from thermodynamic considerations,[3] it can be shown that the inverse piezoelectric effect, represented by the piezoelectric constant e^S, is a consequence of direct piezoelectric effect given by the piezoelectric constant e^E, and thus

$$-e^S = e^E = e \tag{5.5}$$

and Eq. (5.2) can be written as

$$T = c^E S - eE \tag{5.6a}$$

$$D = eS + \varepsilon^S E \tag{5.6b}$$

Using similar arguments, three other systems can also be deduced:

$$S = s^E T + dE \qquad D = dT + \varepsilon^T E, \tag{5.7}$$

$$S = s^D T + gD \qquad E = -gT + \beta^T D \tag{5.8}$$

and

$$T = c^D S - hD \tag{5.9a}$$

$$E = -hS + \beta^S D \tag{5.9b}$$

The derived piezoelectric constant e and d are measured in units of C/m^2 and C/N, respectively. For instance, in very strong piezoelectrics, such as PZT-5H ceramics, $e = 23.3$ C/m^2; in lithium niobate crystal, $e = 2.5$ C/m^2; and in weak piezoelectrics such as gallium arsenide (GaAs), $e = 0.15$ C/m^2.

Any one of the relations given by Eqs. (5.6) to (5.9) can be used. However, in device design one type of piezoelectric constant may be preferable to others. For instance, for PZT-4, $d = 289 \times 10^{-12}$ C/N and $g = 26 \times 10^{-3}$ V · m/N, and for PVF$_2$, $d = 14 \times 10^{-12}$ C/N and $g = 160 \times 10^{-3}$ V · m/N. From Eq. (5.8) it is seen that PVF$_2$, due to its large g constant, would be preferable material for hydrophones (detectors of acoustic waves in liquids), while PZT-4, with its high d constant, would be preferable material for accelerometer applications.

Limiting our consideration to relations given by Eqs. (5.6) and (5.7), the following interpretation can be given, for instance, to Eq. (5.6). Consider the application of the equation

$$T = c^E S - eE \tag{5.10}$$

to a cylindrical piezoelectric body of radius R much larger than its thickness t, where all the field quantities are in the axial direction, Fig. 5.2(a). If the body is rigidly clamped on both sides of the cylinder, then $S = 0$ and

$$T = -eE \tag{5.11}$$

Therefore an applied electric field will cause stress T. For this reason e is

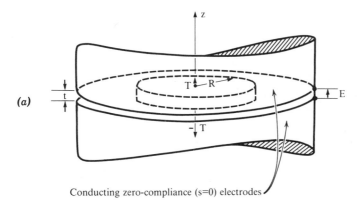

(a)

Conducting zero-compliance (s=0) electrodes

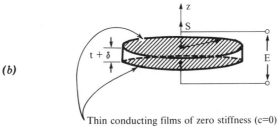

(b)

Thin conducting films of zero stiffness (c=0)

Figure 5.2. A piezoelectric disk with $R \gg t$, (a) Under clamped conditions ($S = 0$), that is, $t = $ const for $E \neq 0$, the disk exhibits stress $T = -eS$. (b) Under zero stress conditions ($T = 0$) for $E \neq 0$, the disk exhibits strain $S = eE/c^E = \delta t/t$.

referred to as the *piezoelectric stress constant*.[†] On the other hand, if all sides of the body are free, $T = 0$, as shown in Fig. 5.2(b), and

$$S = \frac{e}{c^E} E \tag{5.12}$$

Thus an applied electric field will cause an axial strain, that is, an increase in the relative length of the cylinder equal to $1 + \delta t/t$. Similar considerations can be applied to Eq. (5.7), where d is termed the piezoelectric strain constant. For PZT-5H ceramics with $c^E = 11.7 \times 10^{10}$ N/m², $e = 23.3$ C/m², and $E = 5$ kV/m, we obtain $S = 10^{-6}$ from Eq. (5.12) and $T = -1.17 \times 10^5$ N/m² from Eq. (5.11). Although the computed strain is relatively small, the stress is extremely large, qualifying the material for ultrasonic generation applications.

Using Eqs. (5.6) and (5.7) it is easy to show that

$$d = es^E \qquad e = dc^E$$

$$\varepsilon^T = \varepsilon^S \left(1 + \frac{e^2}{c^E \varepsilon^S} \right) \qquad \varepsilon^S = \varepsilon^T - \frac{d^2}{s^E} \tag{5.13}$$

$$s^E c^E = 1$$

Similar relations[3] can be found for the other two sets given by Eqs. (5.8) and (5.9).

Equations (5.13) demonstrate that piezoelectric coupling, manifesting itself through constants e and d, requires two dielectric constants, ε^S and ε^T. The value of ε^S, the dielectric constant at constant strain, is very difficult to measure because of the requirement that the specimen be rigidly clamped. Usually another three constants, for instance ε^T, e, and c^E, are measured and the value of ε^S is found from Eq. (5.13). In strong piezoelectrics, the values of the two dielectric constants can differ by nearly 50%. For instance, for PZT-5H, $\varepsilon^T = 3400\varepsilon_0$ and $\varepsilon^S = 1470\varepsilon_0$, where ε_0 is the free-space permittivity, while for GaAs, $\varepsilon_{11}^T \simeq \varepsilon_{11}^S = 11.1\varepsilon_0$.

5.2   THE PIEZOELECTRIC COUPLING FACTOR

In practice it is often necessary to compare different piezoelectric materials in order to assess their performance for intended application. Parameters such as the efficiency of coupling to mechanical vibrations with an outside electric field and collinearity of phase and group velocity, the directions of the applied electric field, and the resulting particle displacement have to be considered.

[†]The "clamped" condition, requiring ideally rigid electrodes, is difficult to achieve. The response of an ideally rigid (stiff) body would be $S \to 0$ for $T \neq 0$, and thus from Hooke's law, $s \to 0$ or $c \to \infty$.

The efficiency of coupling is usually assessed through the piezoelectric (electro-mechanical) coupling factor. The concept of the coupling factor can be introduced by considering one-dimensional wave propagation in piezoelectric material.

Since there are no free charges in piezoelectric dielectrics, it follows that

$$\nabla \cdot \vec{D} = 0 \qquad (5.14)$$

and

$$\nabla \times \vec{E} = -\frac{\partial \vec{B}}{\partial t} \qquad (5.15)$$

$$\vec{J} = \frac{\partial \vec{D}}{\partial t} \qquad (5.16)$$

Again, the last equation, where $\vec{J}$ is the displacement current density, indicates that only piezoelectric dielectrics will be considered. The magnetic flux density $\vec{B}$ is associated mainly with electromagnetic waves, which in the presence of piezoelectricity are coupled to acoustic waves. Since the velocity of electromagnetic waves is approximately 10^4 to 10^5 times higher than the velocity of acoustic waves, this coupling is very weak; that is, the contribution of magnetic field associated with the acoustic wave to the total magnetic field is very small. In this case, with negligible error, it is proper to use for acoustic waves

$$\nabla \times \vec{E} \approx 0 \qquad (5.17)$$

which results by similarity with electrostatics, in a potential ϕ, termed the quasi-static potential, from which the electric field is obtained as[†]

$$\vec{E} = -\nabla\phi \qquad (5.18)$$

For one-dimensional variations with propagation along the z axis, Eqs. (5.14) and (5.18) reduce to

$$\frac{\partial D_z}{\partial z} = 0 \qquad (5.19)$$

and

$$E_z = -\frac{\partial \phi}{\partial z} \qquad (5.20)$$

Assuming that the wave propagation direction is along a pure mode axis,

[†] Thus the electric field of the piezoelectric wave is always in the direction of propagation. The transverse electric field is associated with the electromagnetic wave, which has been neglected.

defined in Chap. 1, we can consider either longitudinal or shear wave propagation.

For longitudinal waves, Eq. (5.6) is written as

$$T_3 = c^E \frac{\partial u_z}{\partial z} - eE_z \tag{5.21a}$$

$$D_z = e\frac{\partial u_z}{\partial z} + \varepsilon^S E_z \tag{5.21b}$$

and Eq. (1.4) as

$$\frac{\partial T_3}{\partial z} = \rho_m \frac{\partial^2 u_z}{\partial t^2} \tag{5.22}$$

Combining Eqs. (5.19), (5.21), and (5.22), we obtain the wave equation in the form

$$\rho_m \frac{\partial u_z^2}{\partial t^2} = c^E\left(1 + \frac{e^2}{c^E \varepsilon^S}\right)\frac{\partial^2 u_z}{\partial z^2} \tag{5.23}$$

The harmonic solution of the wave equation is given by

$$u_z = M\exp[j(\omega t - \beta_a z)] + N\exp[j(\omega t + \beta_a z)] \tag{5.24}$$

where M and N are constants and either $M = 0$ or $N = 0$ in a piezoelectric of infinite extent. Using Eqs. (5.23) and (5.24), the propagation constant is obtained as

$$\beta_a = \frac{\omega}{v_L^D} \tag{5.25}$$

where

$$v_L^D = \left(\frac{c^E}{\rho_m}\right)^{1/2}(1 + K^2)^{1/2} = v_L(1 + K^2)^{1/2} \tag{5.26}$$

and

$$K^2 = \frac{e^2}{c^E \varepsilon^S} \simeq -\frac{2\Delta v_L}{v_L} \tag{5.27}$$

Since all the fields vary as $\exp[j(\omega t \pm \beta_a z)]$, it is easily shown using Eqs. (5.20) to (5.22) and (5.25) to (5.27) that

$$D_z = 0 \tag{5.28a}$$

causing $J = 0$ in Eqs. (5.16).

Using $D_z = 0$, Eq. (5.21b) can be written as

$$\phi = \frac{e}{\varepsilon^S} u_z = \frac{e v_z}{j\omega \varepsilon^S} \qquad (5.28b)$$

and therefore a longitudinal piezoelectric wave propagating along a pure mode axis, coinciding with the $+z$ direction, in a piezoelectric of infinite extent, is characterized by the following fields:

$$\left.\begin{aligned} T_3 &= -\rho_m v_L^D v_z \\ v_z &= j\omega M \exp\left[j(\omega t - \beta_a z) \right] \\ \phi &= \frac{e v_z}{j\omega \varepsilon^S} \\ D_z &= 0 \end{aligned}\right\} \qquad (5.29)$$

Note that the wave equation (5.23), except for the constants, is similar to Eq. (3.93) for longitudinal waves in an isotropic solid. Thus, it is possible to define a constant

$$c^D = c^E\left(1 + \frac{e^2}{c^E \varepsilon^S}\right) = c^E(1 + K^2) \qquad (5.30)$$

termed the *piezoelectrically stiffened elastic constant*, applicable only when $D_z = 0$, for the wave propagation in the given direction (z direction). The factor K^2 defined by Eq. (5.27) is termed the *piezoelectric (electromechanical) coupling factor (coefficient)* and indicates that acoustic vibrations can be set up when an outside RF electric field is applied to the material. Higher values of K^2 indicate, in general, a stronger piezoelectric from the point of view of wave excitation efficiency. It is seen from Eq. (5.30) that, effectively, the piezoelectricity increases the stiffness constant c^E by an amount equal to K^2 percent, and c^D is termed the "stiffened" elastic constant. Similarly, ε^T in Eq. (5.13) is called the *stiffened dielectric constant*, and Eq. (5.26) gives the *stiffened longitudinal velocity* v_L^D in a piezoelectric medium. This velocity is always larger than the velocity v_L of an acoustically equivalent nonpiezoelectric medium. In strong piezoelectrics such as PZT-5H, $K = 0.75$; in lithium niobate crystal, $K = 0.16$; and in the weak piezoelectric GaAs, $K = 0.065$. Therefore, in computing the longitudinal velocity in strong piezoelectrics, the stiffening must be taken into account since its relative contribution may be as high as 56% (PZT-5H). In weak piezoelectrics such as GaAs, that is not the case, the contribution being only 0.4%, and therefore $v_L^D \approx v_L$. For PVF$_2$, $e \approx -0.165$ C/m^2, $K = 0.2$, and $v_L^D \approx v_L = 2150$ m/s.

5.3 BULK PIEZOELECTRIC TRANSDUCERS

In the preceding section, it has been demonstrated that the piezoelectric effect can be used to convert mechanical energy into electrical energy and vice versa.

In practice, devices using the piezoelectric effect are termed piezoelectric transducers. Piezoelectric transducers are used from DC to microwave frequencies, and the current record[10] for high-frequency excitation of bulk acoustic waves is 114 GHz.[†] The DC applications of piezoelectrics are, for instance, different pressure transducers and strain gauges; low-frequency applications are in sonars, phonograph pickups, loudspeakers, and ultrasonic cleaning machines. In a medium frequency range, approximately from 1 to 200 MHz, piezoelectric transducers are used extensively in frequency control systems as resonators and also in nondestructive testing (NDT) applications. In the 0.001 to 2 GHz range they are used in acousto-optical modulators, and in the microwave range (up to 18 GHz) they are used as delay lines.

5.4 PIEZOELECTRIC THIN-DISK TRANSDUCERS

In the simplest possible applications, bulk transducers are made of thin piezoelectric disks and have either a single electric port as shown in Fig. 5.3(a) when used as resonators because of their high Q factor, or an electric port and an acoustic port as shown in Fig. 5.3(b) when used as generators of acoustic waves in liquids, or again an electric and an acoustic port with the acoustic port bonded to a solid, when used as generators of acoustic waves in solids. In NDT applications, transducers are usually acoustically coupled to the specimen under investigation by means of a liquid couplant or gel, and in delay-line applications they are firmly bonded to the solid. In the resonator applications, Fig. 5.3(a), if the transducer material M_b is assumed lossless and the two surrounding media (M_a and M_c) are air, the electric input impedance is a pure reactance. In applications (b) and (c) in Fig. 5.3, one of the important parameters is the ratio of the available electric power to the total radiated acoustic power. Depending on the ratio of the impedances of media M_a and M_c, the piezoelectric disk M_b acts as a low-Q resonator in these two cases as well. In order to achieve a broad-band operation, the Q factor must be lowered either with lossy backing material of appropriate impedance or with additional matching layers, like M_c in Fig. 5.3(c).

In all cases, the thickness l of the piezoelectric usually between $\lambda_a/4$ and $\lambda_a/2$, is much smaller (at least by an order of magnitude) than the lateral dimensions. The operating frequency of thin-disk transducers is determined by the thickness l and the boundary conditions, this being the reason they are sometimes called thickness-mode transducers. Shear modes are usually used in resonator applications, while both shear and longitudinal (also called dilatational or expander) modes are used for other applications.

Since the two lateral dimensions of the material are very large compared with the thickness and the wavelength, the thin-disk geometry is, in essence, an

---

[†]Some later work was done in which the electric field of light around 10^{12} Hz excited bulk acoustic waves directly in a piezoelectric (private communication).

one-dimensional model. The metallic electrodes are assumed to be very thin, providing no mass loading. In Fig. 5.4(a), if the z axis is a pure-mode axis, one type of mode can be excited, either longitudinal or shear, assuming that the lateral dimensions of the disk (in the x and y directions) are infinite. In practice the latter is not satisfied, and thus a certain amount of excitation of other modes occurs as well. Sometimes z is not a pure-mode axis. In spite of these limitations, the one-dimensional theory to be presented agrees well with experimental results and is routinely used by transducer engineers.

In what follows, only longitudinal mode excitation is considered, with the wave propagating in the z direction with a collinear electric field E_z. All the field quantities have components only in the z direction and are functions of either z or time t or both. Thus, the nonzero field quantities are $S_{zz} = S_3$, E_z, v_z, u_z, and $T_{zz} = T_3$. The object of the computation which follows is to develop an electrical analogue, shown in Fig. 5.4(b) of the thin-disk transducer shown

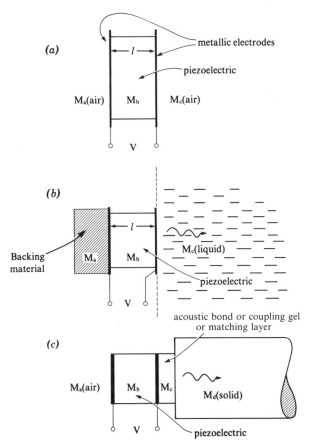

Figure 5.3. Schematic of thin-disk transducers. (a) Resonator. (b) Generator of acoustic waves in liquids. (c) Generator of acoustic waves in solids.

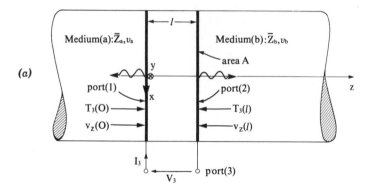

(a)

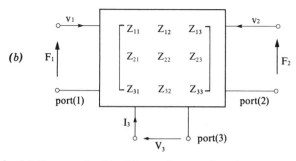

(b)

Figure 5.4. (a) Geometry for thin-disk transducer studies. (b) Electrical equivalent of (a) with acoustic ports (1 and 2) and electric port (3).

in Fig. 5.4(a). Comparing Fig. 5.4(a) and (b) it follows that

$$v_z(0) = v_1 \quad \text{and} \quad v_z(l) = -v_2 \tag{5.31}$$

In transducer design, it is customary to use the force acting on a surface of the transducer rather than the stress. Remembering that the traction force is acting from the piezoelectric (Sect. 1.1), $T_3(0)$ and $T_3(l)$ must be chosen as shown in Fig. 5.4(a). On the other hand, by circuit theory convention, in Fig. 5.4(b) the traction forces are so directed that the power is flowing toward the piezoelectric. These considerations lead to

$$T_{(1)} = -T_3(0) \quad \text{and} \quad T_{(2)} = -T_3(l) \tag{5.32}$$

where the subscripts in parentheses indicate the circuit ports rather than the axes. It follows that the traction forces are

$$F_1 = AT_{(1)} = -AT_3(0) \quad \text{and} \quad F_2 = AT_{(2)} = -AT_3(l) \tag{5.33}$$

where A is the cross-sectional area of the transducer, $A = w\tau$, Fig. 5.5(a).

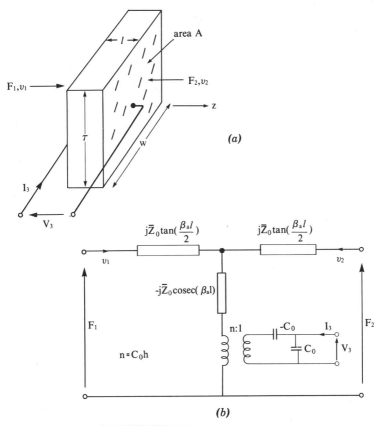

Figure 5.5. (*a*) Piezoelectric thin-disk transducer. (*b*) Mason's equivalent circuit (electric field parallel with particle velocity).

The current I_3 in Fig. 5.4(*a*), given by Eq. (5.16), is found to be

$$I_3 = j\omega D_z A = J_c A \qquad (5.34)$$

where $j\omega D_z$ is the displacement current density in the piezoelectric and J_c is the conduction current density in the metal electrodes. Thus D_z must be nonzero and independent of coordinate z in the region of the piezoelectric. The solution, Eq. (5.29), of the wave equation (5.23) has to be modified to take this into account. Since the piezoelectric region is of finite thickness, both terms in Eq. (5.24) are now required. Suppressing the time dependence, this can be written as

$$v_z(z) = (M'\sin \beta_a z + N'\cos \beta_a z) j\omega \qquad (5.35)$$

and

$$D_z = \text{constant} = \frac{I_3}{j\omega A} \qquad (5.36)$$

It is easy to show that Eqs. (5.35) and (5.36) are also a solution of the wave equation with unknowns M' and N'. From Eq. (5.31), $v_z(0) = v_1$, and on using Eq. (5.35) it follows that

$$v_1 = N'j\omega \tag{5.37}$$

and for $v_z(l) = -v_2$,

$$-v_2 = j\omega M' \sin \beta_a l + v_1 \cos \beta_a l \tag{5.38}$$

resulting in

$$j\omega M' = -v_2 \csc \beta_a l - v_1 \cot \beta_a l \tag{5.39}$$

From Eqs. (5.35), (5.37), and (5.39), the solution for the velocity inside the piezoelectric is thus obtained, while D_z is given by Eq. (5.36). All other field quantities can be derived once $v_z(z)$ and D_z are determined. The electric field is obtained from Eq. (5.21b) as

$$E_z = \frac{1}{\varepsilon^S}\left[D_z - \frac{e}{j\omega}\frac{\partial v_z}{\partial z}\right] \tag{5.40}$$

and on substitution of Eq. (5.40) into (5.21b) it follows that

$$T_3(z) = -\frac{e}{\varepsilon^S}D_z + \frac{1}{j\omega}\frac{\partial v_z}{\partial z}\left[c^E + \frac{e^2}{\varepsilon^S}\right] \tag{5.41}$$

Using Eqs. (5.30) and (5.36) and introducing

$$h = \frac{e}{\varepsilon^S} \tag{5.42}$$

the notation used in Eq. (5.9a), it follows that

$$T_3(z) = \frac{-hI_3}{j\omega A} + \frac{c^D}{j\omega}\frac{\partial v}{\partial z} \tag{5.43}$$

and, on using Eq. (5.35), the expressions for stresses at the boundaries are

$$T_3(0) = \frac{-hI_3}{j\omega A} + \frac{c^D \beta_a M' j\omega}{j\omega}$$

$$T_3(l) = \frac{-hI_3}{j\omega A} + \frac{c^D \beta_a}{j\omega}[M'\cos \beta_a l - N'\sin \beta_a l]j\omega \tag{5.44}$$

Using Eqs. (5.37) and (5.39), this can be written as

$$T_3(0) = -\frac{h}{j\omega A}I_3 - \frac{\left(\rho_m c^D\right)^{1/2}}{j}\left[v_2\csc\beta_a l + v_1\cot\beta_a l\right]$$

$$T_3(l) = -\frac{h}{j\omega A}I_3 - \frac{\left(\rho_m c^D\right)^{1/2}}{j}\left[v_1\csc\beta_a l + v_2\cot\beta_a l\right] \qquad (5.45)$$

The voltage V_3 across the plate is found from

$$V_3 = \int_0^l E_z \, d_z \qquad (5.46)$$

where E_z, obtained from Eqs. (5.35), (5.36), and (5.40), is given by

$$E_z(z) = \frac{1}{\varepsilon^S}\left[\frac{I_3}{j\omega A} - \frac{e}{j}\left(\frac{\rho_m}{c^D}\right)^{1/2} j\omega\left(M'\cos\beta_a z - N'\sin\beta_a z\right)\right] \qquad (5.47)$$

yielding

$$V_3 = \frac{1}{\varepsilon^S}\left[\frac{I_3 l}{j\omega A} - \frac{e}{j\omega}\left(M'\sin\beta_a l + N'\cos\beta_a l - N'\right)j\omega\right] \qquad (5.48)$$

and on using Eqs. (5.37) and (5.39) it follows that

$$V_3 = \frac{I_3}{j\omega C_0} + \frac{e}{j\omega\varepsilon^S}(v_1 + v_2) \qquad (5.49)$$

where $C_0 = \varepsilon^S A/l$ is the clamped (zero strain) capacitance of the transducer. Using Eqs. (5.33), (5.45), and (5.49), it is now possible to formulate the Z matrix of the circuit shown in Fig. 5.4(b) as

$$\begin{bmatrix} F_1 \\ F_2 \\ V_3 \end{bmatrix} = -j\begin{bmatrix} \bar{Z}_0\cot\beta_a l & \bar{Z}_0\csc\beta_a l & h/\omega \\ \bar{Z}_0\csc\beta_a l & \bar{Z}_0\cot\beta_a l & h/\omega \\ h/\omega & h/\omega & 1/\omega C_0 \end{bmatrix}\begin{bmatrix} v_1 \\ v_2 \\ I_3 \end{bmatrix} \qquad (5.50)$$

where

$$\bar{Z}_0 = A Z_0 = A\sqrt{\rho_m c^D}$$

$$\beta_a = \omega\left(\frac{\rho_m}{c^D}\right)^{1/2} = \frac{\omega}{v_L^D}$$

$$c^D = c^E + \frac{e^2}{\varepsilon^S} \qquad (5.51)$$

$$h = \frac{e}{\varepsilon^S}$$

$$C_0 = \frac{\varepsilon^S A}{l}, \qquad A = w\tau$$

The impedance $\bar{Z}_0$ defined by Eq. (5.51) is measured in units of kg/s. Using Eq. (5.50) with appropriate impedances at ports 1 to 3, it is now possible to compute all the parameters relevant to the transducer design. An alternative way would be to devise an electrically equivalent circuit describing the performance of the transducer in terms of lumped or distributed parameters. The problem of finding an equivalent circuit from a set of equations is not unique, since a very large number of circuits could be devised. Mason's equivalent circuit,[11] suggested in 1948 and shown in Fig. 5.5, has been used and is presently being used by many engineers. The transformer ratio of this circuit, given by

$$n = hC_0 = \frac{eA}{l} \tag{5.52}$$

depends on dimensions of the piezoelectric disk and the piezoelectric constant e. Using the relation $\csc x = \cot(x/2) - \cot x$, it is easy to show the equivalence between the equivalent circuit and Eq. (5.50). In a nonpiezoelectric medium, $e = 0$ and the transformer is short-circuited, the resulting equivalent network being applicable to longitudinal wave propagation in nonpiezoelectrics. If the electric port is open-circuited, $I_3 = 0$, the sum of the two capacitances is zero and the transformer is again short-circuited. An interesting case occurs when both acoustic ports are rigidly clamped. This condition corresponds to "open" acoustic ports ($v_1 = v_2 = 0$). To achieve this condition, the acoustic ports must be terminated with very high acoustic impedances. The highest acoustic impedance available (Table 1, Sect. 1.2) is that of tungsten, whose value is 105×10^6 in SI units, and should be compared with values 30×10^6 to 36×10^6 in SI units being typical for impedances of ceramic disks. Thus it is seen that the clamped condition is impossible to realize properly. Nevertheless, if the clamped condition is applied to Eq. (5.50) with $I_3 = 0$, it follows that

$$\frac{F_1}{V_3} = \frac{F_2}{V_3} = n \tag{5.53}$$

which means that if the forces are applied on the acoustic ports from terminations having very large impedances, all other impedances in the circuit can be neglected, the circuit being reduced to the transformer only.

Depending on the application, various boundary conditions can now be imposed and the relevant parameters found. For instance, $F_1 = F_2 = 0$ corresponds to the resonator, that is, unloaded transducer, shown in Fig. 5.3(a), and the electric input impedance can be found as $Z_3 = V_3/I_3$. The configuration shown in Fig. 5.3(b) corresponds to

$$F_1 = -\bar{Z}_a v_1 \quad \text{and} \quad F_2 = -\bar{Z}_c v_2 \tag{5.54}$$

where $\bar{Z}_a$ and $\bar{Z}_c$ are the acoustic characteristic impedances of media M_a and M_c, respectively. The device in Fig. 5.3(c) corresponds to $F_1 = 0$ and $F_2 = -\bar{Z}'_c v_2$ in Eq. (5.50) where $\bar{Z}'_c$ is the equivalent acoustic impedance to the right of the piezoelectric (medium M_b).

Another equivalent circuit, more recently (1970) proposed by Krimholtz et al.,[12] is shown in Fig. 5.6. The parameters of the circuit are

$$C_0 = \frac{\varepsilon^S A}{l}$$

$$v^D = v_L^D = \left(\frac{c^D}{\rho_m} \right)^{1/2}$$

$$\overline{Z}_0 = A Z_0 = A \rho_m v^D = A \left(\rho_m c^D \right)^{1/2} \qquad (5.55)$$

$$m = \frac{\omega \overline{Z}_0}{2h} \csc \left(\frac{\omega l}{2 v^D} \right)$$

$$Z_1 = j \overline{Z}_0 \left(\frac{h}{\omega \overline{Z}_0} \right)^2 \sin \left(\frac{\omega l}{v^D} \right)$$

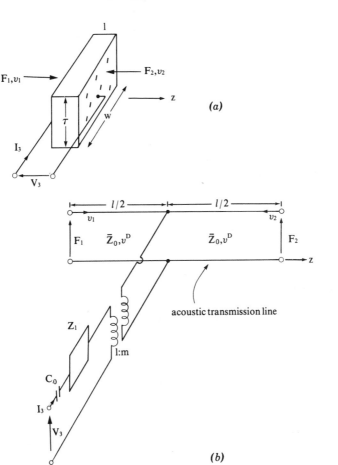

Figure 5.6. (*a*) Piezoelectric thin-disk transducer. (*b*) Equivalent circuit (electric field parallel with particle velocity) proposed by Krimholtz et al.[12]

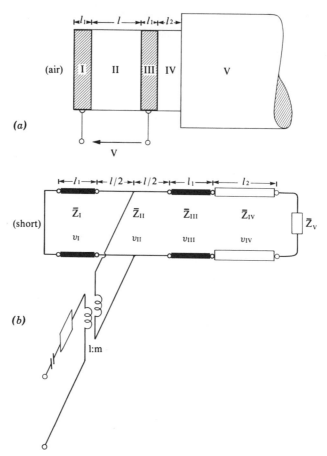

Figure 5.7. (*a*) A more realistic configuration of the thin-disk transducer: I and III, metallic layers; II, piezoelectric layer; IV, adhesive bond layer; V, nonpiezoelectric material. (*b*) Equivalent circuit of the structure.

In this equivalent circuit, $\bar{Z}_0$ and v^D are the acoustic characteristic impedance, measured in units of kg/s, and the longitudinal velocity, respectively, associated with the acoustic transmission line. The line length is equal to the thickness of the transducer. Thus, it would appear that this equivalent circuit is closer to physical reality, since a distinction is drawn between the lumped-element electrical behavior and the distributed acoustical behavior of the transducer. To demonstrate this point, consider the more realistic type of transducer construction shown in Fig. 5.7. By inspection of physical configuration, the equivalent circuit can be immediately drawn.

5.5 TIME DOMAIN ANALYSIS OF THE TRANSDUCER

An additional physical insight of the transducer operation can be gained by considering its behavior in the time domain. For the sake of clarity, the

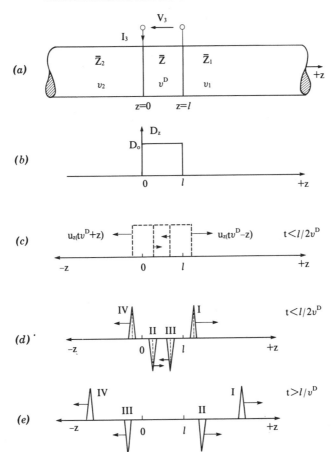

Figure 5.8. (*a*) One-dimensional geometry used in time-domain studies. (*b*) Electric displacement. (*c*) Mechanical displacement waves. (*d*), (*e*) Stress waves.

boundary effects will be eliminated at first by assuming that the transducer terminations, shown in Fig. 5.8(*a*), are of infinite extent and are made out of material whose acoustic properties are identical with those of the thin-disk transducer itself, that is,

$$v_1 = v_2 = v^D \quad \text{and} \quad \bar{Z}_1 = \bar{Z}_2 = \bar{Z} \tag{5.56}$$

In this case, if the conductor layers are assumed infinitely thin, there will be no reflections from the boundaries at $z = 0$ and $z = l$, since the system is acoustically homogeneous in the z direction.

In order to investigate the origin of acoustic generation in the transducer when voltage $V_3(t)$ is applied to port 3 of Fig. 5.5(*b*),[†] consider Eqs. (5.9b),

---

[†]It is a simple matter to express Eqs. (5.50) in terms of v_1, v_2, and V_3 as independent variables.

(5.10), and (5.22), which are repeated here for convenience:

$$E_z = -hS_3 + \beta^S D_z \tag{5.9b}$$

$$T_3 = c^E S_3 - eE_z \tag{5.10}$$

$$\frac{\partial T_3}{\partial z} = \rho_m \frac{\partial^2 u_z}{\partial t^2} \tag{5.22}$$

When E_z is eliminated from the first two equations and then T_3 substituted in the last equation, it follows that

$$\rho_m \frac{\partial^2 u_z}{\partial t^2} - (c^E + eh)\frac{\partial^2 u_z}{\partial z^2} = -e\beta^S \frac{\partial D_z}{\partial z} \tag{5.57}$$

For the one-dimensional case, it follows that h, the piezoelectric transmitting constant, can be expressed as $h = e/\varepsilon^S$, and β^S, the dielectric impermeability, as $\beta^S = 1/\varepsilon^S$. Using Eq. (5.30), this can be written in more compact form as

$$\frac{\partial^2 u_z}{\partial z^2} - \frac{1}{v^{D2}} \frac{\partial^2 u_z}{\partial t^2} = \frac{e}{\varepsilon^S c^D} \frac{\partial D_z}{\partial z} = \frac{1}{e} \frac{K^2}{1 + K^2} \frac{\partial D_z}{\partial z} \tag{5.58}$$

For convenience, a constant k_T, termed *the piezoelectric coupling constant for transversely clamped material* (no motion in the direction transverse to the electric field, that is, $v_x = v_y = 0$), is defined as

$$k_T^2 \equiv \frac{K^2}{1 + K^2} \tag{5.59}$$

where K, the piezoelectric coupling factor, is given by Eq. (5.27). The solution of Eq. (5.58) is given by (Ref. 4, Chap. 3)

$$u_z(t, z) = -\frac{v^D k_T^2}{2e} \int\int d\tau \, d\xi \, \chi\left[\frac{(t - \tau)}{T} - \frac{|z - \xi|}{l}\right] \frac{\partial D_z(\xi/l, \tau/T)}{\partial \xi} \tag{5.60}$$

where the integration is performed over the space–time volume occupied by the source region, and where the normalization constant is $T = l/v^D$. To find the impulse response, we set $D_z = D_z(\xi/l)\delta(\tau/T)$, where δ is the Dirac delta function, $D_z(z)$ is given in Fig. 5.8(b) with $D_0 = lI_3/v^D A$ (Exercise 1), and I_3 is the peak value of the RF current.

The solution of Eq. (5.60) (Exercise 1) is given by

$$
u_z(t, z) = -\frac{lk_T^2}{2e} \times
\begin{cases}
D_z\left(\dfrac{tv^D - z}{l}\right), & z > l \\[3mm]
D_z\left(\dfrac{tv^D + z}{l}\right), & z < 0
\end{cases}
\tag{5.61}
$$

indicating that, on impulse excitation of electric displacement, two mechanical displacement waves will be excited, as shown in Fig. 5.8(c). The particle velocity of the waves is

$$
v_z(t, z) = -\frac{v^D k_T^2}{2e} \times
\begin{cases}
\dot{D}_z\left(\dfrac{tv^D - z}{l}\right), & z > l \\[3mm]
\dot{D}_z\left(\dfrac{tv^D + z}{l}\right), & z < 0
\end{cases}
\tag{5.62}
$$

where the dot indicates differentiation with respect to the total argument. The corresponding stresses are

$$
T_3(t, z) = -Zv_z(t, z) = \frac{v^D k_T^2 Z}{2e} \dot{D}_z\left(\frac{tv^D - z}{l}\right), \qquad z > l \tag{5.63}
$$

for waves propagating in the $+z$ direction, and

$$
T_3(t, z) = Zv_z(t, z) = -\frac{v^D k_T^2 Z}{2e} \dot{D}_z\left(\frac{tv^D + z}{l}\right), \qquad z < 0 \tag{5.64}
$$

for waves propagating in the $-z$ direction. It follows from Fig. 5.8(b) that the electric displacement can be written as

$$
D_z\left(0, \frac{z}{l}\right) = D_0\left[\chi\left(\frac{z}{l}\right) - \chi\left(\frac{z-l}{l}\right)\right] \tag{5.65}
$$

where χ is the step function, equal to unity for $t \geq 0$ and zero otherwise. The derivative of D_z with respect to the total argument is

$$
\dot{D}_z = D_0\left[\delta\left(\frac{z}{l}\right) - \delta\left(\frac{z-l}{l}\right)\right] \tag{5.66}
$$

and it follows from Eqs. (5.63) and (5.64) that the gradient of the electric displacement is the origin of the acoustic waves in the transducer. It is now evident that the matched transducer, when excited with impulse-type electric displacement, will produce four stress impulses, two propagating in the $+z$ direction and two propagating in the $-z$ direction, as shown in Fig. 5.8(d).

These impulses originate in the conductor region, assumed infinitely thin, at the interfaces between the piezoelectric and the matched terminations.

Concentrating exclusively on the progressive wave (the one propagating in the $+z$ direction), it follows from Eqs. (5.62), (5.63), and (5.66) that

$$T_3(t, z) = -Zv_z(t, z)$$

$$v_z(t, z) = -v_0 \left[\delta\left(\frac{tv^D - z}{l} \right) - \delta\left(\frac{tv^D - z + l}{l} \right) \right] \qquad (5.67)$$

where

$$v_0 = \frac{v^D k_T^2 D_0}{2e} \qquad (5.68)$$

It is now possible to discuss the transient response of the transducer by following the rule that the shortest impulse response corresponds to the largest possible bandwidth. When the system is acoustically homogeneous in the z direction, the impulse response for $z > l$ (or $z < 0$) consists of two impulses only, yielding the largest possible bandwidth but not the best power transfer, since one-half of the power is lost to the regressive wave (propagating in the $-z$ direction). Therefore the transducer will have at least 3 dB conversion loss. In order to improve the conversion loss, it is tempting to terminate the plane $z = 0$ with zero impedance, corresponding to an air-backed transducer. Note that terminating the plane $z = 0$ with a very large (ideally infinite) impedance is not technologically feasible, due to a limited range of available acoustic impedances (see Sect. 1.2). In the case of an air-backed transducer, the pulse propagating from the $z = l$ plane toward the $z = 0$ plane is totally reflected at $z = 0$ with a 180° change of phase, see Eq. (1.23), thereafter propagating in the $+z$ direction and arriving at time $v^D/2l$ after the first pulse at the plane $z = l$. This is shown in Fig. 5.9. A lossless air-backed transducer thus provides maximum power transfer but not the maximum bandwidth, since the duration of the impulse response time has been extended by 50%.

In a more general case, when both planes $z = 0$, l have certain reflection and transmission coefficients given by

$$r_1 = \frac{\overline{Z}_1 - \overline{Z}}{\overline{Z}_1 + \overline{Z}} \qquad \tau_1 = \frac{2\overline{Z}_1}{\overline{Z}_1 + \overline{Z}} \qquad (5.69)$$

$$r_2 = \frac{\overline{Z}_2 - \overline{Z}}{\overline{Z}_2 + \overline{Z}} \qquad \tau_2 = \frac{2\overline{Z}_2}{\overline{Z}_2 + \overline{Z}} \qquad (5.70)$$

and

$$\tau_3 = \frac{2\overline{Z}}{\overline{Z}_1 + \overline{Z}} \qquad \tau_4 = \frac{2\overline{Z}}{\overline{Z}_2 + \overline{Z}} \qquad (5.71)$$

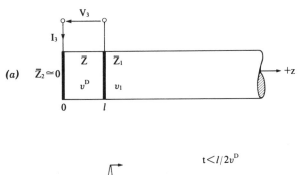

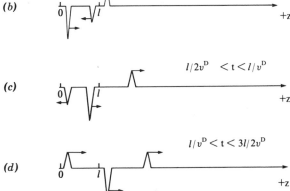

Figure 5.9. (*a*) An air-backed transducer. (*b*), (*c*), (*d*) Snapshots of stress waves upon impulse excitation.

where usually $r_1 < 0$ and $r_2 < 0$ for the lossless case; for $l/2v^D < t$, the amplitude of the impulse I in Fig. 5.8(*e*) should be multiplied by τ_1, and the amplitude of impulse III by τ_3.

For $l/2v^D < t$, the amplitude of impulse II should be multiplied by τ_4 and the amplitude of impulse IV by τ_2. For $l/v^D < t < 2l/v^D$, two impulses, not shown in the figure, are propagating in the region $0 < z < l$. They will produce a series of impulses in the regions $z < 0$ and $z > l$ for $t > 2l/v^D$. Their amplitudes are easy to find using Eqs. (5.69) through (5.71). It is evident that the response, for instance, in the region $z > l$ will consist of an alternate series of decaying pulses (ringing), their exact nature being dependent on the reflection and transmission coefficient and the length l of the piezoelectric region. Under these circumstances, the transducer behaves as an imperfect resonator. Due to a long impulse response, the bandwidth is reduced and the efficiency is low, since power is lost due to radiation into the $z < 0$ region. In practice this situation prevails, since the impedance of piezoelectric materials used for the thin disk is typically in the range $30–36 \times 10^6$ SI units, while the impedance of water is 1.5×10^6 SI units, leading to a mismatch ratio of

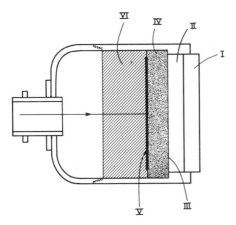

Figure 5.10. Schematic of a transducer construction. I and II, matching layers; III and V, conducting layers; IV, piezoelectric disk; VI, backing material.

20–23. This is a much higher mismatch ratio than the one encountered in optics, where, for instance, for a glass-to-air interface the mismatch ratio is approximately 1.5. In acoustics, due to a high mismatch ratio, more than one quarter-wave layer is needed for the perfect match of a single frequency. At present, the accuracy required in constructing such layers is very difficult to achieve. For instance, a glass quarter-wave plate at 3 MHz has a thickness of approximately 300 μm, and a quarter-wave epoxy resin layer, about 150 μm. However, a production error of, say, 20% in the thickness of the epoxy bond will cause substantial deformation in the amplitude of the transfer function, rendering the device almost useless. In a more classical design of the transducers, the matching of the emitting (receiving) face of the transducer is not attempted, and only the loading of the back face by a (lossy) material is made. In practice,[13] both methods are presently used as shown in Fig. 5.10. For broadband transducers, an important design parameter is the thickness of the disk, since for a monochromatic wave only a quarter-wave transformer provides a perfect match. In broadband design,[14] for instance, $\lambda/4$ and $3\lambda/4$ transformers do not give the same impulse response, and the thickness of the disk l is usually optimized in the region $\lambda/4 < l < \lambda/2$.

Finally, it is worth mentioning that the geometric configuration of neither equivalent circuit, described in Sect. 5.4, suggests the complex generation process, dependent on the gradient of the electric displacement, presented in this section.

5.6   THE THIN-DISK RESONATOR

For the thin-disk resonator, shown in Fig. 5.3(a), with infinitely thin metallization layers and air-loading on both faces, it follows that $F_1 = F_2 = 0$ in Eq. (5.50). Under these conditions the fields inside the resonator are found from

Eqs. (5.35), (5.38), (5.39), and (5.50) and are given by

$$E_z = j\frac{e}{\varepsilon^S}\frac{v}{v_L^D}\frac{[(1 + K^2)/K^2]\cos(\beta_a l/2) - \cos[\beta_a(l/2 - z)]}{\sin(\beta_a l/2)} \quad (5.72)$$

$$S_3 = j\frac{v}{v_L^D}\frac{\cos[\beta_a(l/2 - z)]}{\sin(\beta_a l/2)} \quad (5.73)$$

and

$$v_z(z) = j\omega u_z(z) = v\frac{\sin[\beta_a(l/2 - z)]}{\sin(\beta_a l/2)} \quad (5.74)$$

with $v_1 = v_2 = v$ and $D_z = I_3/j\omega A$,

$$I_3 = -\frac{\overline{Z}_0\omega v}{h}\cot\left(\frac{\beta_a l}{2}\right) \quad (5.75)$$

and

$$V_3 = -\frac{2jhv}{\omega} - \frac{jI_3}{\omega C_0} \quad (5.76)$$

Combining Eqs. (5.75) and (5.76), the expression for electrical input impedance is obtained as

$$Z_3 = \frac{V_3}{I_3} = \frac{1}{j\omega C_0}\left[1 - k_T^2\frac{\tan(\beta_a l/2)}{\beta_a l/2}\right] \quad (5.77)$$

where k_T^2 is the piezoelectric coupling constant for transversely clamped material given by Eq. (5.59). The appearance of this constant is to be expected, since the transverse dimensions of the resonator are much larger than its thickness, thus resulting in negligible transverse motion. It is convenient for the purpose of the analysis to write Eq. (5.77) in the form

$$Z_3 = \frac{1}{j\omega C_0} + Z_m \quad (5.78)$$

where the first term is the clamped capacitive reactance of the transducer, and the second term is the equivalent motional impedance, which can be either inductive or capacitive in character. In practice, the inductive character of this impedance is important. The series equivalent circuit corresponding to Eq. (5.78) and the admittance $Y_3 = 1/Z_3$ are shown in Fig. 5.11.

The frequencies of parallel resonance (antiresonance) are obtained from Eq. (5.77) for $Y_3 = 0$. This results in $\beta_a l = \pi(2n + 1)$, and using $\omega = v_L^D\beta_a$ it

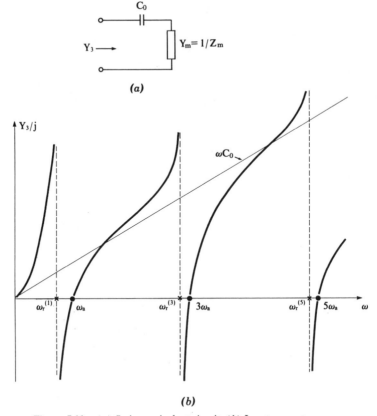

Figure 5.11. (*a*) Series equivalent circuit. (*b*) Input susceptance.

follows that the antiresonance frequency of the nth mode is given by

$$\omega_a^{(n)} = \frac{\pi(2n + 1)v_L^D}{l} = (2n + 1)\omega_a \tag{5.79}$$

where

$$\omega_a = \frac{\pi v_L^D}{l} \tag{5.80}$$

is the lowest antiresonance frequency corresponding to a transducer thickness of half the acoustic wavelength. Equation (5.80) can be written as

$$f_a = \frac{v_L^D}{2l} = \frac{K_F}{l} \tag{5.80a}$$

where K_F is the frequency constant. This parameter is usually expressed in units of kHz · mm or MHz · μm. For instance, for AT-cut quartz, $K_F = 1664$ kHz · mm, and for X-cut quartz $K_F = 2849$ kHz · mm. Thus a 1-mm-thick AT-cut quartz plate will resonate with a frequency of 1.664 MHz. It is to be noted that the frequencies of harmonic modes $n = 1, 2, 3, \ldots$ in Eq. (5.79) in real resonators are not precisely integral multiples of the fundamental and, depending on the method of construction, may be either slightly smaller or larger than $(2n + 1)f_a$.

The resonant frequencies $\omega_r^{(n)}$ are obtained from $Z_3 = 0$ and are given by the solution of the equation

$$\frac{\tan\left[(\pi/2)\left(\omega_r^{(n)}/\omega_a\right)\right]}{(\pi/2)\left(\omega_r^{(n)}/\omega_a\right)} = \frac{1}{k_T^2} \tag{5.81}$$

where Eq. (5.80) and the dispersion relation $\omega = v_a\beta_a$ were used.

In order to develop an equivalent circuit for Z_3, the relation

$$\tan y = \sum_{n=0}^{\infty} \frac{2y}{\left[(2n + 1)\pi/2\right]^2 - y^2} \tag{5.82}$$

is used in the expression for impedance Z_m, Eq. (5.78), and

$$Z_m = -\frac{1}{j\omega C_0} \sum_{n=0}^{\infty} \frac{k_{T,n}^2}{1 - \omega^2/\omega_a^{(n)2}} \tag{5.83}$$

is obtained, where $k_{T,n}^2$ is the piezoelectric coupling constant for the nth mode given by

$$k_{T,n}^2 = \frac{8k_T^2}{\left[(2n + 1)\pi\right]^2}. \tag{5.84}$$

Combining Eqs. (5.78) and (5.83) (see Exercise 2), it follows that

$$Z_3 = \frac{1}{j\omega C_0}\left[1 - k_T^2 - \sum_{n=0}^{\infty} \frac{\omega k_{T,n}^2}{\omega_a^{(n)2} - \omega^2}\right] \tag{5.85}$$

which is useful in finding the equivalent circuit. The latter is shown in Fig. 5.12(a). It is seen that the total clamped capacitance is $C_0/(1 - k_T^2)$, while the other elements' values are given by

$$C_n = \frac{C_0}{k_{T,n}^2} \qquad \text{and} \qquad L_n = \frac{1}{C_n\omega_a^{(n)2}} \tag{5.86}$$

For operation in the vicinity of the harmonic (antiresonance) frequency $\omega_a^{(n)}$, the equivalent circuit can be approximated by the circuit shown in Fig.

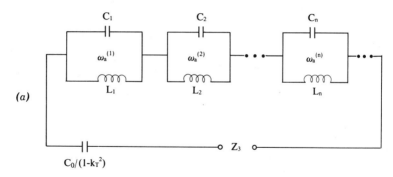

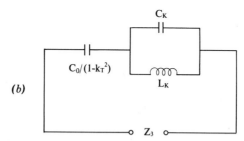

Figure 5.12. (*a*) Exact series equivalent circuit of the resonator. (*b*) Approximate equivalent circuit for operation near frequency $\omega_a^{(k)}$.

5.12(*b*).

Piezoelectric plates thinner than 60 μm are too fragile to handle in the routine operations required for piezoelectric resonator production. On the other hand, operation at higher harmonic frequencies is achieved because of the lower coupling constant, as given by Eq. (5.84). In frequency control applications, a compromise is reached by operating up to the fifth, and sometimes the seventh, harmonic frequency, resulting in the frequency coverage up to 300 MHz. This is necessary in order to avoid the thin disks that would be required for fundamental mode operation at higher frequencies, Eq. (5.80). Two resonators are shown in Fig. 5.13.

In the crystal filter design, where the resonators are used as circuit elements, the fundamental mode is usually employed. In this application the useful characteristic of the resonator is in the frequency interval $\omega_r^{(1)} \equiv \omega_s$ to $\omega_a \equiv \omega_p$, where the susceptance is inductive, as shown in Fig. 5.11(*b*). In this case, higher order terms in Eq. (5.85) are usually neglected, resulting in the approximation

$$Z_3 \approx \frac{1}{j\omega C_0}\left[1 - k_T^2 - \frac{8k_T^2}{\pi^2}\frac{\omega^2}{\omega_p^2 - \omega^2}\right] \qquad (5.87)$$

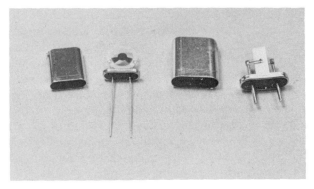

Figure 5.13. Crystal resonators. (*Left*) A 20-MHz 3rd overtone resonator. (*Right*) A 500-kHz fundamental frequency resonator. (*Courtesy of J. F. Gehrels, Leigh Instruments Ltd., Don Mills, Ontario.*)

with the equivalent circuit shown in Fig. 5.12(b) having the antiresonance (parallel resonance) at ω_p and the series resonance at ω_s. The latter is determined from Eq. (5.87) for $Z_3 = 0$ and is given by

$$\frac{\omega_p - \omega_s}{\omega_s} \approx \frac{4}{\pi^2} k_T^2 \tag{5.88}$$

Thus, the coupling constant k_T limits the relative frequency interval ω_s to ω_p, which is useful in crystal filter design. In some applications, such as frequency synthesizers, LiTaO$_3$ or LiNbO$_3$ crystals must be used instead of quartz in order to obtain a larger frequency interval (ω_s, ω_p). Another equivalent circuit often used in practice is shown in Fig. 5.14(b) and can be derived from Eq. (5.87) by setting $Z_3(s) = X(s)$, where

$$X(s) = \frac{1 + L_s C_s s^2}{s\left(s^2 L_s C_s C_p + C_s + C_p\right)} \tag{5.89}$$

and where C_s and L_s are motional capacitance and inductance, respectively, C_p is the static capacitance, $s = j\omega$, and $r = 0$ in Fig. 5.14(b). The series and parallel resonance are obtained for $X(s) = 0$ and $X(s) = \infty$, respectively, and are given by

$$\omega_s^2 = \frac{1}{L_s C_s} \quad \text{and} \quad \omega_p^2 = \omega_s^2 \left[1 + \frac{C_s}{C_p}\right] \tag{5.90}$$

It is easy to show that

$$\frac{C_s}{C_p} = \frac{f_p^2 - f_s^2}{f_s^2} = \frac{\Delta f}{2 f_s} = \frac{f_p - f_s}{2 f_s} \tag{5.91}$$

which should be compared with Eq. (5.88). The impedance variation versus frequency for the approximate equivalent circuit with losses included is shown in Fig. 5.14(a). For instance, for AT-cut quartz operating in the shear mode, $\Delta f/f_s \approx 0.2\%$ ($k_T^2 \approx 0.005$). This results in the allowable values of C_p for circuit design of $C_p/C_s \geqslant 232$. The minimum value of C_s that can be routinely achieved is $C_s = 0.0194$ pF. The Q factor of inductance L_s for crystal resonators is in the range 10^4 to 1.5×10^5 and is dependent on the method of encapsulation of the resonator. In practice, the value of k_T is usually determined from Eq. (5.88) by measuring the frequencies ω_s and ω_p. However, care must be taken not to excite other modes that are present, because of the finite transverse dimensions and because the propagation direction of the cut chosen may not coincide with a pure-mode axis.

Recently the experimental work by Toki et al.[15] on resonators made with piezoelectric ceramics has demonstrated the inadequacy of the equivalent circuit

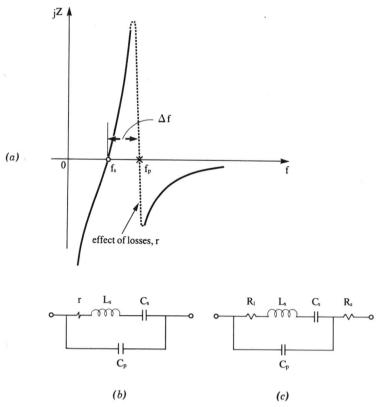

Figure 5.14. (a) Impedance variations of the approximate equivalent circuit in vicinity of $\omega_a = \omega_p$. (b) Approximate equivalent circuit for the fundamental mode of operation. (c) Improved equivalent circuit for the fundamental mode of operation.[15]

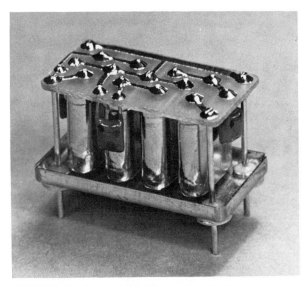

Figure 5.15. A 10.7-MHz 8-pole monolithic filter made in bulk resonator technology. (*Courtesy of J. F. Gehrels, Leigh Instruments Ltd., Don Mills, Ontario.*)

shown in Fig. 5.14(b). It was found that in the frequency range (f_s, f_p), Fig. 5.14(a), the resistance r is strongly frequency dependent, $r = r(f)$. For instance, for a variety of resonators with k_T in the range 0.15 to 0.60 and Q factors in the range 98 to 1030, where Q is defined as $Q = \omega_p L_s / r(f_p)$, the ratio $r(f_s)/r(f_p)$ varied from 1.05 to 1.42. Furthermore, Toki et al. demonstrated that the ratio

$$\frac{R_1 + R_s}{R_1} = \frac{r(f_s)}{r(f_p)}$$

is a resonator constant and showed that the equivalent circuit given in Fig. 5.14(c) with $R_1 = r(f_p)$ and $R_s = r(f_s) - r(f_p)$ is a far better representation of the resonator performance. In the new equivalent circuit, R_1 and R_s are constant in the frequency range of interest, or if the equivalent circuit shown in Fig. 5.14(b) is to be used, then

$$r = R_1 + R_s \left(\frac{f_p^2 - f_s^2}{f_p^2 - f_s^2} \right)^2 .$$

The new equivalent circuit takes into account the total dielectric loss of the encapsulated resonators. Physically, part of this dielectric loss is due to dielectric (and acoustic) losses in the piezoelectric material itself. With the increasing thickness l of the resonator, the inductance L_s and the resistances R_1 and R_s all increase in value. The inductance L_s is independent of the resonator diameter for typical commercially available resonators, while both R_1 and R_s are inversely

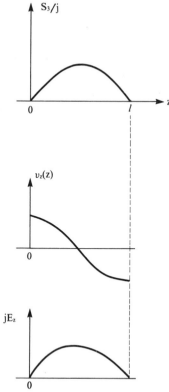

Figure 5.16.   Fields of the resonator operating at $\beta_a \pi = l$.

proportional to the diameter. A monolithic filter made with bulk resonator technology is shown in Fig. 5.15.

Using Eqs. (5.72) to (5.76), it is possible to study the fields of the resonator. For instance, fundamental mode operation ($\beta_a l = \pi$) results in the fields shown in Fig. 5.16.

5.7   TRANSDUCER AS EXCITER OF BULK WAVES

The basic trade-off in the design of thin-disk transducers is between bandwidth (duration of impulse response) and efficiency (sensitivity when the transducer is used as the receiver) as described in Sect. 5.5. In practice, the design[13, 14, 16–18] is carried out numerically by optimizing in some way configurations similar to the one shown in Fig. 5.10, also taking parasitic effects into account whenever possible.

It is generally accepted that below 100 MHz the thickness of metallic electrodes (usually on the order of 0.1 μm) can be neglected, as well as the associated mass-loading effects. At 100 MHz, for the acoustic velocity of 5 km/s, $\lambda/8 = 6\ \mu$m and the ratio of $\lambda/8$ to electrode thickness is about 60. The resulting reflection coefficient can be considered negligible. However, even below 100 MHz the RF resistance and inductance (skin effect) of the electrodes are certainly in evidence. The former increases the noise figure and competes with the radiation resistance of the transducer for its share of the input electric power, thus contributing to transducer losses. For this reason it is difficult to achieve a transducer input loss less than 3–5 dB.

Design parameters of interest are usually the electric input impedance Z_3, the input electric power P_3, and the output acoustic power P_2 as shown in Fig. 5.17(a) for an air-backed transducer. At present there is no unique way to achieve optimization of the circuit, for instance, shown in Fig. 5.7. However, a certain insight into the design can be gained by considering a much simpler case of an air-backed transducer, as shown in Fig. 5.17(a), where the metallization layers are assumed infinitely thin. Under these circumstances $F_1 = 0$ and $F_2 = -\overline{Z}_2 v_2$ in Eq. (5.50), where it is assumed that both $\overline{Z}_0$ and $\overline{Z}_2$ are real. It follows, after some algebra, that

$$Z_3 = \frac{V_3}{I_3} = \frac{1}{j}\left\{\frac{1}{\omega C_0} + \frac{k_T^2}{\omega C_0(\beta_a l)}\left[\frac{-j\overline{Z}_2/\overline{Z}_0\sin\beta_a l + 2(1 - \cos\beta_a l)}{-\sin\beta_a l + j\overline{Z}_2/\overline{Z}_0\cos\beta_a l}\right]\right\}$$

$$(5.92)$$

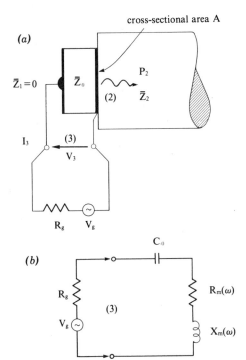

Figure 5.17. (a) An air-backed transducer radiating into material of acoustic impedance $\overline{Z}_2$. (b) Its input equivalent circuit.

or

$$Z_3 = \frac{1}{j\omega C_0} + Z_m \tag{5.93}$$

where Z_m is the motional acoustic impedance

$$P_3 = \frac{V_3 I_3^*}{2} = \left(\frac{1}{j\omega C_0} + Z_m\right)\frac{|I_3|^2}{2} = \frac{|I_3|^2}{2\,j\omega C_0} + P_m \tag{5.94}$$

where P_3 is the complex power delivered to port 3 in Fig. 5.17, P_m is the complex power delivered to the motional acoustic impedance Z_m, and

$$I_3 = \frac{V_g}{R_g + Z_m + 1/j\omega C_0} \tag{5.95}$$

The complex power delivered to medium $\bar{Z}_2$ is given by

$$P_2 = \tfrac{1}{2}(-v_2^* F_2) = \frac{\bar{Z}_2 |v_2|^2}{2} \tag{5.96}$$

Since $\bar{Z}_2$ is assumed real, the complex power is equal to the average power in this case. Furthermore, since the transducer is assumed to be lossless, it follows that

$$\mathrm{Re}\{P_3\} = \mathrm{Re}\{P_m\} = P_2 \tag{5.97}$$

where

$$P_m = \frac{Z_m |V_g|^2}{2\left[(R_g + R_m)^2 + (X_m - 1/\omega C_0)^2\right]} \tag{5.98}$$

and

$$Z_m = R_m + jX_m \tag{5.99}$$

with the input electrical circuit as shown in Fig. 5.17(b). A transducer used for longitudinal wave excitation in solids is shown in Fig. 5.18.

5.7.1 Unmatched Transducer at Resonance

At the parallel resonant frequency $\omega = \omega_a$, that is, at $\beta_a l = \pi$, the motional acoustic impedance obtained from Eq. (5.92) is real and is given by

$$Z_m(\omega_a) = R_m(\omega_a) = R_{m0} = \frac{4k_T^2}{\pi\omega_a C_0}\frac{\bar{Z}_0}{\bar{Z}_2} \tag{5.100}$$

Figure 5.18. A bulk transducer used for longitudinal wave excitation in solids. (*Courtesy of KB-Aerotech, Division of Krautkramer-Branson, Inc.*)

The resistance R_{m0} is called the *radiation resistance* of the resonant transducer radiating into material of acoustic impedance $\bar{Z}_2$. The power dissipated in the radiation resistance is obtained from Eq. (5.98) and is given for $\omega = \omega_a$ by

$$P_m(\omega_a) = \frac{R_{m0}|V_g|^2}{2\left[(R_g + R_{m0})^2 + (1/\omega_a C_0)^2\right]} \qquad (5.101)$$

The power available from the source is

$$P_g = \frac{|V_g|^2}{8R_g} \qquad (5.102)$$

The input power loss is then given by

$$\text{Input power loss} = \frac{P_m(\omega_a)}{P_g} = \frac{4R_g R_{m0}}{(R_g + R_{m0})^2 + (1/\omega_a C_0)^2} \qquad (5.103)$$

Since in practice R_g is typically 50 Ω, and R_{m0} is a few ohms, it follows that $R_g \gg R_{m0}$, and the ratio given by Eq. (5.103) is maximum for $R_g = 1/\omega_a C_0$. It is seen from Eqs. (5.100) and (5.51) that for a chosen material and operating frequency this condition determines the area A of the transducer. In the case of matched output $\bar{Z}_0 = \bar{Z}_2$, since $k_T^2 \ll 1$, it follows from Eq. (5.100) that $R_{M0} \ll 1/\omega_a C_0$.

For $R_g = 1/\omega_a C_0$, the minimum power loss (maximum power input), in dB, is given by

$$10 \log\left(\frac{P_m(\omega_a)}{P_g}\right) = 10 \log\left(\frac{2R_{m0}}{R_g}\right) \tag{5.104}$$

which is equal to $10 \log(2.5 k_T^2)$ in the case of matched output, $\bar{Z}_0 = \bar{Z}_2$. For $k_T = 0.3$ this results in the minimum input power loss of -6.5 dB. For the case when the capacitance C_0 is tuned out at resonance by inductance $L = 1/\omega_a^2 C_0$, the expression for input power loss, Eq. (5.103), depends only on resistances. If $R_g = R_{m0}$, there will be no input power loss. However, this is a highly idealized situation, since the lossy inductance, resistive losses in the electrodes, and resistance in the leads will always cause losses of a few decibels. At resonant frequency, the equivalent acoustic admittance $\bar{Y}_A$, looking into the transducer from the loading medium, as well as the equivalent acoustic force F_A caused by emf V_g can also be found. They are given by

$$F_A = \frac{2j(h/\omega_a)V_g}{R_g + 1/j\omega_a C_0} \tag{5.105}$$

and

$$\bar{Y}_A = \bar{G}_A + j\bar{B}_A = \frac{\pi}{4k_T^2}\left[\frac{R_g \omega_a C_0}{\bar{Z}_0} - j\frac{1}{\bar{Z}_0}\right] \tag{5.106}$$

where $\bar{G}_A$ is the acoustic conductance and $\bar{B}_A$ is the acoustic (inductive) susceptance. Equation (5.106) can also be written as

$$\bar{Y}_A = \frac{1}{4}\left(\frac{\omega_a}{h}\right)^2\left[R_g + \frac{1}{j\omega_a C_0}\right] \tag{5.107}$$

with the equivalent acoustic circuit shown in Fig. 5.19(a). Matching considerations, using the equivalent acoustic circuit, result in conclusions similar to those obtained with the equivalent electric circuit shown in Fig. 5.17(b).

5.7.2 Frequency Dependence

The case of the simplest matching circuit, the series inductance, has been considered in the preceding section. In the majority of cases, however, due to a high mismatch between R_g and R_{m0}, a transformer must be added to the matching circuit, as shown in Fig. 5.19(b). The inductance L is used to tune out the capacitance C_0 at resonance, and the transformer ratio is adjusted to be $n^2 R_{m0} = R_g$. This procedure results in the minimum power loss at frequency ω_a but also, due to the series resonant circuit L, C_0, R_{m0}, contributes to

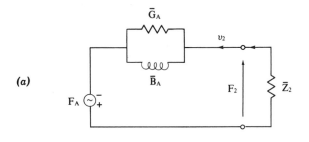

(a)

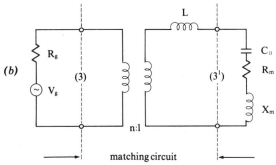

(b)

Figure 5.19. (a) Equivalent acoustic circuit at the interface with the load $\bar{Z}_2$. (b) Electric matching network consisting of a transformer and an inductance L.

reduction of the bandwidth. Under these circumstances it is customary to define a Q factor, Q_e, as an approximate measure of the electric bandwidth. The frequency dependence of the motional acoustic resistance R_m, Eq. (5.99), also contributes to the reduction of bandwidth and another quality factor, termed acoustic, Q_a is defined to take this effect into account. To clarify these points, consider the power being delivered to the load, Eq. (5.97), for the transducer shown in Fig. 5.17(a). Using Eq. (5.98), it follows that

$$P_2 = \frac{R_m(\omega)|V_g|^2}{2\left[(R_g + R_m(\omega))^2 + (X_m(\omega) - 1/\omega C_0)^2\right]} \tag{5.108}$$

The power impulse response of the transducer is obtained for $v_g(t) = \delta(t)$, that is, for $V_g = V_g(\omega) = 1$, in which case Eq. (5.108) represents the square of the unnormalized transfer function $H_u(\omega)$. Under these circumstances it follows that

$$P_2(\omega) = |H_u(\omega)|^2 = \frac{R_m(\omega)}{2\left[(R_G + R_m(\omega))^2 + (X_m(\omega) - 1/\omega C_0)^2\right]} \tag{5.109}$$

Similarly, the unnormalized transfer function of the impulse power available from the source is obtained from Eq. (5.102) and is given by

$$P_g(\omega) = |H_g(\omega)|^2 = \frac{1}{8R_g} \tag{5.110}$$

The normalized power transfer function $H_p(\omega)$ is obtained by division of Eqs. (5.109) and (5.110) and is given by

$$H_p(\omega) = \frac{P_2(\omega)}{P_g(\omega)} = \left|\frac{H_u(\omega)}{H_g(\omega)}\right|^2 = \frac{4R_g R_m(\omega)}{[R_g + R_m(\omega)]^2 + [X_m(\omega) - 1/\omega C_0]^2} \tag{5.111}$$

Using Eqs. (5.92) and (5.99), it can be shown that

$$R_m = R_{m0}\left(\frac{\omega_a}{\omega}\right)^2 \left[\frac{(\bar{Z}_2/\bar{Z}_0)^2 \sin^4(\beta_a l/2)}{1 + [(\bar{Z}_2/\bar{Z}_0)^2 - 1]\cos^2\beta_a l}\right] \tag{5.112}$$

where R_{m0} is given by Eq. (5.100). It is now possible to write the normalized power transfer function, Eq. (5.111), as

$$H_p(\omega) = H_e(\omega)H_a(\omega) \tag{5.113}$$

where H_e and H_a are the electric and acoustic power transfer functions, respectively, given by

$$H_e(\omega) = \frac{4R_g R_{m0}(\omega_a/\omega)^2}{[R_g + R_m(\omega)]^2 + [X_m(\omega) - 1/\omega C_0]^2} \tag{5.114}$$

and

$$H_a(\omega) = \frac{(\bar{Z}_2/\bar{Z}_0)^2 \sin^4(\pi\omega/2\omega_a)}{1 + [(\bar{Z}_2/\bar{Z}_0)^2 - 1]\cos^2(\pi\omega/\omega_a)} \tag{5.115}$$

where the relation

$$\beta_a l = \frac{\omega l}{v_a} = \pi\frac{\omega}{\omega_a} \tag{5.116}$$

has been used.

Using Eqs. (5.114) and (5.115), it is now possible to discuss the effect of Q_e associated with H_e and of Q_a associated with H_a on the bandwidth under various conditions. The function H_a is plotted in Fig. 5.20. In what follows the function H_a will be evaluated for three different cases of acoustic loading conditions.

1. When $\bar{Z}_2 = \bar{Z}_0$, it follows from Eq. (5.115) that

$$H_a(\omega) = \sin^4\left(\frac{\pi}{2}\frac{\omega}{\omega_a}\right) \tag{5.117}$$

with 3 dB points at $\omega_- = 0.64\omega_a$ and $\omega_+ = 1.36\omega_a$, resulting in the relative bandwidth of 72%, or $Q_a \approx 1.39$.

2. When $\bar{Z}_2 \ll \bar{Z}_0$, using the approximation $\cos \pi\omega/\omega_a \approx -1$, Eq. (5.115) can be written as

$$H_a(\omega) \approx \frac{1}{(\bar{Z}_0/\bar{Z}_2)^2 \sin^2\left[\pi(\omega - \omega_a)/\omega_a\right] + 1} \tag{5.118}$$

and on using $\sin x \approx x$ it follows that

$$\left|\frac{\omega_\pm - \omega_a}{\omega_a}\right| = \frac{1}{\pi}\frac{\bar{Z}_2}{\bar{Z}_0} \quad \text{and} \quad Q_a = \frac{\pi\bar{Z}_0}{2\bar{Z}_2} \tag{5.119}$$

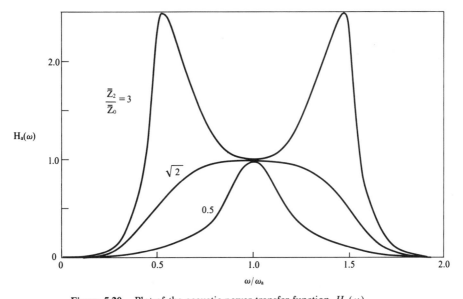

Figure 5.20. Plot of the acoustic power transfer function, $H_a(\omega)$.

Thus in this case a very narrow bandwidth is obtained. This was expected, since, due to a high-impedance mismatch, the transducer is acting as $\lambda/2$ resonator at the fundamental resonance ω_a.

3. When $\bar{Z}_2 \gg \bar{Z}_0$, the denominator in Eq. (5.115) is minimized at frequencies ω_1 and ω_2 given by

$$\pi \left| \frac{\omega_{1,2} - \omega_a}{\omega_a} \right| = \frac{\pi}{2} \tag{5.120}$$

and two peaks become prominent, as shown in Fig. 5.20, at frequencies $\omega_1 = 0.5\omega_a$ and $\omega_2 = 1.5\omega_a$. At these two frequencies the transducer acts as a resonator of length $\lambda/4$ and $3\lambda/4$, respectively.

4. When $\bar{Z}_2/\bar{Z}_0 = \sqrt{2}$, the transfer function is maximally flat, and from Eq. (5.115) it follows that $\omega_- = 0.5\omega_a$ and $\omega_+ = 1.5\omega_a$, resulting in a relative bandwidth of 100% and $Q_a = 1$.

An estimate of electric Q, Q_e is obtained using Eq. (5.114) for the matching circuit shown in Fig. 5.19(b), where the motional reactance can be found from Eqs. (5.92), (5.93), and (5.99). In practice, when k_T^2 is small, it follows that $1/\omega C_0 \gg X_m$, and the design procedure is usually carried out by neglecting X_m in Eq. (5.114). For a voltage source with resistance R_g, as in Fig. 5.19(b), an estimate for Q_e is given by

$$Q_e \approx \frac{\omega_a L}{R_{m0} + R_g/n^2} = \frac{1}{2\omega_a C_0 R_{m0}} = \frac{\pi}{8k_T^2} \frac{\bar{Z}_2}{\bar{Z}_0} \tag{5.121}$$

since under optimum matching for high-efficiency operation $R_{m0} = R_g/n^2$. The two Q factors given by Eqs. (5.119) and (5.121) are inversely proportional, and it is expected that the optimum bandwidth will be obtained for $Q_e \simeq Q_a$, or

$$Q_e = Q_a = \frac{\pi}{4k_T} = Q \tag{5.122}$$

and

$$\frac{\bar{Z}_2}{\bar{Z}_0} = 2k_T \tag{5.123}$$

Taking into account the multiplicative operation in Eq. (5.113), the relative bandwidth is found as

$$\frac{\omega_+ - \omega_-}{\omega_a} = \frac{1}{Q\sqrt{2}} = \frac{2\sqrt{2}}{\pi} k_T \tag{5.124}$$

and is dependent solely on the coupling coefficient k_T. For high coupling materials such as PZT-5H with $k_T = 0.56$, the relative bandwidth is 50%, $\bar{Z}_2 \approx \bar{Z}_0$, and $Q \approx 1.4$. A larger bandwidth may be obtained by using synthetic backing material, wideband tuning networks, and $\lambda/4$ matching layers between transducer and loaded medium $\bar{Z}_2$. These design approaches usually require extensive numerical studies of the power transfer function, Eq. (5.113).

Current acousto-optic bulk-wave technology[19] requires transducers for operation well into the gigahertz range. By bonding a piezoelectric plate to the substrate and then using ion milling, a 0.25-μm-thick transducer operating at 11 GHz has been made.[20] Sputter-deposited ZnO transducers are presently used in the 0.2-to-18 GHz band. Below 0.2 GHz, the necessary ZnO layers are more than 10 μm thick, requiring long deposition time. Transducers having center frequencies in the 2-to-6 GHz range, with bandwidths of 1 GHz, have been made using sputtered ZnO layers.[19]

The ZnO layers are also used in piezoelectric diffused metal oxide semiconductor (PI-DMOS) strain transistors, which consist of a ZnO thin-film capacitor and an insulator-gate field-effect transistor† capable of acting as strain gauges for the detection of acoustic waves in the gigahertz range.[21] The thin-film ZnO-MOS transistors can also be designed to have essentially DC response.[22]

5.8 ENERGY CONSERVATION THEOREM

For a plane longitudinal wave propagating along the z axis, coinciding with a pure-mode direction, in piezoelectric material of infinite cross section, the one-dimensional energy conservation theorem is obtained by combining Eqs. (1.34) and (5.6), which are repeated here for convenience:

$$-\frac{\partial(T_3 v_z)}{\partial z} = -\left[\frac{\partial}{\partial t}\left(\tfrac{1}{2}\rho_m v_z^2\right) + T_3 \frac{\partial S_3}{\partial t}\right] + F_z^b v_z \qquad (1.34)$$

$$T_3 = c^E S_3 - e E_z \qquad (5.6a)$$

$$D_z = e S_3 + \varepsilon^S E_z \qquad (5.6b)$$

Equation (5.6a) is modified using Eq. (1.27) to take viscous losses into account:

$$T_3 = c^E S_3 + \eta \frac{\partial S_3}{\partial t} - e E_z \qquad (5.125)$$

Equation (5.6b) is solved for E_z, which is then substituted in Eq. (5.125). The resulting expression for T_3 is substituted in Eq. (1.34), which is then integrated

---

† Note that silicon NPN planar transistors with their emitter-base junction mechanically coupled to a diaphragm are available commercially.

over a volume V, as shown in Fig. 1.5(a), resulting in[3]

$$\iint_{A_2} (-T_z v_z)\, dA - \iint_{A_1} (-T_3 v_z)\, dA$$

$$= -\iiint_V \frac{\partial}{\partial t}\left[\tfrac{1}{2}\rho_m v_z^2\right] dV - \iiint_V \frac{\partial}{\partial t}\left[\tfrac{1}{2}\left(c^E + \frac{e^2}{\varepsilon^S}\right)S_3^2\right] dV$$

$$- \iiint_V \eta\left[\frac{\partial S_3}{\partial t}\right]^2 dV + \iiint_V \frac{e}{\varepsilon^S} D_z \frac{\partial S_3}{\partial t}\, dV + \iiint_V F_z^b v_z\, dV$$

$$\text{(5.126)}$$

Many terms in Eq. (5.126) have already been identified in Sect. 1.5, Eq. (1.36). It is seen that the strain energy density has been augmented by the presence of piezoelectricity and is now given by

$$u_s^D = \frac{1}{2}\left(c^E + \frac{e^2}{\varepsilon^S}\right)S_3^2 = \frac{c^D S_3^2}{2} \tag{5.127}$$

The term

$$P_s^D = \iiint_V \frac{e}{\varepsilon^S} D_z \frac{\partial S_3}{\partial t}\, dV \tag{5.128}$$

represents the power supplied by piezoelectric sources. Denoting the total kinetic and strain energy by U^D, where the superscript D means that the increase in the strain energy due to the piezoelectric effect has been taken into account, Eq. (5.126) can be written as

$$\iint_{A_2} \Pi\, dA - \iint_{A_1} \Pi\, dA + \frac{\partial U^D}{\partial t} + P_d = P_s + P_s^D \tag{5.129}$$

where Π, P_d, and P_s are given by Eqs. (1.37), (1.38), and (1.39), respectively. According to the energy conservation theorem in piezoelectrics, Eq. (5.129), the power supplied by both mechanical and piezoelectric sources is partly radiated into the surroundings, partly stored as field energy, and partly dissipated in the material. The expression for the instantaneous radiated acoustic power remains unchanged and is given by Eq. (1.41).

An additional insight into the meaning of Eq. (5.129) can be gained by applying Eq. (5.129) to waves propagating in material that extends infinitely the z direction as well. In this case, Eq. (5.29) is applicable, and $D_z = 0$ in Eqs. (5.6b) and (5.129). From Eq. (5.6b) it follows that

$$S_3 = -\frac{\varepsilon^S}{e} E_z \tag{5.130}$$

Denoting the strain energy density at constant electric field and constant electric displacement by

$$u_s^E = \frac{c^E S_3^2}{2} \quad \text{and} \quad u_s^D = \frac{c^D S_3^2}{2} \tag{5.131}$$

respectively, and electric field energy densities at constant strain and constant stress by

$$u_e^S = \frac{\varepsilon^S E_z^2}{2} \quad \text{and} \quad u_e^T = \frac{\varepsilon^T E_z^2}{2} \tag{5.132}$$

respectively, it follows that

$$\frac{u_e^S}{u_s^E} = \frac{u_e^T}{u_s^D} = K^2 \tag{5.133}$$

$$\frac{u_e^S}{u_s^D} = \frac{K^2}{1 + K^2} = k_T^2 \tag{5.134}$$

and

$$\frac{u_e^T}{u_s^E} = K^2(1 + K^2) \tag{5.135}$$

where K^2 and k_T^2 are given by Eqs. (5.27) and (5.59), respectively. An important point here is that Eqs. (5.133) to (5.135) are point relations applicable to any point in the piezoelectric of infinite extent. The strain energy is proportional to electric field energy, the constant of proportionality being dependent on K^2. Since $D_z = 0$, it follows that $P_s^D = 0$ in Eq. (5.129), meaning that in a piezoelectric of infinite extent the acoustic waves cannot be set up by application of an external electric field, body forces being the only means of initiating the generation of acoustic waves. However, if the piezoelectric material is of finite extent in the z direction, as is the case considered in Sect. 5.4, where

$$I_3 = A \frac{\partial D_z}{\partial t} \tag{5.34}$$

the piezoelectric source term may play a major role. In this case, the relations given by Eqs. (5.133) to (5.135) are no longer applicable, since $D_z \neq 0$.

As shown in Sect. 1.5 for time-harmonic excitation, the complex form of the energy conservation theorem is obtained by multiplying Eq. (1.33) by v_z^* and multiplying the complex conjugate of Eq. (1.7) by T_3. It follows, then, by addition that

$$\frac{\partial}{\partial z}\left(\frac{-v_z^* T_3}{2}\right) = \tfrac{1}{2}\left[-j\omega\rho_m v_z v_z^* + j\omega T_3 S_3^* + F_z^b v_z^*\right] \qquad (5.136)$$

The field E_z is first eliminated from Eqs. (5.6b) and (5.125), and the resulting expression for T_3 is then substituted in Eq. (5.136), giving

$$\frac{\partial}{\partial z}\left(\frac{-v_z^* T_3}{2}\right) = \frac{1}{2}\left\{-j\omega\rho_m v_z v_z^* + j\omega\left[c^E + \frac{e^2}{\varepsilon^S}\right]S_3 S_3^*\right\} + \frac{(j\omega)^2 \eta S_3 S_3^*}{2}$$

$$+ \frac{F_z^b v_z^*}{2} - \frac{j\omega e}{2\varepsilon^S} D_z S_3^* \qquad (5.137)$$

After integration over the elementary volume $dA\,dz$ it follows that

$$\iint_{A_2}\left[\frac{-v_z^* T_3}{2}\right] dA - \iint_{A_1}\left[\frac{-v_z^* T_3}{2}\right] dA - j\omega\left[U_s^D - U_v\right] + \langle P_d\rangle = P_s + P_s^D$$

$$(5.138)$$

where, as in Sect. 1.5, the first two terms represent the complex acoustic power, and U_v, $\langle P_d\rangle$, and P_s are as given by Eqs. (1.45), (1.46), and (1.47), respectively. Due to piezoelectricity the peak strain energy is increased and is now

$$U_s^D = \iiint_V \frac{c^D S_3 S_3^*}{2}\, dV \qquad (5.139)$$

where c^D has been given by Eq. (5.30), and

$$P_s^D = -\frac{j\omega e}{2\varepsilon^S}\iiint_V D_z S_3^*\, dV \qquad (5.140)$$

is the complex power supplied by piezoelectric sources.

In a piezoelectric of infinite extent, $D_z = 0$ and therefore $P_s^D = 0$. If the piezoelectric is also lossless, then $\eta = 0$, that is, $\langle P_d\rangle = 0$, and for wave propagation far away from sources, $P_s = 0$. With all these assumptions, Eq. (5.138) becomes

$$\iint_{A_2}\Pi\, dA - \iint_{A_1}\Pi\, dA = j\omega\left[U_s^D - U_v\right] = 0 \qquad (5.141)$$

which means that the wave energy is conserved. The Q factor of the medium, calculated in the manner shown in Sect. 1.5, is given by

$$Q^D = \frac{c^D}{\omega\eta} = Q^E(1 + K^2)$$ (5.142)

where $Q^E = c^E/\omega\eta$.

5.9 PIEZOELECTRIC SEMICONDUCTORS

In piezoelectric semiconductors,[23-27] in addition to the piezoelectric effect already described, there are mobile electrons and/or holes. In the quiescent regime, electrical neutrality exists, $\nabla \cdot \vec{D}_0 = 0$, where $\vec{D}_0$ is the DC electric displacement vector. Such a state of electrons and holes is defined as the solid-state plasma.[23] Application of an external field, such as an RF electric field, can cause deviations from neutrality. For instance, for an N-type semiconductor,

$$\nabla \cdot \vec{D}_1 = \rho_1 = -qn_1$$ (5.143)

where $\vec{D}_1$ and ρ_1 are the RF electric displacement and RF space-charge density, respectively, q is the magnitude of the electronic charge, and n_1 is the RF electron number density. The periodic potential of the crystal lattice causes the electron mass to be lower than the free-electron mass m_0. This effective mass, m^*, can range from $0.01m_0$ (Bi) to $\simeq m_0$ (most metals). If the application of the outside field is suddenly terminated and the plasma is left to itself, the neutrality will be regained by collective oscillations of electrons with plasma frequency ω_p, $\omega_p^2 = n_0q^2/\varepsilon m^*$, where ε is the lattice permittivity. The positive charges, being associated with the crystal lattice, can be considered immobile, and electrons in the process of collective oscillations will be bunched and debunched, the restoring force being the Coulomb force. For typical semiconductors, $\omega_p \simeq 10^{12}$ rad/s. The Coulomb field associated with an electron is shielded at a distance by the presence of surrounding charges. This is characterized by the Debye length λ_D, which gives a measure of the distance over which the fields of an electron can be observed. Specifically, $\lambda_D \simeq 2\pi v_\theta/\omega_p$, where $v_\theta = 2k_BT/m^*$ is the average thermal velocity of electrons, where k_B is the Boltzmann constant and T is the absolute temperature. The Debye length can also be regarded as giving the minimum size of plasma in which the collective effects can be observed. For typical semiconductors, $\lambda_D = 1$ μm. In solids, electrons undergo collisions with positive charges, with crystal lattices, and among themselves. This effect can be taken into account by defining the scattering frequency ν. For solids, ν ranges from 10^{10} to 10^{13} rad/s. The presence of scattering and finite electron temperature causes the diffusion effects to become important whenever the electrons are bunched. The diffusion constant D_N is defined as $D_N = v_\theta^2/2\nu$. In most solids, $D_N \simeq 10^{-1}$

m^2/s. The finite conductivity of semiconductors can be taken into account by defining the dielectric relaxation frequency, $\omega_R = \omega_p^2 \varepsilon_0/\nu\varepsilon = \sigma/\varepsilon$, where ε_0 is the free-space permittivity. The electric field of a piezoelectric wave of frequency ω will be screened quickly by redistribution of electrons if $\omega \ll \omega_R$. However, for $\omega \geqslant \omega_R$ the redistribution of electrons is not fast enough, and the electric field will be strong, causing the material to behave as a good dielectric.

For instance, the motion of electron charge in an N-type semiconductor can be described mathematically as follows. The equation of continuity is

$$\frac{\partial \rho}{\partial t} + \nabla \cdot \vec{J} = 0 \tag{5.144}$$

where the total space-charge density $\rho = \rho_0 + \rho_1 = -qn$, where n is the total electron number density and $\rho_0 = -qn_0$ is the equilibrium space-charge density. Furthermore, $\vec{J}$ is the space-charge current density, $\vec{J} = \rho\vec{v}$, where $\vec{v}$ is the electron velocity. With these qualifications, Eq. (5.144) can be written as

$$\frac{\partial n}{\partial t} + \nabla \cdot (n\vec{v}) = 0. \tag{5.145}$$

and the rate of change of momentum of an electron[23] as

$$m^* \frac{d\vec{v}}{dt} = -q\vec{E} - m^*\nu\vec{v} + m^*\nu D_N \frac{\nabla\rho}{\rho} \tag{5.146}$$

Consider an infinite piezoelectric semiconductor biased by a DC electric field $\vec{E}_0$ in the $+z$ direction as shown in Fig. 5.21. The applied field will cause electrons to move in the $-z$ direction with drift velocity $\vec{v}_0 = -\mu\vec{E}_0$, where μ is the electron mobility, and will also generate electric displacement $\vec{D}_0$. If an RF signal is superimposed on this DC state, an RF space charge will result. Denoting by subscript 0 the DC quantities and by subscript 1 the RF quantities, it follows that

$$n = n_0 + n_1(z, t)$$

$$\vec{E} = [E_0 + E_1(z, t)]\hat{z}$$

$$\vec{D} = [D_0 + D_1(z, t)]\hat{z}$$

$$\vec{v} = [v_0 + v_1(z, t)]\hat{z} \tag{5.147}$$

where all the RF quantities are assumed to be much smaller than the DC quantities. Assuming that all RF quantities vary as $\exp j(\omega t - kz)$, for instance, $E_1(z, t) = E_1 \exp j(\omega t - kz)$, it follows from Eq. (5.145) that

$$n_1 = \frac{n_0 k v_1}{\omega - v_0 k} \tag{5.148}$$

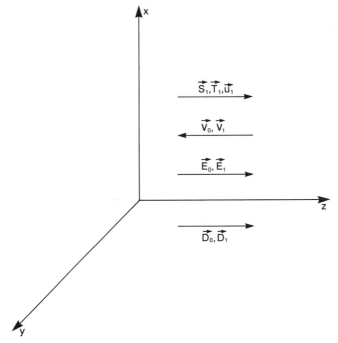

Figure 5.21.   The geometry for analysis of the acoustoelectric effect. See text for details.

and from Eq. (5.143) that

$$D_1 = \frac{-jqn_1}{k}$$

Combining the last two equations, it is found that

$$D_1 = \frac{-jqn_0v_1}{\omega - v_0k} \tag{5.149}$$

In Eq. (5.146), the total differential is given by

$$\frac{dv}{dt} = \frac{\partial v}{\partial t} + \frac{\partial v}{\partial z}\frac{\partial z}{\partial t} = \frac{\partial v_1}{\partial z} + \frac{\partial v_1}{\partial t}v_0$$

$\nabla = -jk\hat{z}$, and after neglecting the DC and nonlinear terms and combining with Eq. (5.148), it follows that

$$v_1 = \frac{jqE_1}{m^*\left[\omega - kv_0 - j\nu - k^2\nu D_N/(\omega - v_0k)\right]} \tag{5.150}$$

Finally, by combining Eqs. (5.149) and (5.150), the electric displacement is found to be

$$D_1 = \frac{\omega_p^2 \varepsilon^S E_1}{(\omega - kv_0)(\omega - kv_0 - j\nu) - k^2 \nu D_N} \tag{5.151}$$

where $\omega_p^2 = n_0 q^2 / m^* \varepsilon^S$.

The RF fields (index 1) describing a plane longitudinal piezoelectric wave propagating in the $+z$ direction, Eqs. (5.6), can be written as[†]

$$T_1 = c^E S_1 - eE_1 \quad \text{and} \quad D_1 = eS_1 + \varepsilon^S E_1$$

and from Eq. (5.22) as

$$\frac{\partial T_1}{\partial z} = \rho_m \frac{\partial^2 u_1}{\partial t^2} \tag{5.152}$$

with the associated longitudinal strain $S_1 = \partial u_1 / \partial z$, where u_1 is the RF displacement in the z direction, Fig. 5.21. For harmonic waves of the form $\exp j(\omega t - kz)$ it follows from the last three equations that

$$D_1 = \varepsilon^S E_1 \left[1 + \frac{K^2 k^2 v_L^2}{k^2 v_L^2 - \omega^2} \right] \tag{5.153}$$

where K^2 is the electromechanical coupling coefficient, Eq. (5.27), and v_L is the acoustic longitudinal wave velocity, $v_L^2 = c^E / \rho_m$.

Combining Eqs. (5.153) and (5.151) results in a relation determining the existence of various wave types (dispersion relation),

$$(\omega^2 - k^2 v_L^2) \left[1 - \frac{\omega_p^2}{(\omega - kv_0)(\omega - kv_0 - j\nu) - D_N \nu k^2} \right] = K^2 k^2 v_L^2 \tag{5.154}$$

Piezoelectric semiconductors are collision dominated, $\nu \gg kv_0$, $\nu \gg \omega$, and therefore the last equation becomes

$$(\omega^2 - k^2 v_L^2) \left[1 + \frac{\omega_R}{D_N k^2 + j(\omega - kv_0)} \right] = K^2 k^2 v_L^2 \tag{5.155}$$

The solution of Eq. (5.155) can be obtained by using the approximation

$$k \simeq \frac{\omega(1 + j\alpha)}{v_L}, \quad \alpha \ll 1$$

[†] The subscript 1 here and in Eqs. (5.152) and (5.153) indicates the RF quantities oriented in the z direction, Fig. 5.21.

which, if the gain α is larger than zero, will result in spatial wave amplification in the $+z$ direction, $\exp(\omega \alpha z / v_L) \exp j(\omega t - \omega z / v_L)$. It is found that[25]

$$\alpha \simeq \frac{\frac{1}{2} K^2 (\omega_R / \omega) \gamma}{(\omega_R / \omega)^2 (1 + \omega^2 / \omega_R \omega_D)^2 + \gamma^2} \tag{5.156}$$

where $\omega_D = v_L^2 / D_N$ is defined as the diffusion frequency, and $\gamma = v_0 / v_L - 1$. It is seen from Eq. (5.156) that if $\gamma > 0$, that is, if $v_0 / v_L > 1$, the wave will be amplified. This is often referred to as the *acoustoelectric effect*. The applied DC electric field necessary to cause gain is on the order of 1 kV/cm. For instance, for CdS the relevant parameters[24] are $v_L = 1$ km/s, $\varepsilon^S \simeq 5.4\varepsilon_0$, $n_0 \simeq 10^{19}$ m^{-3}, and $\mu \simeq 3 \times 10^{-2}$ m^2/V $\cdot$ s, $\gamma \simeq 1.7$, and $v_0 \simeq 2.7 v_L$.

The acoustoelectric effect in piezoelectric semiconductors, such as CdS, ZnO, GaAs, InSb, and many others,[23-26,28] has been observed in many other circumstances for both longitudinal and shear waves. The DC quantities, such as the drift velocity, in finite samples have important ramifications for the gain.[29,30]

The theory presented here shows that when the electron-drift velocity exceeds the acoustic velocity, acoustoelectric amplification (gain) results. This is the well-known traveling-wave amplification effect.[31,32] From Eq. (5.156) it can be shown that the gain per unit length α / λ_a, $\lambda_a = v_L / f$, is maximum when $\omega^2 = \omega_R \omega_D$. The electron collective oscillation requires that $\lambda_a > \lambda_D$, for otherwise the RF electric field of the piezoelectric wave will not cause appreciable bunching of the electrons, thus resulting in less gain. In other words, if $\tau_R = 1 / \omega_R$ and τ_D is the time necessary for electrons to diffuse one wavelength, $\tau_D \simeq \lambda_a^2 / D_N$, it is required that $\tau_D > \tau_R$; if the diffusion is fast compared to the dielectric relaxation time, the full bunching of the electrons will not occur.[23]

REFERENCES

1. P. Curie, *C. R.*, **91**, 294 (1880); **91**, 383 (1880).

2. L. Brillouin, *Wave Propagation in Periodic Structures*, Dover, New York, 1953, Chap. 2.

3. B. A. Auld, *Acoustic Fields and Waves in Solids*, Vol. I, Wiley, New York, 1973, Chap. 8.

4. R. A. Heising, *Quartz Crystals for Electrical Circuits*, Electronic Industries Association, Washington, 1978, Chap. 1.

5. W. L. Bond, *Crystal Technology*, Wiley, New York, 1976, Chap. 1.

6. A. Holden and P. Singer, *Crystals and Crystal Growing* (*Sci. Study Ser.*), Doubleday, Garden City, 1960.

7. O. E. Mattiat, Ed., *Ultrasonic Transducer Materials*, Plenum, New York, 1971, Chaps. 2 and 3.

8. H. Kawai, "The Piezoelectricity of Poly(vinylidene) Fluoride," *Jap. J. Appl. Phys.*, **8**, 975–976 (1969).

9. L. N. Bui, H. J. Shaw, and L. T. Zitelli, "Study of Acoustic Wave Resonance in Piezoelectric PVF$_2$ Film," *IEEE Trans. Sonics Ultrason.*, **SU-24**(5), 331–336 (1977).

10. J. Ilukor and E. H. Jacobsen, "Generation and Detection of Coherent Elastic Waves at 114,000 Mc/sec," *Science*, **153**, 1113–1114 (1966).

11. W. P. Mason, *Electromechanical Transducers and Wave Filters*, Van Nostrand, New York, 1948.

12. R. Krimholtz, D. A. Leedom, and G. L. Matthaei, "New Equivalent Circuits for Elementary Piezoelectric Transducers," *Electron. Lett.*, **6**, 338–389 (1970).

13. T. R. Meeker, "Thickness Mode Piezoelectric Transducers," *Ultrasonics*, **1972**, 26–36.

14. C. S. DeSilets, J. D. Fraser, and G. S. Kino, "The Design of Efficient Broad-Band Piezoelectric Transducers," *IEEE Trans. Sonics Ultrason.*, **SU-25**(3), 115–125 (1978).

15. M. Toki, Y. Tsuzuki, and O. Kawano, "A New Equivalent Circuit for Piezoelectric Ceramic Disk Resonators," *Proc. IEEE*, **68**(8), 1032–1033 (1980).

16. T. M. Reeder and D. K. Winslow, "Characteristics of Microwave Acoustic Transducers for Volume Wave Excitation," *IEEE Trans. Microwave Theory and Tech.*, **MTT-17**, 927–941 (1969).

17. W. P. Mason and R. N. Thurston, Eds., *Physical Acoustics*, Vol. 14, Academic, New York, 1979, Chap. 4.

18. Reference 3, Vol. II, Chap. 11.

19. E. K. Kirchner, "Deposited Transducer Technology for Use with Acousto-Optic Bulk Wave Devices," *Proc. SPIE*, **214**, 102–109 (1979).

20. H. C. Huang, J. D. Knox, Z. Turski, R. Wargo, and J. J. Hanak, "Fabrication of Submicron LiNbO$_3$ Transducers for Microwave Acoustic (Bulk) Delay Lines," *Appl. Phys. Lett.*, **24**, 109–111 (1974).

21. K. W. Yeh and R. S. Muller, "Piezoelectric D-MOS Transducers," *Appl. Phys. Lett.*, **29**(9), (1976).

22. P. L. Chen, R. S. Muller, R. M. White, and R. Jolly, "Thin Film ZnO-MOS Transistors with Virtually DC Response," *1980 Ultrason. Symp. Proc.*, 945–948 (1980).

23. M. C. Steele and B. Vural, *Wave Interaction in Solid State Plasmas*, McGraw-Hill, New York, 1969, Chaps. 1 and 8.

24. A. R. Hutson, J. H. McFee, and D. L. White, "Ultrasonic Amplification in CdS," *Phys. Rev. Lett.*, **7**, 237–239 (1961).

25. D. L. White, "Amplification of Ultrasonic Waves in Piezoelectric Semiconductors," *J. Appl. Phys.* **33**, 2547–2554 (1962).

26. K. Blötekjaer and C. F. Quate, "The Coupled Modes of Acoustic Waves and Drifting Carriers in Piezoelectric Crystals," *Proc. IEEE*, **52**, 360–377 (1964).

27. H. Berger, R. I. Harrison, and S. P. Denker, "Power Flow and Energy Storage in Piezoelectric Semiconductor Devices," *IEEE Trans. Microwave Theory Techniq.*, **MTT-18**(2), 105–111 (1970).

28. F. S. Hickernell and N. G. Sakiotis, "An Electroacoustic Amplifier with Net Electrical Gain," *Proc. IEEE*, **52**, 194–195 (1964).

29. R. M. White, "Some Effects of Heating and Dispersion in the Ultrasonic Amplifier," *IEEE Trans. Sonics Ultrason.*, **SU-13**, 69–73 (1966).

30. C. W. Turner and K. P. Weller, "DC Flow Considerations for Crossed-Field Acoustic Amplifiers and Other Solid-State Traveling-Wave Devices," *IEEE Trans. Electron Devices*, **ED-16**, 787–797 (1969).

31. See, for instance, J. R. Pierce, *Almost All About Waves*, MIT Press, Cambridge, MA, 1974, Chap. 11; also C. C. Johnson, *Field and Wave Electrodynamics*, McGraw-Hill, New York, 1965, Chap. 9.

32. A. Scott, *Active and Nonlinear Wave Propagation in Electronics*, Wiley, New York, 1970, Chap. 2.

EXERCISES

1. Using Eq. (5.60) with $D_z = D_z(\xi/l)\delta(\tau/T)$, where δ is the Dirac delta function and $D_z(\xi/l) = D_0\{\chi(\xi/l) - \chi[(\xi - l)/l]\}$, show (a) that the solution of Eq. (5.60) is given by

$$u_z(t, z) = \frac{-k_T^2 D_0 l}{2e}\left\{\chi\left[\frac{tv^D - |z|}{l}\right] - \chi\left[\frac{tv^D - |z - l|}{l}\right]\right\}$$

and (b) that $D_0 = lI_3/v^D A$.

2. Derive Eq. (5.85) and the equivalent circuit shown in Fig. 5.12(a) by combining Eqs. (5.78) and (5.83).

3. Derive a shunt equivalent circuit for the impedance Z_m in Eq. (5.78). *Note*:

$$\cot x = \frac{1}{x} + \sum_{n=1}^{\infty} \frac{2x}{x^2 - n^2\pi^2}$$

4. (a) Show that for a lossy piezoelectric, $\eta \neq 0$ in Eq. (1.28), the Z matrix, Eq. (5.50), can be written as

$$\begin{bmatrix} F_1 \\ F_2 \\ V_3 \end{bmatrix} = -j\begin{bmatrix} \tilde{Z}_0\cot\gamma l & \tilde{Z}_0\csc\gamma l & h/\omega \\ \tilde{Z}_0\csc\gamma l & \tilde{Z}_0\cot\gamma l & h/\omega \\ h/\omega & h/\omega & 1/\omega C_0 \end{bmatrix}\begin{bmatrix} v_1 \\ v_2 \\ v_3 \end{bmatrix}$$

where $\tilde{Z}_0 = \bar{Z}_0(1 + j/2Q)$, $\gamma = \beta_a - j\alpha$, $\beta_a \gg \alpha$, $\alpha \approx \omega\eta\beta_a/2c^D$, and $Q \approx c^D/\omega\eta$.

(b) Show that in this case the electric input impedance, Eq. (5.77), of a lossy resonator can be written as

$$Z_3 = \frac{1}{j\omega C_0}\left[1 - k_T^2\frac{\tan(\gamma l/2)}{(\gamma^* l/2)}\right] = R(\omega) + jX(\omega)$$

(c) Find explicit expressions for $R(\omega)$ and $X(\omega)$ near resonance and show that for $\omega \approx \omega_0$

$$\frac{X(\omega)}{R(\omega)} \approx 2Q\frac{\sin\Omega}{\Omega + \sin\Omega} \approx -2Q\delta$$

where $\Omega = \pi(\omega/\omega_0) = \pi(1 + \delta)$. Comment on the possibility of using the above equation for measuring the Q factor.

5. Consider a slug of lossy backing material, as shown in Fig. 5.22, characterized with impedance $\bar{Z}_0$ (real) and $\gamma = \beta_a - j\alpha$, $\alpha \ll \beta_a$.

(a) Using Eq. (E1.5) of Exercise 3, Chap. 1, show that

$$\bar{Z}_a = \bar{Z}_0\left(\frac{\alpha l + j\tan\beta_a l}{1 + j(\alpha l)\tan\beta_a l}\right)$$

where $\beta_a = 2\pi/\lambda_a$.

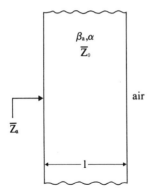

Figure 5.22.

(b) Show that for $l = \lambda_0/2$ near resonance, $\beta l \approx \pi$, the above expression can be written as

$$\bar{Z}_a \simeq \bar{Z}_0(\alpha l + j \tan \beta_a l) \simeq \bar{Z}_0 \left(\alpha l + j \frac{\pi \lambda_0 \delta}{\lambda_a} \right)$$

where

$$\delta = (\omega - \omega_0)/\omega_0, \quad \omega_0 = v_a \pi/l.$$

(c) Compare the expression in (b) with the impedance of a series RLC circuit near resonance,

$$Z = R + j \left(\omega L - \frac{1}{\omega C} \right) \approx R + j 2 \omega_0 \delta L, \qquad \omega_0^2 = \frac{1}{LC}$$

and identify R, L, and C in terms of acoustic quantities. Show that the Q factor at resonance is $Q = \pi/2\alpha l$.

(d) Find the length l needed to achieve the acoustic $Q = 20$ for a material with $\alpha = 4.34$ dB/cm. If the acoustic velocity of the material is $v_a = 3140$ m/s, find the resonant frequency.

6. Repeat (a) to (d) in Exercise 5 for $l = \lambda_0/4$.
    (a) Show in this case that the input acoustic admittance is given by

$$\bar{Y} = \bar{Y}_a \left[\alpha l + j \frac{\pi}{2} \frac{\lambda_0 \delta}{\lambda_a} \right]$$

(b) Compare the above expression with the admittance of a shunt RLC circuit near resonance,

$$Y = G + j 2 \delta \omega_0 C, \qquad \omega_0^2 = \frac{1}{LC}$$

and identify G, L, and C in terms of acoustic quantities.

(c) Show that the Q factor at resonance is $Q = \pi/4\alpha l$.

(d) Find the Q factor for rutile ($v_a = 7900$ m/s, $\alpha = 3.5$ dB/cm) at frequency 3 GHz.

7. Consider the equivalent circuit of the transducer in Fig. 5.6 with $l = \lambda_0/2$ terminated with a backing impedance $F_1 = -\bar{Z}_B v_1$ and the load impedance $F_2 = -\bar{Z}_L v_2$ both assumed real.

   (a) The acoustic transmission line in the equivalent circuit, Fig. 5.6, can be regarded in this case to consist of two $\lambda/4$ acoustic lines. Using Eq. (E1.5) of Exercise 3 of Chap. 1, show that near resonance, $\omega \approx \omega_0 = \pi v^D/l$, the equivalent acoustic admittance of the two quarter-wave lines at the output of the transformer in Fig. 5.6 is given by

   $$\bar{Y}_a \simeq \frac{\bar{Z}_L + \bar{Z}_B}{\bar{Z}_0^2} + \frac{2}{j\bar{Z}_0 \tan \beta l}$$

   (b) Using Eq. (5.82) for $n = 0$, represent the imaginary part of the computed admittance by an LC shunt resonant circuit and show that the acoustic Q factor of the circuit is given by $Q_a = \pi \bar{Z}_0/2(\bar{Z}_B + \bar{Z}_L)$.

   (c) Find the electrical input impedance at resonance of the equivalent circuit of Fig. 5.6 and, assuming that a series tuning inductance has been used at the input, show that the electrical Q factor is given by $Q_e = \pi(\bar{Z}_B + \bar{Z}_L)/4k_T^2\bar{Z}_0$.

   (d) From the condition for optimum bandwidth, $Q_e = Q_a$, find a relation between Z_B, Z_L, Z_0, and k_T. Consider $Z_L = 1.5 \times 10^6$ SI units (water), $Z_0 = 33.7 \times 10^6$ SI units (PZT-5A with $k_T = 0.5$), and comment regarding the choice of the backing material, Fig. 1.6.

8. Consider Fig. 5.9. Present an argument as to why the impulse generated at $z = 0$ for $t < l/2v^D$ has twice the magnitude of the impulses generated at $z = l$.

9. Consider a composite transducer consisting of M thin piezoelectric transducers of thickness l spaced a distance $B \gg l$ apart as shown in Fig. 5.23. The nonpiezoelectric material in which the transducers are firmly bonded has identically the same acoustical properties (impedance, velocity) as the piezoelectric material. Using Eqs. (5.67) and the theory presented in Chap. 3, find the impulse response and the transfer function [the equivalent of Eqs. (3.13) and (3.14)] of the composite transducer at a plane $z > a$.

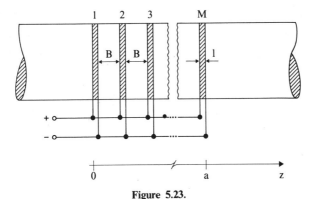

Figure 5.23.

10. (a) Using Eq. (5.50) for a lossy nonpiezoelectric ($h = 0$) material, show that the transmission line equation for the case can be written as

$$F_1 = F_2 \cosh(\beta l) - v_2 \overline{Z}_0 \sinh(\beta l)$$

$$v_1 = \frac{F_2}{\overline{Z}_0} \sinh(\beta l) - v_2 \cosh(\beta l)$$

where $\beta = \alpha + j\beta_a$, as shown in Fig. 5.24.

(b) If the line is loaded with acoustic load $\overline{R}_L$ such that $F_2 = -\overline{R}_L v_2 = \overline{R}_L v_L$, show that the force and the velocity at distance z are given by

$$F = v_L \{ \overline{R}_L \cosh[\beta(l - z)] + \overline{Z}_0 \sinh[\beta(l - z)] \}$$

$$v = v_L \left\{ \cosh[\beta(l - z)] + \frac{\overline{R}_L}{\overline{Z}_0} \sinh[\beta(l - z)] \right\}$$

(c) The dissipative loss is defined as $L = 10 \log(P_i/P_L)$, where P_i is the input power at $z = 0$ and P_L is the power delivered to the load. Defining the series acoustic resistance per unit length of the transmission line (see Exercise 5) as $R = \overline{Z}_a/l = \overline{Z}_0 \alpha$, it follows that

$$\frac{P_i}{P_L} = 1 + \frac{P_t}{P_L} = 1 + \frac{R \int_0^l v \cdot v^* \, dz}{v_L \cdot v_L^* \overline{R}_L}$$

where P_t is the power dissipated in the transmission line between $z = 0$ and $z = l$.

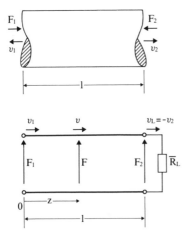

Figure 5.24.

By combining the last three equations, show that

$$L = 10 \log \left\{ 1 + \alpha l \left[\left(\frac{1 + SWR^2}{SWR} \right) \frac{\sinh \alpha l}{2 \alpha l} \pm \left(\frac{1 - SWR^2}{SWR} \right) \frac{\sin(2\beta_a l)}{2\beta_a l} \right. \right.$$

$$\left. \left. + 2 \left(\frac{\cos(2\alpha l) - 1}{2\beta_a l} \right) \right] \right\}$$

where $\beta_a = 2\pi/\lambda_a$, SWR is the standing wave ratio, the plus sign is used when $\overline{R}_L > \overline{Z}_0$, and the negative sign applies when $\overline{R}_L < \overline{Z}_0$.

(d) Show that for $\alpha l \ll 1$ and $SWR^2 \approx 1$, the one-way attenuation of the acoustic transmission line is given by $\alpha_t \approx \alpha[(1 + SWR^2)/2SWR]$, that is, the existence of the standing waves increases the attenuation per unit length.

11. Consider an air-backed transducer matched to water with two acoustic $\lambda_a/4$ transformers. The materials used are PZT-5A, fused quartz, and Lucite. The diameter of the contact area is 2.4 cm. The PZT-5A material is half-wave resonant at $f_a = 1.5$ MHz. Assuming that both transformers are $\lambda_a/4$ at f_a, compute the normalized power transfer function, Eq. (5.113), under the assumption that a simple series inductance has been used for electrical matching purposes. Plot the magnitude and phase of the normalized power transfer function for $0 \leqslant f/f_a \leqslant 2$.

12. Repeat Exercise 11 under the assumption that the acoustic transformers are $\lambda_a/4$ at $0.9f_a$.

Chapter 6

Piezoelectric Materials

It has been pointed out in Chapter 5 that piezoelectricity is intimately related to crystallographic properties of materials and that piezoelectricity cannot exist in materials possessing central symmetry. This suggests that the symmetry properties of solids determine, to a large extent, their macroscopic behavior. It is therefore of interest to consider certain aspects of the internal symmetry of solids.[1-5]

A solid in which different atoms occur in a repetitive three-dimensional (or two-dimensional) array is called a *crystal*, and the properties of such a solid are said to be *crystalline* in nature. The repetitive three-dimensional array, the *motif* of the crystal, is dependent on the chemical structure of the crystal and may consist of a group of molecules (rutile), a group of atoms (zinc), a single molecule (sodium chloride), or a single atom (copper). The distance between adjacent motifs is typically on the order of 3×10^{-10} m, and the volume associated with each motif is approximately equal to 3×10^{-29} m^3. The motif in a crystal is repeated according to a certain scheme of repetition, the *space lattice*. Thus the crystal structure of any crystalline solid is completely specified by the space lattice and the motif.

In the space lattice, the motif is represented by a single point, resulting in an infinite three-dimensional array of points such that there is no distinction between any two points. A *unit cell*, made of a combination of lattice points, is chosen to be the representative part of the lattice. The space lattice can be divided into an infinite number (long-range order) of identical unit cells. The most general type of unit cell, shown in Fig. 6.1, is the parallelepiped with axes a, b, and c obeying the right-hand rule and completely specified by edges a, b, and c and angles α, β, and γ. The labeling of edges, angles, and faces shown in Fig. 6.1 has been adopted by convention.

If a solid is composed of many crystals (also called grains), it is termed *polycrystal* and its properties are said to be *polycrystalline* in nature. In the polycrystal, the grains are in mutual contact, forming a mass that may or may not be void-free.

In studies of crystals, the symmetry elements of the crystal necessary to describe its external shape are termed *macroscopic symmetry elements*, and those that concern the structure at an atomic level and have no influence on external shape are termed *microscopic symmetry elements*. Macroscopic symme-

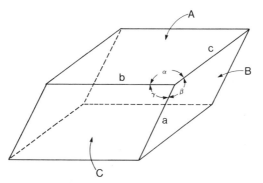

Figure 6.1. The parallelepiped unit cell completely specified by its edges, a, b, c and angles α, β, and γ.

try elements[†] are sufficient to describe the phenomenon of piezoelectricity and must be fully specified for a particular crystal; that is, they must be specified for any point of the crystal lattice. A *crystal class* (point group of symmetry) is a collection of all symmetry elements at any point of the lattice.

6.1 CRYSTAL SYSTEMS AND CLASSES

The complete set of symmetry elements that may exist in a space lattice consists of 10 elements, namely, *rotational axes* $(1, 2, 3, 4, 6)$, *center of symmetry* $(\bar{1})$, *rotational-inversion axes* $(\bar{3}, \bar{4}, \bar{6})$, and the *mirror plane* (m).

The rotational axes can be onefold, twofold, threefold, fourfold, and six-fold, corresponding to numerical symbols 1, 2, 3, 4, and 6, respectively, and are illustrated in Fig. 6.2. It is seen that a one-fold rotational axis means no symmetry at all. For an n-fold rotational axis, the rotation angle θ between successive positions of self-coincidence is equal to $360°/n$.

The *center-of-symmetry operation*, shown in Fig. 6.2(f), is called *inversion*, and the point p, denoted by a dot or numerical symbol $\bar{1}$, is called the *center of symmetry*. Thus the inversion operation requires that every point be projected through the center, the resulting point being equidistant from the center but on the opposite side.

The numerical symbols $\bar{3}, \bar{4}, \bar{6}$ correspond to rotation followed by inversion or vice versa, as shown in Fig. 6.3, and can be expressed symbolically as

$$3 + \bar{1} \rightarrow \bar{3}, \qquad 4 + \bar{1} \rightarrow \bar{4}, \qquad 6 + \bar{1} \rightarrow \bar{6} \qquad (6.1)$$

Space lattices can also be symmetrical about a plane that divides a body into halves that are mirror images, as illustrated in Fig. 6.3(d). The mirror

[†]Symmetry elements determine the dependence of other macroscopic properties, such as electric conductivity, viscosity, and magnetic susceptibility on direction in the crystal.

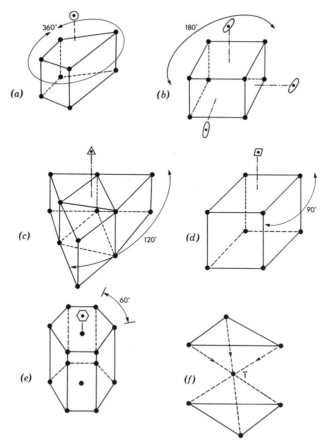

Figure 6.2. (a–e) One-, two-, three-, four-, and sixfold rotational axes, respectively. (f) Center-of-symmetry operation.

plane is denoted symbolically by m. The mirror symmetry operation is equivalent to a twofold rotation-inversion axis, $m \rightarrow \bar{2} \rightarrow 2 + \bar{1}$. The sixfold rotation-inversion axis is equivalent to a threefold rotation axis and mirror reflection across a plane orthogonal to the axis, $\bar{6} \rightarrow 3 + m$.

Depending on the degree of symmetry, crystals are classified into *seven crystal systems*, each system in turn being divided into classes (points groups). The seven systems listed in ascending order of rotational symmetry are

> *Triclinic* (2 classes), one onefold axis
> *Monoclinic* (3 classes), one twofold axis
> *Orthorombic* (3 classes), three mutually perpendicular twofold axes
> *Rhombohedral* [also called *trigonal*] (5 classes), one threefold axis
> *Tetragonal* (7 classes), one fourfold axis

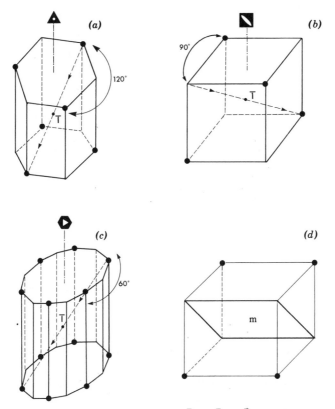

Figure 6.3. Illustration of rotation-inversion. (*a*) $\bar{3}$. (*b*) $\bar{4}$. (*c*) $\bar{6}$. (*d*) Mirror operation, *m*.

Hexagonal (7 classes), one sixfold axis,
Cubic (5 classes), four threefold axes at 70°32′ to each other and three twofold axes.

Each of the seven systems can have more than one unit cell.[†] There are four basic types of unit cells. A *primitive unit cell*, denoted by *P*, contains a single lattice point, while a *nonprimitive unit cell* (a base-centered unit cell, a face-centered unit cell, or a body-centered unit cell) contains two or more lattice points. A base-centered unit cell, denoted by *A*, *B*, or *C*, since there may be three kinds, has lattice points at the cell corners and also at the center of two opposite faces. A body-centered unit cell, denoted by *I*, has lattice points at each corner and additional points at the geometrical center. A face-centered unit cell, denoted by *F*, has lattice points at each corner and additional points at the geometrical

---

[†]Unit cells, that is, Bravais lattices, are useful in understanding various coupling mechanisms such as piezoelectricity at the atomic level. A description of the crystal structure requires that the positions of all atoms be stated relative to the Bravais lattice.

center of each of the six faces. Since there is one motif at each point, the four types of cells contain one motif for a *P*-cell, two motifs for an *A*, *B*, or *C* cell, four motifs for an *F* cell, and two motifs for an *I* cell. The existence of four kinds of unit cells and seven systems suggest that there may be $4 \times 7 = 28$ different lattices. Of these, only 14 differ from one another. These 14 lattices are termed Bravais lattices,[1,4] and every crystal has a lattice that must be one of the 14 Bravais lattices. The 14 Bravais lattices are the triclinic *P*, the monoclinic *P* and *C*, the orthorhombic *P*, *C*, *I*, and *F*, the rhombohedral *P*, the hexagonal *P*, the tetragonal *P* and *I*, and the cubic *P*, *I*, and *F*. In all cases, the *A* and *B* cells are either unnecessary because of redundancy or are impossible because of symmetry arguments.

Each class within the system is denoted by a combination of numerical symbols associated with symmetry operation.[1] For instance, a class denoted by the symbol $4/m$ has a fourfold rotation axis with a mirror plane at right angles to it. This notation is known as *full Hermann-Mauguin notation*. In some cases, the full notation tends to be cumbersome, and often the *abbreviated Hermann-Mauguin notation*,[1] is used instead. For instance, a class denoted in full notation by $4/m2/m2/m$ is denoted in abbreviated notation as $(4/mmm)$, the parentheses being the characteristic of the abbreviated notation. For all classes, both types of notations are shown in Table 6, where the distinction is made between piezoelectric and nonpiezoelectric classes. It is seen that each system has at least one piezoelectric class.

The common symmetry component of each system imposes certain restrictions regarding the parameters a, b, c, α, β, and γ of the unit cell shown in Fig. 6.1. Thus each system can also be characterized in terms of the parameters of the unit cell. For instance, in the cubic system $a = b = c$, and $\alpha = \beta = \gamma = 90°$.

The system of axes a, b, c and the associated angles is termed the *natural coordinate system* of the crystal. The unit vectors $\hat{a}$, $\hat{b}$, and $\hat{c}$ specify the directions of unit translations which can be of value a, b, and c, respectively. To identify a specific direction or plane in the crystal, the *Miller system of indices* is used.

For directions, the Miller indices are derived as follows. For instance, a direction determined by a line passing through the origin of the unit cell and a point $3a$, $-2b$, $0c$, upon division by unit translations is denoted by $[3\ \ \bar{2}\ \ 0]$. For a line passing through a point $-3a/2$, $2b/3$, c, after division by unit translations, the rationalization is done multiplying by 6 (the lowest common denominator), which results in the Miller indices of $[\bar{9}\ \ 4\ \ 6]$. Another notation that is also used is of the type $\langle spq \rangle$, which denotes a particular family of directions in the Bravais lattices of the particular crystal system. For instance, in the cubic system the family $\langle 100 \rangle$ includes the directions $[100]$, $[\bar{1}00]$, $[010]$, $[0\bar{1}0]$, $[001]$, and $[00\bar{1}]$. The acoustic wave propagation along all these directions has the same properties, and it is thus much simpler to state the family of directions rather than to list all the equivalent directions.

TABLE 6 Hermann-Mauguin Notation for Various Crystal Classes

Crystal System	Full Notation		Abbreviated Notation	
	Piezo-electrics	Nonpiezo-electrics	Piezo-electrics	Nonpiezo-electrics
Triclinic	1	$\bar{1}$	(1)	$(\bar{1})$
Monoclinic	m		(m)	
		$2/m$		$(2/m)$
	2		(2)	
Orthorhombic	222		(222)	
		$2/m2/m2/m$		(mmm)
	$2mm$		(mm)	
Rhombohedral	3	$\bar{3}$	(3)	$(\bar{3})$
(trigonal)	$3m$		$(3m)$	
		$\bar{3}2/m$		$(\bar{3}m)$
	32		(32)	
Tetragonal	$4mm$	$4/m2/m2/m$	$(4mm)$	$(4/mmm)$
	$\bar{4}2m$		$(\bar{4}2m)$	
	422		(42)	
		$4/m$		$(4/m)$
	4		(4)	
	$\bar{4}$		$(\bar{4})$	
Hexagonal	$\bar{6}$	$6/m$	$(\bar{6})$	$(6/m)$
	$\bar{6}2m$		$(\bar{6}2m)$	
	6		(6)	
	$6mm$		$(6mm)$	
	622	$6/m2/m2/m$	(62)	$(6/mmm)$
Cubic	23	432	(23)	(43)
		$2/m\bar{3}$		$(m3)$
	$\bar{4}3m$		$(\bar{4}3m)$	
		$4/m\bar{3}2/m$		$(m3m)$

Miller indices of a plane are determined by the following procedure:

1. Determine the intercepts that the plane makes on the axes of the lattice.
2. Divide the intercepts by appropriate translations a, b, and c.
3. Invert the dividends.

4. Rationalize.

5. Place the rationalized numbers in parentheses.

For instance, a plane ∞, $-2b$, ∞ after division becomes ∞, -2, ∞, and after inversion, 0, $-\frac{1}{2}$, 0, resulting after rationalization in $(0\bar{1}0)$. For a plane $-2a$, $3b$, $-4c$, the Miller notation would be $(\bar{6}4\bar{3})$.

The Miller system of indices requires that similar planes have the same interplanar spacing and be composed of an identical array of atoms, and that similar directions have the same unit translations. These requirements are satisfied in all crystal systems except the hexagonal system, where there are difficulties due to the sixfold axis of rotation.[5] In this case, *Miller-Bravais indices* are used. The relationships between the three Miller indices $[SPQ]$ and the Miller-Bravais indices $[spqr]$ are as follows:

$$s = \tfrac{1}{3}(2S - P), \quad p = \tfrac{1}{3}(2P - S)$$

$$q = -(s + p), \quad r = Q \tag{6.2}$$

and

$$S = s - q, \quad P = p - q, \quad Q = r \tag{6.3}$$

Thus a direction $[010]$ corresponds to $[\bar{1}2\bar{1}0]$ in Miller-Bravais notation or to $[\bar{1}2.0]$ in shorter Miller-Bravais notation, where the index q is often omitted and replaced by a decimal point. Similarly for planes, (110) corresponds to $(1\bar{1}20)$ or to (11.0). It is to be noted that the hexagonal system is the only system where Miller-Bravais notation is necessary, the Miller system being adequate for the other six systems.

The theoretical treatment of elasticity and electricity has been developed in terms of the rectangular coordinate system, termed the *crystal* (or *crystallographic*) system and denoted by X, Y, Z or x_1, x_2, x_3, respectively, the latter notation being easier to use in tensor calculus. The relationship between the natural axes a, b, c and the crystallographic axes X, Y, Z must be known in order to use the proper constants. These relationships, for each crystal system, have been adopted by convention. Various piezoelectric, elastic, and other constants of a particular crystal specimen are evaluated in terms of X, Y, Z axes. The latter may or may not be different from the *coordinate axes* x, y, z that one chooses at will in considering, for instance, a particular wave propagation problem. However, in order to use the appropriate material constants, such as permittivity and stiffnesses in the problem at hand, the relationship between x, y, z and X, Y, Z must also be known. It is important to note that in practice when denoting crystal cuts, both Miller and Miller-Bravais indices are used in the X, Y, Z system of coordinates.

The crystal cuts for transducers can also be specified without Miller or Miller-Bravais notation. According to *IEEE Standard on Piezoelectricity*,[6] the first two letters denote the initial plate orientation in which the thickness and length fall along the X, Y or Z axes, respectively. Thus the first two letters completely specify the unrotated cut, that is, the first letter (X, Y, or Z) indicates the direction of the plate thickness before any rotation is made and the second letter indicates the plate length before any rotations. For instance, a YZ cut is shown in Fig. 6.4. The remaining three symbols are used to indicate the plate edges used as axes of rotation, followed by a list of rotation angles ϕ, θ, ψ. For instance,[5] YZ *twl* 30°/15°/40° means the YZ plate rotated by $\phi = 30°$ about thickness t, then by $\theta = 15°$ about width w and $\psi = 40°$ about length l.

The ferroelectric ceramics,[7] which are isotropic polycrystals before poling, are isotropic after poling only in the plane transverse to the poling axis. The poling axis is taken to be the Z axis by convention, and the material is said to be acoustically uniaxial having an ∞-fold axis of symmetry and denoted in Hermann-Mauguin notation by ∞m. Since the highest axis of symmetry is natural crystals in a sixfold axis, the poled ferroelectric ceramics are usually included in the hexagonal class $6mm$. In another synthetic piezoelectric, PVF$_2$ (polyvinylidene fluoride),[8] which is available in flexible sheet form, two axes must be distinguished, the poling and stretching axes. The poling axis is taken to be the Z axis, and the stretching axis is taken to be the X axis, the material belonging to orthorhombic class $2mm$. An important class[7] of piezoelectrics belonging to hexagonal class $6mm$, where the c axis of the natural coordinate system is taken to be the sixfold axis of symmetry and parallel to the Z axis of the crystallographic coordinate system, can be deposited in crystalline layers by such techniques as molecular beam epitaxy[9] or in polycrystalline layers by sputter deposition. Materials such as CdS, ZnO, CdSe, AlN, and BeO all belong to hexagonal class $6mm$. The crystalline layers exhibit exactly the same properties as the bulk crystals, and they are used mainly for optical applications. Polycrystalline layers of these materials have been grown with the Z axis normal

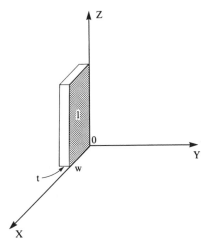

Figure 6.4. A YZ cut.

or parallel to the plane of the film.[7] In particular, when grown with the Z axis normal to the specimen surface, the individual grains need only have their Z axes parallel to one another (transverse isotropy) in order to exhibit a strong piezoelectric effect for longitudinal acoustic waves. The resulting coupling factors can be as much as 90% of the bulk single-crystal value, making these materials important for microwave acoustic transducers.

It has been pointed out in Chap. 5 that all crystal classes not possessing a center of symmetry except one (cubic class 432) are piezoelectric. For cubic class 432, the symmetry requirement, due to threefold axes, reduces the piezoelectric effect to zero.

In isotropic materials the choice of axes is unimportant, the materials often being classified in the cubic crystal system or in the *isotropic* (*crystal*) *system*, which is added to the existing seven crystal systems shown in Table 6.

6.2   PIEZOELECTRIC CONSTITUTIVE RELATIONS[†]

The symmetry of the crystal affects various constants in the stiffness (compliance), dielectric, and piezoelectric tensors. The latter can be defined through generalization of the one-dimensional constitutive relations, Eq. (5.6), and is given in the full tensor notation as[10]

$$T_{jk} = c^E_{jklm}S_{lm} - e_{jki}E_i$$

$$D_i = e_{ijk}S_{jk} + \varepsilon^S_{ij}E_j \tag{6.4}$$

where the tensors T_{jk}, c_{jklm}, and S_{jk} have been defined in Chap. 3, and the electric displacement D_i, electric field E_j, and dielectric tensor ε_{ij} are familiar quantities known from the theory of electromagnetism. The latter is a second-order symmetry tensor, $\varepsilon_{ij} = \varepsilon_{ji}$, having at most six independent components

$$\left[\varepsilon^S_{ij}\right] = \begin{bmatrix} \varepsilon^S_{xx} & \varepsilon^S_{xy} & \varepsilon^S_{xz} \\ \varepsilon^S_{xy} & \varepsilon^S_{yy} & \varepsilon^S_{yz} \\ \varepsilon^S_{xz} & \varepsilon^S_{yz} & \varepsilon^S_{zz} \end{bmatrix} \tag{6.5}$$

Both D_i and E_j are column vectors given by

$$[D_i] = \begin{bmatrix} D_x \\ D_y \\ D_z \end{bmatrix}, \qquad [E_j] = \begin{bmatrix} E_x \\ E_y \\ E_z \end{bmatrix} \tag{6.6}$$

In Eq. (6.4), e_{ijk} is a third-rank tensor, containing at most 27 independent

[†]Unless otherwise stated, in all the derivations that follow it will be assumed that the coordinate x, y, z and the crystallographic X, Y, Z axes coincide.

components, such that

$$e_{ijk} = \left(\frac{\partial D_i}{\partial S_{jk}}\right)_E = -\left(\frac{\partial T_{jk}}{\partial E_i}\right)_S \tag{6.7}$$

The tensor e_{ijk} is symmetric with respect to the two last indices,

$$e_{ijk} = e_{ikj} \tag{6.8}$$

which is a consequence of the symmetry in the strain and stress tensors. Equation (6.8) reduces the number of independent components from 27 to 18. The last two indices can have at most six different values, allowing for the abbreviated notation, Eq. (3.5), with the conversion table

jk	J
xx	1
yy	2
zz	3
$yz = zy$	4
$xz = zx$	5
$xy = yx$	6

$$\tag{6.9}$$

and with

$$e_{iJ} = e_{ijk}, \qquad i = x, y, z; J = 1 \text{ to } 6 \tag{6.10}$$

The latter expression can be represented by a matrix

$$[e_{iJ}] = \begin{vmatrix} e_{x1} & e_{x2} & e_{x3} & e_{x4} & e_{x5} & e_{x6} \\ e_{y1} & e_{y2} & e_{y3} & e_{y4} & e_{y5} & e_{y6} \\ e_{z1} & e_{z2} & e_{z3} & e_{z4} & e_{z5} & e_{z6} \end{vmatrix} \tag{6.11}$$

termed the *piezoelectric stress matrix*. Therefore, Eq. (6.4) can be written in abbreviated form as

$$T_J = c^E_{JK}S_K - e_{Ji}E_i$$

$$D_i = e_{iJ}S_J + \varepsilon^S_{ij}E_j \tag{6.12}$$

where $i, j = x, y, z$ and $J, K = 1$ to 6, or in the matrix form as

$$[T] = [c^E][S] - [\tilde{e}][E]$$

$$[D] = [e][S] + [\varepsilon^S][E] \tag{6.13}$$

where the column vectors $[T]$, $[S]$, $[E]$, and $[D]$ are given by Eqs. (3.5), (3.18), and (6.6), the matrices $[c^E]$, $[\varepsilon^S]$, and $[e]$ by Eqs. (3.42), (6.5), and (6.11), and where $\tilde{e}$ indicates the transpose of matrix e. Equation (6.4) can also be written in symbolic notation as

$$\overset{\leftrightarrow}{T} = \overset{\leftrightarrow}{c}^E : \vec{S} - \overset{\leftrightarrow}{e} \cdot \vec{E}$$

$$\vec{D} = \overset{\leftrightarrow}{e} : \vec{S} + \overset{\leftrightarrow}{\varepsilon}^S \cdot E \tag{6.14}$$

where

$$\overset{\leftrightarrow}{e} = \overset{\sim}{\overset{\leftrightarrow}{e}} \tag{6.15}$$

is the transpose of the third-rank tensor.

In general, the full tensor notation, Eq. (6.4), is useful in problems involving coordinate transformations, while the matrix notation, Eq. (6.13), is useful in carrying out numerical computations.[†]

In analogy with Eq. (6.13), Eq. (5.7) can be written as

$$[S] = [s^E][T] + [\tilde{d}][E]$$

$$[D] = [d][T] + [\varepsilon^T][E] \tag{6.16}$$

where $[d]$ is 3×6 matrix termed the *piezoelectric strain matrix*.

The other forms of the constitutive relations and relationships between various constants can also be defined.[10,11] Equation (6.6) can be written in a single matrix form as

$$
\begin{bmatrix} S_1 \\ S_2 \\ S_3 \\ S_4 \\ S_5 \\ S_6 \\ D_x \\ D_y \\ D_z \end{bmatrix} =
\begin{bmatrix}
s_{11}^E & s_{12}^E & s_{13}^E & s_{14}^E & s_{15}^E & s_{16}^E & d_{x1} & d_{y1} & d_{z1} \\
s_{12}^E & s_{22}^E & s_{23}^E & s_{24}^E & s_{25}^E & s_{26}^E & d_{x2} & d_{y2} & d_{z2} \\
s_{13}^E & s_{23}^E & s_{33}^E & s_{34}^E & s_{35}^E & s_{36}^E & d_{x3} & d_{y3} & d_{z3} \\
s_{14}^E & s_{24}^E & s_{34}^E & s_{44}^E & s_{45}^E & s_{46}^E & d_{x4} & d_{y4} & d_{z4} \\
s_{15}^E & s_{25}^E & s_{35}^E & s_{45}^E & s_{55}^E & s_{56}^E & d_{x5} & d_{y5} & d_{z5} \\
s_{16}^E & s_{26}^E & s_{36}^E & s_{46}^E & s_{56}^E & s_{66}^E & d_{x6} & d_{y6} & d_{z6} \\
d_{x1} & d_{x2} & d_{x3} & d_{x4} & d_{x5} & d_{x6} & \varepsilon_{xx}^T & \varepsilon_{xy}^T & \varepsilon_{xz}^T \\
d_{y1} & d_{y2} & d_{y3} & d_{y4} & d_{y5} & d_{y6} & \varepsilon_{xy}^T & \varepsilon_{yy}^T & \varepsilon_{yz}^T \\
d_{z1} & d_{z2} & d_{z3} & d_{z4} & d_{z5} & d_{z6} & \varepsilon_{xz}^T & \varepsilon_{yz}^T & \varepsilon_{zz}^T
\end{bmatrix}
\begin{bmatrix} T_1 \\ T_2 \\ T_3 \\ T_4 \\ T_5 \\ T_6 \\ E_x \\ E_y \\ E_z \end{bmatrix}
$$

defining a symmetric 9×9 *elasto-piezo-dielectric matrix*. By using the 9×9 matrix it is now possible to express the characteristic of all crystal classes.[‡] Similar matrices can also be written using the three other forms of constitutive relations. In order to simplify the tabulation the following convention has been adopted:[6]

[†]However, the Bond transformation technique is applicable to matrices, see Sect. 6.4.
[‡]The following ten classes are pyroelectric: (1), (2), (m), (mm), (4), (4mm), (3), (3m), (6), and (6mm).

●   Indicates a nonzero constant
○   Indicates the negative of ●
×   Indicates $2(s_{11} - s_{12})$
—   A line between any two nonzero constants indicates numerical equality
◼   Indicates twice the numerical value of ●
◻   Indicates twice the numerical value of ○
NIE   Number of independent elastic constants
NIP   Number of independent piezoelectric constants
NID   Number of independent dielectric constants

6.2.1   Triclinic

Class (1)   Class ($\bar{1}$)

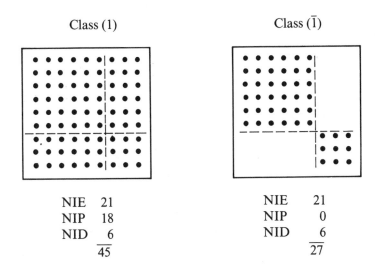

NIE   21   NIE   21
NIP   18   NIP   0
NID   6   NID   6
      $\overline{45}$ $\overline{27}$

6.2.2   Monoclinic

Class (2)   Class (m)   Class ($2/m$)

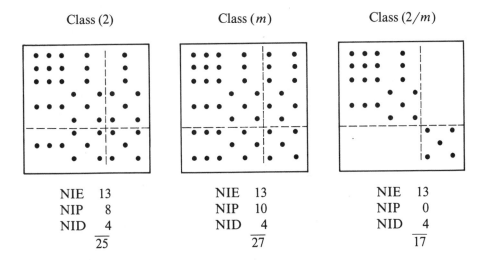

NIE   13   NIE   13   NIE   13
NIP   8   NIP   10   NIP   0
NID   4   NID   4   NID   4
     $\overline{25}$ $\overline{27}$ $\overline{17}$

6.2.3   Orthorhombic

Class (222)   Class (*mm*)   Class (*mmm*)

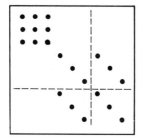   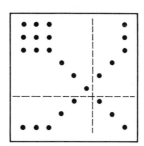   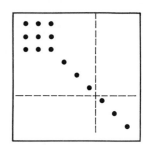

NIE	9		NIE	9		NIE	9
NIP	3		NIP	5		NIP	0
NID	3		NID	3		NID	3
	15			17			12

6.2.4   Rhombohedral (Trigonal)

Class ($\bar{3}$)   Class ($\bar{3}m$)

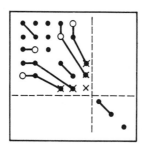   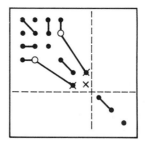

NIE	7		NIE	6
NIP	0		NIP	0
NID	2		NID	2
	9			8

Class (3)   Class (3m)   Class (32)

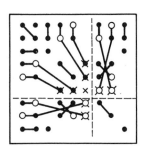   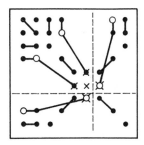   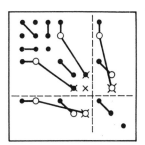

NIE   7   NIE   6   NIE   6
NIP   6   NIP   4   NIP   2
NID   2   NID   2   NID   2
        ──   ──   ──
        15   12   10

6.2.5   Tetragonal

Class (4)   Class ($\bar{4}$)   Class (4mm)

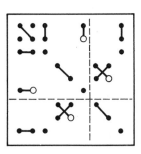   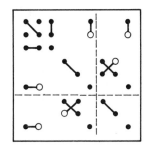   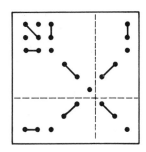

NIE   7   NIE   7   NIE   6
NIP   4   NIP   4   NIP   3
NID   2   NID   2   NID   2
        ──   ──   ──
        13   13   11

Class ($\bar{4}2m$)   Class (42)

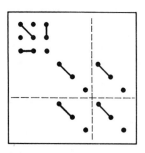
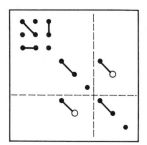

NIE   6   NIE   6
NIP   2   NIP   1
NID   2   NID   2
      $\overline{10}$ $\overline{9}$

Class ($4/m$)   Class ($4/mmm$)

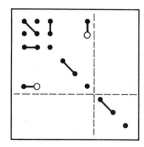
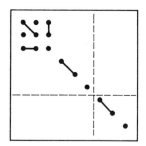

NIE   7   NIE   6
NIP   0   NIP   0
NID   2   NID   2
      $\overline{9}$ $\overline{8}$

6.2.6 Hexagonal

Class (6)   Classes (6mm), (∞m)   Class ($\bar{6}$)

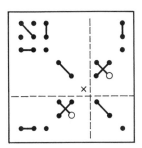 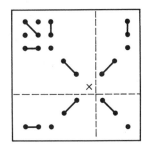 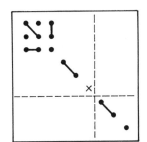

NIE	5
NIP	4
NID	2
	$\overline{11}$

NIE	5
NIP	3
NID	2
	$\overline{10}$

NIE	5
NIP	2
NID	2
	$\overline{9}$

Class ($\bar{6}2m$)   Class (62)   Classes (6/m), (6/mmm)

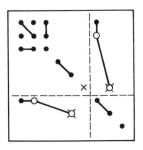 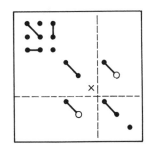

NIE	5
NIP	1
NID	2
	$\overline{8}$

NIE	5
NIP	1
NID	2
	$\overline{8}$

NIE	5
NIP	0
NID	2
	$\overline{7}$

6.2.7   Cubic

Classes (23), ($\bar{4}3m$)   Classes (43), ($m3$), ($m3m$)

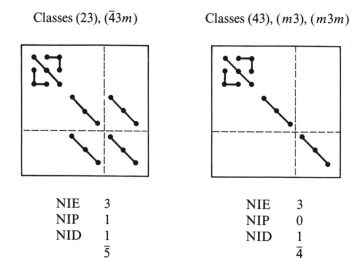

NIE	3		NIE	3
NIP	1		NIP	0
NID	1		NID	1
	$\bar{5}$			$\bar{4}$

6.2.8   Isotropic

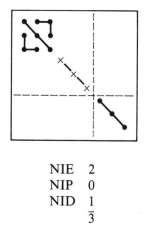

NIE   2
NIP   0
NID   1
      $\bar{3}$

For nonpiezoelectric materials the superscript E in $[s^E]$ matrices and the superscript T in $[\varepsilon^T]$ matrices should be removed from Eq. (6.17).

6.2.9   Piezoelectric Strain Constants of Some Materials

Various constants associated with the constitutive relation, Eq. (6.16), for several materials are given in Tables 7 and 8. Hard materials such as diamonds

TABLE 7 Compliance and Other Constants of Some Piezoelectric Materials[a]

Constant	Units	1 SiO₂	2 LiNbO₃	3 LiTaO₃	4 TeO₂	5 PZT-5A	6 ZnO	7 Bi₁₂GeO₂₀	8 GaAs
s_{11}^E	pm²/N	12.77	5.78	4.87	115	16.4	7.86	9.9	12.64
s_{33}^E		9.60	5.02	4.36	10.3	18.8	6.94		
s_{12}^E		−1.79	−1.01	−0.58	−104.8	−5.74	−3.43	−2.43	−4.234
s_{13}^E		−1.22	−1.47	−1.25	−2.11	−7.72	−2.206		
s_{44}^E		20.04	17.0	10.8	37.7	47.5	23.57	40.0	18.6
s_{14}^E		4.50	−1.02	0.64					
s_{66}^E					15.2				
d_{x1}	pC/N	2.31							
d_{x4}		−0.723			8.13			28	2.6
d_{x5}			68	26		584	−11.34		
d_{y2}			21	7					
d_{z1}			−1	−2		−171	−5.43		
d_{z3}			6	8		374	11.67		
$\varepsilon_{xx}^T/\varepsilon_0$	—	4.52	84	51	22.9	1730	9.16	40	12.5
$\varepsilon_{zz}^T/\varepsilon_0$		4.68	30	45	24.7	1700	12.64		
ρ_m	g/cm³	2.65	4.65	7.46	6.00	7.75	5.68	9.20	5.31
System:		Trig.	Trig.	Trig.	Tetrag.	Hexag.	Hexag.	Cubic	Cubic
Class:		(32)	(3m)	(3m)	(42)	(∞m)	(6mm)	(23)	(4̄3m)

[a] 1, Quartz (right-handed); 2, Lithium niobate; 3, Lithium tantalate; 4, Paratellurite; 5, Lead titanate zirconate (5A); 6, Zinc oxide; 7, Bismuth germanium oxide; and 8, Gallium arsenide.

TABLE 8 Compliance and Other Constants of Some Nonpiezoelectric Materials[a]

Constants	Units	1 Sapphire	2 Rutile	3 Aluminum	4 Gold	5 Silver	6 Titanium	7 Diamond	8 Silicon
s_{11}	pm²/N	2.38	6.788	15.9	23.3	23.2	9.59	1.12	7.68
s_{33}		2.20	2.592				6.99		
s_{12}		−0.70	−4.017	−5.8	−10.7	−9.93	−4.62	−0.22	−2.14
s_{13}		−0.38	−0.799				−1.90		
s_{44}		7.05	8.072	35.2	23.8	22.9	21.4	2.07	12.56
s_{14}		0.49							
s_{66}			5.302						
$\varepsilon_{xx}/\varepsilon_0$		9.34	89					5.67	11.7
$\varepsilon_{zz}/\varepsilon_0$		11.54	173						
ρ_m	g/cm³	3.986	4.26	2.695	19.3	10.49	4.5	3.512	2.332
System:		Trig.	Tetrag.	Cubic	Cubic	Cubic	Hexag.	Cubic	Cubic
Class:		$(\bar{3}m)$	$(4/mmm)$	$(m3m)$	$(m3m)$	$(m3m)$	$(6/mmm)$	$(m3m)$	$(m3m)$

[a] 1, Al_2O_3; 2, TiO_2; 3, Al; 4, Au; 5, Ag; 6, Ti; 7, C; and 8, Si.

and rutile have small compliances, whereas weak piezoelectrics such as GaAs and quartz have very small piezoelectric strain constants, and strong piezoelectrics such as PZT-5A and LiNbO$_3$ possess very large piezoelectric strain constants.

6.3 PIEZOELECTRIC STRESS CONSTANTS

In many practical situations such as in transducer studies and design, it is convenient to use the piezoelectric stress constitutive relation,[6, 10, 11] Eq. (6.13),

$$[T] = [c^E][S] - [\tilde{e}][E]$$
$$[D] = [e][S] + [\varepsilon^S][E] \qquad (6.13)$$

In what follows, the $[c^E]$, $[e]$, and $[\varepsilon^S]$ matrices are shown for each class.

6.3.1 Triclinic System

Classes (1), ($\bar{1}$):

$$
\begin{bmatrix}
c_{11}^E & c_{12}^E & c_{13}^E & c_{14}^E & c_{15}^E & c_{16}^E \\
 & c_{22}^E & c_{23}^E & c_{24}^E & c_{25} & c_{26}^E \\
 & & c_{33}^E & c_{34}^E & c_{35}^E & c_{36}^E \\
 & & & c_{44}^E & c_{45}^E & c_{46}^E \\
 & & & & c_{55}^E & c_{56}^E \\
\text{symmetric} & & & & & c_{66}^E
\end{bmatrix}
\begin{bmatrix}
\varepsilon_{11}^S & \varepsilon_{12}^S & \varepsilon_{13}^S \\
 & \varepsilon_{22}^S & \varepsilon_{23}^S \\
\text{symmetric} & & \varepsilon_{33}^S
\end{bmatrix}
$$

Class (1):

$$
\begin{bmatrix}
e_{x1} & e_{x2} & e_{x3} & e_{x4} & e_{x5} & e_{x6} \\
e_{y1} & e_{y2} & e_{y3} & e_{y4} & e_{y5} & e_{y6} \\
e_{z1} & e_{z2} & e_{z3} & e_{z4} & e_{z5} & e_{z6}
\end{bmatrix}
$$

6.3.2 Monoclinic System

Classes (2), (m), ($2/m$):

$$
\begin{bmatrix}
c_{11}^E & c_{12}^E & c_{13}^E & 0 & c_{15}^E & 0 \\
 & c_{22}^E & c_{23}^E & 0 & c_{25}^E & 0 \\
 & & c_{33}^E & 0 & c_{35}^E & 0 \\
 & & & c_{44}^E & 0 & c_{46}^E \\
 & & & & c_{55}^E & 0 \\
\text{symmetric} & & & & & c_{66}^E
\end{bmatrix}
\begin{bmatrix}
\varepsilon_{xx}^S & 0 & \varepsilon_{xz}^S \\
 & \varepsilon_{yy}^S & 0 \\
\text{symmetric} & & \varepsilon_{zz}^S
\end{bmatrix}
$$

Class (2):

$$\begin{bmatrix} 0 & 0 & 0 & e_{x4} & 0 & e_{x6} \\ e_{y1} & e_{y2} & e_{y3} & 0 & e_{y5} & 0 \\ 0 & 0 & 0 & e_{z4} & 0 & e_{z6} \end{bmatrix}$$

Class (m):

$$\begin{bmatrix} e_{x1} & e_{x2} & e_{x3} & 0 & e_{x5} & 0 \\ 0 & 0 & 0 & e_{y4} & 0 & e_{y6} \\ e_{z1} & e_{z2} & e_{z3} & 0 & e_{z5} & 0 \end{bmatrix}$$

6.3.3   Orthorhombic System

Classes (222), (mm), (mmm):

$$\begin{bmatrix} c_{11}^E & c_{12}^E & c_{13}^E & 0 & 0 & 0 \\ & c_{22}^E & c_{23}^E & 0 & 0 & 0 \\ & & c_{33}^E & 0 & 0 & 0 \\ & & & c_{44}^E & 0 & 0 \\ & & & & c_{55}^E & 0 \\ \text{symmetric} & & & & & c_{66}^E \end{bmatrix} \quad \begin{bmatrix} \varepsilon_{xx}^S & 0 & 0 \\ & \varepsilon_{yy}^S & 0 \\ \text{symmetric} & \varepsilon_{zz}^S \end{bmatrix}$$

Class (222):

$$\begin{bmatrix} 0 & 0 & 0 & e_{x4} & 0 & 0 \\ 0 & 0 & 0 & 0 & e_{y5} & 0 \\ 0 & 0 & 0 & 0 & 0 & e_{z6} \end{bmatrix}$$

Class (mm):

$$\begin{bmatrix} 0 & 0 & 0 & 0 & e_{x5} & 0 \\ 0 & 0 & 0 & e_{y4} & 0 & 0 \\ e_{z1} & e_{z2} & e_{z3} & 0 & 0 & 0 \end{bmatrix}$$

6.3.4 Rhombohedral (Trigonal) System

Classes (3), $(\bar{3})$:

$$
\begin{bmatrix}
c_{11}^E & c_{12}^E & c_{13}^E & c_{14}^E & -c_{25}^E & 0 \\
 & c_{11}^E & c_{13}^E & -c_{14}^E & c_{25}^E & 0 \\
 & & c_{33}^E & 0 & 0 & 0 \\
 & & & c_{44}^E & 0 & c_{25}^E \\
 & & & & c_{44}^E & c_{14}^E \\
\text{symmetric} & & & & & c_{66}^E
\end{bmatrix}
\qquad
\text{All classes:}
\begin{bmatrix}
\varepsilon_{xx}^S & 0 & 0 \\
 & \varepsilon_{xx}^S & 0 \\
\text{symmetric} & & \varepsilon_{zz}^S
\end{bmatrix}
$$

$$c_{66}^E = \tfrac{1}{2}(c_{11}^E - c_{12}^E)$$

Classes $(3m)$, (32), $(3m)$:

$$
\begin{bmatrix}
c_{11}^E & c_{12}^E & c_{13}^E & c_{14}^E & 0 & 0 \\
 & c_{11}^E & c_{13}^E & -c_{14}^E & 0 & 0 \\
 & & c_{33}^E & 0 & 0 & 0 \\
 & & & c_{44}^E & 0 & 0 \\
 & & & & c_{44}^E & c_{14}^E \\
\text{symmetric} & & & & & c_{66}^E
\end{bmatrix}
$$

$$c_{66}^E = \tfrac{1}{2}\left(c_{11}^E - c_{12}^E\right)$$

Class (3):

$$
\begin{bmatrix}
e_{x1} & -e_{x1} & 0 & e_{x4} & e_{x5} & -e_{y2} \\
-e_{y2} & e_{y2} & 0 & e_{x5} & -e_{x4} & -e_{x1} \\
e_{z1} & e_{z1} & e_{z3} & 0 & 0 & 0
\end{bmatrix}
$$

Class (32):

$$
\begin{bmatrix}
e_{x1} & -e_{x1} & 0 & e_{x4} & 0 & 0 \\
0 & 0 & 0 & 0 & -e_{x4} & -e_{x1} \\
0 & 0 & 0 & 0 & 0 & 0
\end{bmatrix}
$$

Class $(3m)$:

$$
\begin{bmatrix}
0 & 0 & 0 & 0 & e_{x5} & -e_{y2} \\
-e_{y2} & e_{y2} & 0 & e_{x5} & 0 & 0 \\
e_{z1} & e_{z1} & e_{z3} & 0 & 0 & 0
\end{bmatrix}
$$

6.3.5   Tetragonal System

Classes $(4mm)$, $(\bar{4}2m)$, (42), $(4/mmm)$:

$$
\begin{bmatrix}
c_{11}^E & c_{12}^E & c_{13}^E & 0 & 0 & 0 \\
 & c_{11}^E & c_{13}^E & 0 & 0 & 0 \\
 & & c_{33}^E & 0 & 0 & 0 \\
 & & & c_{44}^E & 0 & 0 \\
 & & & & c_{44}^E & 0 \\
\text{symmetric} & & & & & c_{66}^E
\end{bmatrix}
$$

All classes:

$$
\begin{bmatrix}
\varepsilon_{xx}^S & 0 & 0 \\
 & \varepsilon_{xx}^S & 0 \\
\text{symmetric} & & \varepsilon_{zz}^S
\end{bmatrix}
$$

All classes:

$$c_{66}^E = \tfrac{1}{2}\left(c_{11}^E - c_{12}^E \right)$$

Classes (4), $(4/m)$, $(\bar{4})$;

$$
\begin{bmatrix}
c_{11}^E & c_{12}^E & c_{13}^E & 0 & 0 & c_{16}^E \\
 & c_{11}^E & c_{13}^E & 0 & 0 & -c_{16}^E \\
 & & c_{33}^E & 0 & 0 & 0 \\
 & & & c_{44}^E & 0 & 0 \\
 & & & & c_{44}^E & 0 \\
\text{symmetric} & & & & & c_{66}^E
\end{bmatrix}
$$

Class $(4mm)$:

$$
\begin{bmatrix}
0 & 0 & 0 & 0 & e_{x5} & 0 \\
0 & 0 & 0 & e_{x5} & 0 & 0 \\
e_{z1} & e_{z1} & e_{z3} & 0 & 0 & 0
\end{bmatrix}
$$

Class (42):

$$
\begin{bmatrix}
0 & 0 & 0 & e_{x4} & 0 & 0 \\
0 & 0 & 0 & 0 & -e_{x4} & 0 \\
0 & 0 & 0 & 0 & 0 & 0
\end{bmatrix}
$$

Class (4):

$$
\begin{bmatrix}
0 & 0 & 0 & e_{x4} & e_{x5} & 0 \\
0 & 0 & 0 & e_{x5} & -e_{x4} & 0 \\
e_{z1} & e_{z1} & e_{z3} & 0 & 0 & 0
\end{bmatrix}
$$

Class $(\bar{4})$:

$$
\begin{bmatrix}
0 & 0 & 0 & e_{x4} & e_{x5} & 0 \\
0 & 0 & 0 & -e_{x5} & e_{x4} & 0 \\
e_{z1} & -e_{z1} & 0 & 0 & 0 & e_{z6}
\end{bmatrix}
$$

Class $(\bar{4}2m)$:

$$
\begin{bmatrix}
0 & 0 & 0 & e_{x4} & 0 & 0 \\
0 & 0 & 0 & 0 & e_{x4} & 0 \\
0 & 0 & 0 & 0 & 0 & e_{z6}
\end{bmatrix}
$$

6.3.6   Hexagonal System

All classes:

$$
\begin{bmatrix}
c_{11}^E & c_{12}^E & c_{13}^E & 0 & 0 & 0 \\
 & c_{11}^E & c_{13}^E & 0 & 0 & 0 \\
 & & c_{33}^E & 0 & 0 & 0 \\
 & & & c_{44}^E & 0 & 0 \\
 & & & & c_{44}^E & 0 \\
\text{symmetric} & & & & & c_{66}^E
\end{bmatrix}
$$

All classes:

$$
\begin{bmatrix}
\varepsilon_{xx}^S & 0 & 0 \\
 & \varepsilon_{xx}^S & 0 \\
\text{symmetric} & & \varepsilon_{zz}^S
\end{bmatrix}
$$

$$c_{66}^E = \tfrac{1}{2}\left(c_{11}^E - c_{12}^E \right)$$

Class $(\bar{6}2m)$:

$$\begin{bmatrix} e_{x1} & -e_{x1} & 0 & 0 & 0 & 0 \\ 0 & 0 & 0 & 0 & 0 & -e_{x1} \\ 0 & 0 & 0 & 0 & 0 & 0 \end{bmatrix}$$

Class $(\bar{6})$:

$$\begin{bmatrix} e_{x1} & -e_{x1} & 0 & 0 & 0 & -e_{y2} \\ -e_{y2} & e_{y2} & 0 & 0 & 0 & -e_{x1} \\ 0 & 0 & 0 & 0 & 0 & 0 \end{bmatrix}$$

Class $(6mm)$, (∞m):

$$\begin{bmatrix} 0 & 0 & 0 & 0 & e_{x5} & 0 \\ 0 & 0 & 0 & e_{x5} & 0 & 0 \\ e_{z1} & e_{z1} & e_{z3} & 0 & 0 & 0 \end{bmatrix}$$

Class (62):

$$\begin{bmatrix} 0 & 0 & 0 & e_{x4} & 0 & 0 \\ 0 & 0 & 0 & 0 & -e_{x4} & 0 \\ 0 & 0 & 0 & 0 & 0 & 0 \end{bmatrix}$$

Class (6):

$$\begin{bmatrix} 0 & 0 & 0 & e_{x4} & e_{x5} & 0 \\ 0 & 0 & 0 & e_{x5} & -e_{x4} & 0 \\ e_{z1} & e_{z1} & e_{z3} & 0 & 0 & 0 \end{bmatrix}$$

6.3.7 Cubic System

All classes:

$$\begin{bmatrix} c_{11}^E & c_{12}^E & c_{12}^E & 0 & 0 & 0 \\ & c_{11}^E & c_{12}^E & 0 & 0 & 0 \\ & & c_{11}^E & 0 & 0 & 0 \\ & & & c_{44}^E & 0 & 0 \\ & & & & c_{44}^E & 0 \\ \text{symmetric} & & & & & c_{44}^E \end{bmatrix}$$

All classes:

$$\begin{bmatrix} \varepsilon_{xx}^S & 0 & 0 \\ & \varepsilon_{xx}^S & 0 \\ \text{symmetric} & & \varepsilon_{xx}^S \end{bmatrix}$$

Classes (23), $(\bar{4}3m)$:

$$\begin{bmatrix} 0 & 0 & 0 & e_{x4} & 0 & 0 \\ 0 & 0 & 0 & 0 & e_{x4} & 0 \\ 0 & 0 & 0 & 0 & 0 & e_{x4} \end{bmatrix}$$

6.3.8 Isotropic System

$$\begin{bmatrix} c_{11} & c_{12} & c_{12} & 0 & 0 & 0 \\ & c_{11} & c_{12} & 0 & 0 & 0 \\ & & c_{11} & 0 & 0 & 0 \\ & & & c_{44} & 0 & 0 \\ & & & & c_{44} & 0 \\ \text{symmetric} & & & & & c_{44} \end{bmatrix}$$

$$\begin{bmatrix} \varepsilon_{xx} & 0 & 0 \\ & \varepsilon_{xx} & 0 \\ \text{symmetric} & & \varepsilon_{xx} \end{bmatrix}$$

$$c_{12} = c_{11} - 2c_{44}$$

TABLE 9 Stiffness and Other Constants of Some Piezoelectric Materials

Constants	Units	SiO₂[a]	PZT-5H	LiNbO₃	ZnO	PZT-2	CdS	LiTaO₃	Bi₁₂GeO₂₀
c_{11}^E	GN/m²	86.74	126	203	209.7	135	90.7	233	128
c_{33}^E		107.2	117	245	210.9	113	93.8	275	
c_{12}^E		6.99	79.5	53	121.1	67.9	58.1	47	30.5
c_{13}^E		11.91	84.1	75	105.1	68.1	51.0	80	
c_{44}^E		57.94	23	60	42.47	22.2	15.04	94	25.5
c_{14}^E		−17.91		9				−1.1	
c_{66}^E									
e_{x1}	C/m²	0.171							
e_{x4}		0.0436							
e_{x5}			17.0	3.76	−0.48	9.8	−0.21	2.6	0.99
e_{y2}				2.45				1.6	
e_{z1}			−6.5	0.23	−0.573	−1.9	−0.24	0.0	
e_{z3}			23.3	1.8	1.32	9.0	0.44	1.9	
$\varepsilon_{xx}^S/\varepsilon_0$		4.5	1700	44	8.55	504	9.02	41	38
$\varepsilon_{zz}^S/\varepsilon_0$		4.6	1470	29	10.2	260	9.53	43	
ρ_m	g/cm³	2.65	7.5	4.64	5.68	7.6	4.82	7.46	9.20
System:		Trig.	Hexag.	Trig.	Hexag.	Hexag.	Hexag.	Trig.	Cubic
Class:		(32)	(∞m)	(3m)	(6mm)	(∞m)	(6mm)	(3m)	(23)

[a] Right-handed quartz.

198

TABLE 10 Stiffness and Other Constants of Some Nonpiezoelectric Materials[a]

Constants	Units	1 Al	1 Al (polycr.)	2 In	3 Ni	3 Ni (polycr.)	4 Ag	4 Ag (polycr.)	5 Fused Silica
c_{11}	GN/m²	108	111	45.35	250	324	119	139.5	78.5
c_{33}				45.15					
c_{12}		61.3		40.06	160		89.4		
c_{13}				41.51					
c_{44}		28.5	25.0	6.46	118.5	80	43.7	27	31.2
c_{14}									
c_{66}				12.07					
$\varepsilon_{xx}/\varepsilon_0$									3.78
$\varepsilon_{zz}/\varepsilon_0$									
ρ_m	g/cm³	2.695	2.695	7.3	8.905	8.905	10.49	10.49	2.2
System:		Cubic	Isotropic	Tetr.	Cubic	Isotropic	Cubic	Isotropic	Isotropic
Class:		($m3m$)		($4/mmm$)	($m3m$)		($m3m$)		

[a]1, Aluminum; 2, indium; 3, nickel; 4, silver; and, 5, SiO$_2$.

In all permittivity matrices, the superscript S should be removed whenever the material under consideration is nonpiezoelectric. The same applies to the superscript E in stiffness matrices. The other two constitutive relations,[10, 11] obtained by generalization of Eqs. (5.8) and (5.9), can be written as

$$[S] = [s^D][T] + [\tilde{g}][D]$$

$$[E] = -[g][T] + [\beta^T][D] \tag{6.18}$$

and

$$[T] = [c^D][S] - [\tilde{h}][D]$$

$$[E] = -[h][S] + [\beta^S][D] \tag{6.19}$$

It is to be noted that d and g matrices are identical in form, as are e and h matrices and β and ε matrices.

Various constants associated with the constitutive relation, Eq. (6.13), for several piezoelectric and nonpiezoelectric materials are shown in Tables 9 and 10.

6.4 TRANSFORMATION LAWS

In the design of various devices, a particular cut of the crystal and a particular wave propagation direction are chosen for one or more of the following reasons: high electromechanical coupling coefficient of a particular mode, temperature stability, acoustic beam diffraction considerations, and low losses. Numerical procedures used in determining the properties of the crystal cut require the formulation of physical laws under coordinate transformation conditions.

Coordinate system transformations in space consist of either a shift or a rotation, or both. In applying coordinate transformations, the underlying principle[12] is that properly formulated physical laws must be invariant under shifts (*mathematical homogeneity*) and rotations (*mathematical isotropy*) of one coordinate system (x', y', z') with respect to another (x, y, z) and must be independent of the choice of units.

Transformations of the shift type are not of interest to us, since they have already been taken into account by crystal symmetry considerations of the Bravais lattices and, in what follows, only rotational coordinate transformations in the rectangular coordinate system will be considered.

For the two right-handed coordinate systems shown in Fig. 6.5, characterized by unit vectors $\hat{x}$, $\hat{y}$, $\hat{z}$, and $\hat{x}'$, $\hat{y}'$, $\hat{z}'$, the angle between the $\hat{r}_j'$ axis and the $\hat{r}_k$ axis, denoted by $\theta_{j'k}$, is found from the directional cosine

$$\hat{r}_j' \cdot \hat{r}_k = \cos \theta_{j'k} \equiv a_{jk} \tag{6.20}$$

where r_j' and r_k take on the values x', y', z' and x, y, z, respectively.

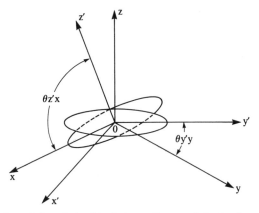

Figure 6.5. Illustration of the relationship between the new (x', y', z') and the old (x, y, z) coordinate system.

It is well known that coordinates of an arbitrary point, or a vector, in the new system (x', y', z') can be specified in terms of the old system (x, y, z) as

$$\begin{bmatrix} x' \\ y' \\ z' \end{bmatrix} = \begin{bmatrix} a'_{xx} & a'_{xy} & a'_{xz} \\ a'_{yx} & a'_{yy} & a'_{yz} \\ a'_{zx} & a'_{zy} & a'_{zz} \end{bmatrix} \begin{bmatrix} x \\ y \\ z \end{bmatrix} \tag{6.21}$$

or

$$[r'] = [a][r] \tag{6.22}$$

which can be written in full tensor notation as

$$r'_i = a_{ij}r_j \tag{6.23}$$

Since the transformation required is to preserve the length of the vector, it follows that

$$[\tilde{r}'][r'] = [\tilde{r}][r] \tag{6.24}$$

and on using Eq. (6.22) it will be found that

$$[\tilde{a}][a] = [I] \tag{6.25}$$

where $[I]$ is the identity matrix, or

$$a_{ik}a_{jk} = \delta_{ij} \tag{6.26}$$

where δ_{ij} is the Kronecker δ symbol. Thus the transformation matrix $[a]$ must be orthogonal.

A tensor of rank one, that is, a vector, an electric field for instance, in the new coordinate system is written as

$$[E'] = [a][E] \tag{6.27}$$

or

$$E'_i = a_{ij}E_j \tag{6.28}$$

and on using Eq. (6.25) it follows that

$$E_j = a_{ij}E'_i \tag{6.29}$$

Thus first-rank tensors transform according to the law given by Eq. (6.28). Similar laws can be derived for tensors of the second rank and higher. For instance, the constitutive relation for zero strain in an anisotropic dielectric can be written from Eq. (6.4) as

$$D_i = \varepsilon_{ij}E_j \tag{6.30}$$

where ε_{ij} is the second-rank (permittivity) tensor. Using Eqs. (6.28), (6.30), and (6.29), the transformation law for second-rank tensors follows from

$$D'_i = a_{ik}D_{kl} = a_{ik}\varepsilon_{kl}E_l = a_{ik}\varepsilon_{kl}a_{jl}E'_j$$

$$= \varepsilon'_{jk}E'_j \tag{6.31}$$

as

$$\varepsilon'_{ij} = a_{ik}a_{jl}\varepsilon_{kl} \tag{6.32}$$

where ε_{kl} are the components of the permittivity tensor with respect to the unprimed (old) system of coordinates.

As an example, consider the permittivity tensor of the tetragonal system, Section 6.3.5, and its transformation into the new system of coordinates obtained by a counterclockwise rotation through an angle ξ about the x axis. For this case, the $[a]$ matrix is given by

$$[a^\xi] = \begin{bmatrix} 1 & 0 & 0 \\ 0 & \cos\xi & \sin\xi \\ 0 & -\sin\xi & \cos\xi \end{bmatrix} \tag{6.33}$$

Since Eq. (6.32) can be written as

$$[\varepsilon'] = [a][\varepsilon][\tilde{a}] \tag{6.34}$$

and it is readily found that

$$\varepsilon'_{xx} = \varepsilon_{xx}$$

$$\varepsilon'_{yy} = \varepsilon_{xx}\cos^2\xi + \varepsilon_{zz}\sin^2\xi \qquad\qquad (6.35)$$

$$\varepsilon'_{zz} = \varepsilon_{xx}\sin^2\xi + \varepsilon_{zz}\cos^2\xi$$

$$\varepsilon'_{zy} = \varepsilon'_{yz} = (\varepsilon_{zz} - \varepsilon_{xx})\cos\xi\sin\xi$$

with the remaining $\varepsilon'_{ij} = 0$. It is seen that a symmetric tensor ε_{ij} under coordinate transformation remains symmetric, which is another general property of tensor transformations.

Using Eqs. (6.28) and (6.32), the transformation laws of higher order tensors are easily found by induction. For instance, for the piezoelectric stress (third-rank) tensor, it follows that

$$e'_{ijp} = a_{ik}a_{jl}a_{pm}e_{klm} \qquad\qquad (6.36)$$

and for the stiffness (fourth-rank) tensor,

$$c'_{ijpt} = a_{ik}a_{jl}a_{pm}a_{ts}c_{klms} \qquad\qquad (6.37)$$

In general, it can be shown that every physical property represented by an even-rank tensor, such as ε_{ij}, c_{ijpt}, is centrosymmetric, and that all physical properties described by an odd-rank tensor, such as e_{ijp}, are zero in crystals with a center of symmetry.

The constitutive relations given by Eqs. (6.12) and (6.16) are conveniently written in terms of abbreviated subscript notation, which is routinely used in engineering practice. It is, therefore, of interest to find a method for performing transformations with abbreviated subscripts, rather than using the full tensor notation. The transformation laws for abbreviated subscripts, known as the *Bond transformation*, is obtained by considering first the transformation law for strain, which, in analogy with Eq. (6.32) is written as[3]

$$S'_{pq} = a_{pi}a_{qj}S_{ij} \qquad\qquad (6.38)$$

Upon examination of various components and their conversion into abbreviated subscripts, this can be written as

$$S'_p = N_{PQ}S_Q \qquad\qquad (6.39)$$

where $P, Q = 1$ to 6, or in matrix form as

$$[S'] = [N][S] \qquad\qquad (6.40)$$

where N is 6×6 matrix. Similarly for stress,

$$T'_{kl} = a_{ki} a_{lj} T_{ij} \tag{6.41}$$

it is found that

$$T'_J = M_{JI} T_I \tag{6.42}$$

where $J, I = 1$ to 6, or

$$[T'] = [M][T] \tag{6.43}$$

Furthermore, it can be shown[3] that

$$[N]^{-1} = [\tilde{M}] \tag{6.44}$$

where $[\ ]^{-1}$ denotes an inverse matrix. Using the derived equations, the transformation laws for stiffness can be obtained from Hooke's law,

$$[T] = [c][S] \tag{3.38}$$

which is written in the new coordinate system as

$$[T'] = [M][T] = [M][c][S] = [M][c][\tilde{M}][S']$$
$$= [c'][S'] \tag{6.45}$$

yielding

$$[c'] = [M][c][\tilde{M}] \tag{6.46}$$

or

$$c'_{PQ} = M_{PI} M_{QJ} c_{IJ} \tag{6.47}$$

with $I, J, P, Q = 1$ to 6. Similarly, for compliances it is found that

$$[s'] = [N][s][\tilde{N}] \tag{6.48}$$

As an example, consider the stiffness tensor of the tetragonal class (42), Sect. 6.3.5, and its transformation into the new system of coordinates obtained by a counterclockwise rotation through an angle ξ about the x axis. Using Eqs. (6.33), (6.41), (6.42), and (3.36), it is found that

$$[M^\xi] = \begin{bmatrix} 1 & 0 & 0 & 0 & 0 & 0 \\ 0 & \cos^2\xi & \sin^2\xi & \sin^2\xi & 0 & 0 \\ 0 & \sin^2\xi & \cos^2\xi & -\sin^2\xi & 0 & 0 \\ 0 & -\frac{1}{2}\sin^2\xi & \frac{1}{2}\sin^2\xi & \cos^2\xi & 0 & 0 \\ 0 & 0 & 0 & 0 & \cos\xi & -\sin\xi \\ 0 & 0 & 0 & 0 & \sin\xi & \cos\xi \end{bmatrix} \tag{6.49}$$

and on using Eq. (6.47) it is found that

$$c'_{11} = c_{11}$$

$$c'_{22} = c_{11}\cos^4\xi + 2c_{13}\sin^2\xi\cos^2\xi + c_{33}\sin^4\xi$$

$$c'_{12} = c'_{21} = c_{12}\cos^2\xi + c_{13}\sin^2\xi$$

$$c'_{33} = c_{11}\sin^4\xi + 2c_{13}\sin^2\xi\cos^2\xi + c_{33}\cos^4\xi$$

$$c'_{23} = c'_{32} = c_{11}\sin^2\xi\cos^2\xi + c_{13}(\sin^4\xi + \cos^4\xi) + c_{33}\sin^2\xi\cos^2\xi \qquad (6.50)$$

$$c'_{31} = c'_{13} = c_{12}\sin^2\xi + c_{13}\cos^2\xi$$

$$c'_{44} = c_{44}\cos^2 2\xi$$

$$c'_{55} = c_{44}\cos^2\xi + c_{66}\sin^2\xi$$

$$c'_{66} = c_{44}\sin^2\xi + c_{66}\cos^2\xi$$

$$c'_{56} = c'_{65} = c_{44}\sin^2\xi - c_{66}\sin\xi\cos\xi,$$

where $c_{66} = \frac{1}{2}(c_{11} - c_{12})$ and where the remaining $c'_{PQ} = 0$.

The piezoelectric stress tensor obtained from Eq. (6.12) for zero strain is

$$T_J = -e'_{Ji}E_i \qquad (6.51)$$

and is written in the new coordinate system as

$$[T'] = [M][T] = -[M][\tilde{e}][E] \qquad (6.52)$$

and on using Eqs. (6.27) and (6.25) as

$$[T'] = -[M][\tilde{e}][\tilde{a}][E'] = -[\tilde{e}'][E'] \qquad (6.53)$$

This results in[10]

$$[\tilde{e}'] = [M][\tilde{e}][\tilde{a}] \qquad (6.54)$$

or

$$[e'] = [a][e][\tilde{M}] \qquad (6.55)$$

which can also be written as

$$e'_{iJ} = a_{ij}e_{jK}M_{JK} \qquad (6.56)$$

where i, $j = x$, y, z and J, $K = 1$ to 6. Similarly, the transformation laws for the piezoelectric strain tensor are obtained from Eq. (6.16) for the zero electric field as

$$[d'] = [a][d][\tilde{N}] \tag{6.57}$$

or

$$d'_{jI} = N_{IK} a_{jl} d_{lK} \tag{6.58}$$

where j, $l = x$, y, z and I, $K = 1$ to 6.

As an example, consider the piezoelectric stress tensor of the tetragonal class (42), Sect. 6.3.5, and its transformation into the new system of coordinates obtained by counterclockwise rotation through an angle ξ about the x axis. Using Eqs. (6.33), (6.49), and (6.54), it is readily found that

$$e'_{x1} = e_{x4}\sin 2\xi$$

$$e'_{x3} = -e_{x4}\sin 2\xi$$

$$e'_{x4} = e_{x4}\cos 2\xi$$

$$e'_{y5} = -e_{x4}\cos^2\xi \tag{6.59}$$

$$e'_{y6} = -e_{x4}\cos\xi \sin\xi$$

$$e'_{z5} = e_{x4}\sin\xi \cos\xi$$

$$e'_{z6} = e_{x4}\sin^2\xi$$

with the remaining $e'_{iJ} = 0$.

For counterclockwise rotation by an angle η about the y axis, the $[a]$ and $[M]$ matrices are given by

$$[a^\eta] = \begin{bmatrix} \cos\eta & 0 & -\sin\eta \\ 0 & 1 & 0 \\ \sin\eta & 0 & \cos\eta \end{bmatrix} \tag{6.60}$$

and

$$[M^\eta] = \begin{bmatrix} \cos^2\eta & 0 & \sin^2\eta & 0 & -\sin 2\eta & 0 \\ 0 & 1 & 0 & 0 & 0 & 0 \\ \sin^2\eta & 0 & \cos^2\eta & 0 & \sin 2\eta & 0 \\ 0 & 0 & 0 & \cos\eta & 0 & \sin\eta \\ \dfrac{\sin^2\eta}{2} & 0 & -\dfrac{\sin^2\eta}{2} & 0 & \cos 2\eta & 0 \\ 0 & 0 & 0 & -\sin\eta & 0 & \cos\eta \end{bmatrix} \tag{6.61}$$

and for counterclockwise rotation by an angle ζ about the z axis

$$[a^\zeta] = \begin{bmatrix} \cos\zeta & \sin\zeta & 0 \\ -\sin\zeta & \cos\zeta & 0 \\ 0 & 0 & 1 \end{bmatrix} \tag{6.62}$$

and

$$[M^\zeta] = \begin{bmatrix} \cos^2\zeta & \sin^2\zeta & 0 & 0 & 0 & \sin 2\zeta \\ \sin^2\zeta & \cos^2\zeta & 0 & 0 & 0 & -\sin 2\zeta \\ 0 & 0 & 1 & 0 & 0 & 0 \\ 0 & 0 & 0 & \cos\zeta & -\sin\zeta & 0 \\ 0 & 0 & 0 & \sin\zeta & \cos\zeta & 0 \\ -\tfrac{1}{2}\sin 2\zeta & \tfrac{1}{2}\sin 2\zeta & 0 & 0 & 0 & \cos 2\zeta \end{bmatrix} \tag{6.63}$$

Using the derived transformation laws, it is possible to relate the symmetry properties of various material tensors to the symmetry operations of the crystal lattice. For instance, if the equations

$$[\varepsilon'] = [a][\varepsilon][\tilde{a}] \tag{6.34}$$

$$[c'] = [M][c][\tilde{M}] \tag{6.46}$$

$$[e'] = [a][e][\tilde{M}] \tag{6.54}$$

result in

$$[\varepsilon'] = [\varepsilon] \tag{6.64}$$

$$[c'] = [c] \tag{6.65}$$

$$[e'] = [e] \tag{6.66}$$

the coordinate transformation performed is said to coincide with a symmetry operation of the crystal lattice, and the matrix $[a]$ is said to be the *generator matrix* of this particular class. Equations (6.34) to (6.66) are the manifestations of Neumann's principle, which states that if a_{ij} is a generator matrix, the components of any tensor property of the crystal to which a_{ij} belongs must be invariant under a_{ij} transformation.

For instance, for the tetragonal class (42), assume that initially the coordinate axes coincide with the crystal axes. Suppose that a new coordinate system is obtained by counterclockwise rotation through an angle 180° about the X crystal axis. In this case, the $[a]$ matrix, Eq. (6.33), becomes

$$\begin{bmatrix} 1 & 0 & 0 \\ 0 & -1 & 0 \\ 0 & 0 & -1 \end{bmatrix} \tag{6.67}$$

Equations (6.35) result in

$$\varepsilon'_{xx} = \varepsilon_{xx}$$

$$\varepsilon'_{yy} = \varepsilon_{xx} \qquad (6.68)$$

$$\varepsilon'_{zz} = \varepsilon_{zz}$$

Eqs. (6.50) in

$$c'_{11} = c_{11}$$

$$c'_{22} = c_{11}$$

$$c'_{12} = c'_{21} = c_{12}$$

$$c'_{33} = c_{33}$$

$$c'_{23} = c'_{32} = c_{13}$$

$$c'_{31} = c'_{13} = c_{13} \qquad (6.69)$$

$$c'_{44} = c_{44}$$

$$c'_{55} = c_{44}$$

$$c'_{66} = c_{66}$$

$$c'_{56} = c'_{65} = 0$$

Eqs. (6.59) result in

$$e'_{x4} = e_{x4}$$

$$e'_{y5} = -e_{x4} \qquad (6.70)$$

These are identical with the material tensors specified for the tetragonal class (42) in Sect. 6.3.5. The result is not surprising, since the operation given by Eq. (6.67) corresponds to the twofold axis of symmetry which is one of the properties characterizing this particular class.

As another example, consider the center of symmetry operator given by

$$[a] = \begin{bmatrix} -1 & 0 & 0 \\ 0 & -1 & 0 \\ 0 & 0 & -1 \end{bmatrix} \qquad (6.71)$$

which reduces Eq. (6.38) to

$$S'_{ij} = S_{ij} \qquad (6.72)$$

or Eq. (6.39) to $[N] = [I] = [\tilde{M}]$. It is seen from Eq. (6.55) that this results in

$$[e'] = -[e] \qquad\qquad (6.73)$$

which can be satisfied only if all tensor components are zero. Thus no crystal class with the center of symmetry can exhibit the phenomenon of piezoelectricity.

Using the derived equations, it is now possible to deduce various material constants in singly, doubly, or triply rotated crystal cuts. (The IEEE standard[6] notation for rotated crystal cuts has been explained at the beginning of the chapter.) Triply rotated crystal cuts are seldom used, and doubly rotated crystal[13] cuts have become of practical importance only recently. This refers in particular to quartz, which is used mostly in the design of precision crystal oscillators. In the past singly rotated crystal cuts were chosen so that the temperature dependence of frequency would be zero at some particular frequency. This resulted in a number of cuts produced by a single rotation around the X axis: the AT cut denoted by YXl-35°, the BT cut denoted by YXl-49°, etc. The performance of these cuts is satisfactory as long as the temperature changes are reasonably slow and small. However, small but rapid changes of temperature cause thermal gradients in the plate, which induce internal stresses in the crystal and thus change the resonant frequency. For this reason the singly rotated cuts of crystals are often referred to as *statistically compensated crystals*. More recently, it has been demonstrated that there exist doubly rotated cuts that are less susceptible to internal as well as external stresses such as those generated by acceleration-induced body forces. These cuts are referred to as SC (*stress-compensated*) cuts, and the resulting crystal plates are said to be dynamically compensated. The SC cut in quartz is approximately determined by $YZwl$ 21.9°/33.9°.

Another method for improving the frequency stability of the oscillators is to use the so-called *enantiomorphous crystals*.[6, 13] These crystals lack a mirror plane of symmetry, center of symmetry, and inversion axes and belong to classes (1), (2), (222), (4), (42), (3), (32), (6), (62), (23), or (43). With the exception of class (43) they are all piezoelectric. Individual crystals of these classes are obtainable in two forms (like left- and right-handed gloves), which are never superposable and are called right-handed and left-handed. Thus enantiomorphs are mirror images of each other. Using two resonators made, for instance, of right-handed and left-handed quartz, it is possible to further improve the frequency stability. This is true not only for singly rotated cuts such as AT or BT, where the pure shear mode is used and consequently the particle motion is strictly in the plane of the plate, but also for SC cuts where the particle motion is neither perpendicular to nor parallel to the plane of the plate.

If e_{iJ} and d_{iJ} are the piezoelectric matrices of the right-handed form, then enantiomorphism requires[6] that the corresponding matrices of the left-handed form be given by $-e_{iJ}$ and $-d_{iJ}$, respectively.

REFERENCES

1. N. F. Kennon, *Patterns in Crystals*, Wiley, New York, 1978.
2. R. D. Bloss, *Crystallography and Crystal Chemistry*, Holt, Rinehart and Winston, New York, 1971.
3. B. A. Auld, *Acoustic Fields and Waves in Solids*, Vol. I, Wiley, New York, 1973, Chaps. 3 and 7.
4. E. Dieulesaint and D. Royer, *Elastic Waves in Solids*, Wiley, New York, 1980, Chap. 2.
5. W. L. Bond, *Crystal Technology*, Wiley, New York, 1976, Chaps. 1 and 2.
6. *IEEE Standard on Piezoelectricity*, ANSI/IEEE Std. 176–1978, Institute of Electrical and Electronics Engineers, New York, 1978.
7. O. E. Mattiat, Ed., *Ultrasonic Transducer Materials*, Plenum, New York, 1971, Chaps. 2 and 3.
8. H. Ohigashi, "Electromechanical Properties of Polarized Polyvinylidene Fluoride Films as Studied by the Piezoelectric Resonance Method," *J. Appl. Phys.*, **47**, 949–955 (1976).
9. A. Y. Cho, "Recent Developments in Molecular Beam Expitaxy (MBE)," *J. Vacuum Sci. Technol.*, **16**, 275 (1979).
10. Ref. 3, Chap. 8.
11. R. Holland and E. P. EerNisse, *Design of Resonant Piezoelectric Devices*, M.I.T. Press, Cambridge, MA, 1969.
12. A. I. Borisenko and I. E. Tarapov, *Vector and Tensor Analysis*, Dover, New York, 1968.
13. A. Ballato, "Crystal Resonators with Increased Immunity to Acceleration Fields," *IEEE Trans. Sonics Ultrason.*, **SU-26**, 195–201 (1980).

EXERCISES

1. Assuming that the coordinate and crystallographic axes coincide, write down in fully developed form the constitutive relations for $LiNbO_3$ using (a) Eq. (6.13), (b) Eq. (6.16), (c) Eq. (6.18), and (d) Eq. (6.19).

2. Using numerical or other methods, find the constants c_{IJ}^E, e_{iJ}, and ε_{ij}^S of AT-cut quarts [YXl-35°].

3. In the design of shear stress gauges, it is of interest to maximize certain piezoelectric constants. Consider Eq. (6.13) written as $D_i = e_{iJ} S_J + \varepsilon_{ij}^S E_j$ with $D_i = \varepsilon_0 E_i + P_i$, where P_i is the electric polarization. If a sample of a piezoelectric material is metallized on all sides, it follows that $P_i = e_{iJ} S_J$. The latter relation, when the metallized sample of a piezoelectric material is rotated through an angle ξ about the x axis can be written as $P_i' = e_{iJ}' S_J'$. By rotating $LiNbO_3$ about the x axis in increments of 1°, show that the ratio e_{y4}'/e_{y2}' has a maximum in the interval $162° < \xi < 164°$. (This defines the 163° Y-cut $LiNbO_3$ plate.)

4. The center of the symmetry operation (inversion), Fig. 6.2(f), transforms a left-handed crystallographic form into the right-handed form. Using the center-of-symmetry operator, Eq. (6.71), find the elastic, piezoelectric, and dielectric constants of left-handed quartz from those given for right-handed quartz in Tables 7 and 9.

5. Using Eqs. (6.13), (6.16), (6.18), and (6.19), show that

$$[g] = [\beta^T][d]$$

$$[s^D] = [s^E] - [\tilde{d}][\beta^T][d]$$

$$[e] = [d][c^E]$$

$$[\varepsilon^S] = [\varepsilon^T] - [d][c^E][\tilde{d}]$$

$$[c^D] = [c^E] + [\tilde{e}][\beta^S][e] = [c^E] + [\tilde{h}][\varepsilon^S][h]$$

$$[h] = [\beta^S][e] = [g][c^D]$$

$$[\beta^S] = [\beta^T] + [h][s^D][\tilde{h}] = [\beta^T] + [g][c^D][\tilde{g}],$$

where $[\beta^T] = [\varepsilon^T]^{-1}$, $[c^E] = [s^E]^{-1}$, $[c^D] = [s^D]^{-1}$, and $[\beta^S] = [\varepsilon^S]^{-1}$.

6. Consider wave propagation along the z axis (poling axis) in the material of infinite extent belonging to class $(6mm)$ or (∞m). (a) Derive the resulting stress constitutive relations under the assumption that only the longitudinal wave is excited.
By comparing the result in (a) with Eqs. (5.21) and (5.27),
(b) find the numerical value of the coupling coefficients for all materials belonging to these classes that are listed in Table 9.
(c) Find the corresponding values of the stiffened longitudinal velocities.

7. Repeat Exercise 6 for class $(3m)$.

Chapter 7

Wave Propagation in Anisotropic Materials

Certain aspects of wave propagation in anisotropic materials have been developed in Chaps. 3, 4, and 6. These concepts are further developed in this chapter to include wave propagation in both piezoelectric and nonpiezoelectric materials.

In the equations that follow, it is assumed that $x_j = x_1, x_2, x_3$ represent the coordinate axes x, y, z, respectively, which coincide with the crystallographic axes X, Y, Z, respectively. The relevant equations are the stress equation of motion, Eq. (2.20),

$$\frac{\partial T_{ij}}{\partial x_j} = \rho_m \ddot{u}_i \tag{7.1}$$

the charge equation of electrostatics, Eq. (5.14),

$$\frac{\partial D_i}{\partial x_i} = 0 \tag{7.2}$$

the strain displacement relation, Eq. (2.15),

$$S_{kl} = \frac{1}{2}\left(\frac{\partial u_k}{\partial x_l} + \frac{\partial u_l}{\partial x_k} \right) \tag{7.3}$$

the quasi-static potential, Eq. (5.18),

$$E_k = -\frac{\partial \phi}{\partial x_k} \tag{7.4}$$

and the constitutive relations, Eq. (6.4),

$$T_{ij} = c_{ijkl}^E S_{kl} - e_{ijk} E_k \tag{7.5}$$

$$D_i = e_{ikl} S_{kl} + \varepsilon_{ik}^S E_k \tag{7.6}$$

Upon substitution of Eqs. (7.3) and (7.4) in Eqs. (7.5) and (7.6), it follows that

$$T_{ij} = c_{ijkl}^{E} \frac{1}{2}\left(\frac{\partial u_k}{\partial x_l} + \frac{\partial u_l}{\partial x_k} \right) + e_{ijk}\frac{\partial \phi}{\partial x_k} \tag{7.7}$$

$$D_i = e_{ikl}\frac{1}{2}\left(\frac{\partial u_k}{\partial x_l} + \frac{\partial u_l}{\partial x_k} \right) - \varepsilon_{ik}^{S}\frac{\partial \phi}{\partial x_k} \tag{7.8}$$

and upon substitution of Eqs. (7.7) and (7.8) into Eqs. (7.1) and (7.2) it follows that

$$c_{ijkl}^{E}\frac{1}{2}\left(\frac{\partial^2 u_k}{\partial x_l \partial x_j} + \frac{\partial^2 u_l}{\partial x_k \partial x_j} \right) + e_{ijk}\frac{\partial^2 \phi}{\partial x_k \partial x_j} = \rho_m \ddot{u}_i \tag{7.9}$$

and

$$e_{ikl}\frac{1}{2}\left(\frac{\partial^2 u_k}{\partial x_l \partial x_i} + \frac{\partial^2 u_l}{\partial x_k \partial x_i} \right) - \varepsilon_{ik}^{S}\frac{\partial^2 \phi}{\partial x_k \partial x_i} = 0 \tag{7.10}$$

If k and l are interchanged in the expressions with parentheses in Eqs. (7.9) and (7.10), the symmetry of c_{ijkl}, Eq. (3.35), with respect to the second pair of subscripts and the symmetry of e_{ikl} with respect to the last two indices, Eq. (6.8), requires the two terms in parentheses to be equal. Therefore,

$$c_{ijkl}^{E}\frac{\partial^2 u_l}{\partial x_k \partial x_j} + e_{ijk}\frac{\partial^2 \phi}{\partial x_k \partial x_j} = \rho_m \ddot{u}_i$$

and

$$e_{ikl}\frac{\partial^2 u_l}{\partial x_k \partial x_i} - \varepsilon_{ik}^{S}\frac{\partial^2 \phi}{\partial x_k \partial x_i} = 0 \tag{7.11}$$

Equations (7.11) represent four coupled equations governing the wave propagation in anisotropic (piezoelectric or nonpiezoelectric) materials. These can be written as

$$c_{1jkl}^{E}\frac{\partial^2 u_l}{\partial x_k \partial x_j} + e_{1jk}\frac{\partial^2 \phi}{\partial x_k \partial x_j} = \rho_m \ddot{u}_1$$

$$c_{2jkl}^{E}\frac{\partial^2 u_l}{\partial x_k \partial x_j} + e_{2jk}\frac{\partial^2 \phi}{\partial x_k \partial x_j} = \rho_m \ddot{u}_2$$

$$e_{3jkl}^{E}\frac{\partial^2 u_l}{\partial x_k \partial x_j} + e_{3jk}\frac{\partial^2 \phi}{\partial x_k \partial x_j} = \rho_m \ddot{u}_3$$

$$e_{ikl}\frac{\partial^2 u_l}{\partial x_k \partial x_i} - \varepsilon_{ik}^{S}\frac{\partial^2 \phi}{\partial x_k \partial x_i} = 0 \tag{7.12}$$

giving for the particular propagation direction the values of the three displacements u_1, u_2, u_3 and the quasistatic potential ϕ. For the most general class of crystals, the triclinic, Eqs. (7.12), upon application of the rules for abbreviated subscript notation, Eq. (6.9), can be written in fully developed form as

$$c_{11}^E \frac{\partial^2 u_1}{\partial x_1^2} + 2c_{16}^E \frac{\partial^2 u_1}{\partial x_1 \partial x_2} + 2c_{15}^E \frac{\partial^2 u_1}{\partial x_1 \partial x_3} + 2c_{56}^E \frac{\partial^2 u_1}{\partial x_2 \partial x_3} + c_{66}^E \frac{\partial^2 u_1}{\partial x_2^2} + c_{55}^E \frac{\partial^2 u_1}{\partial x_3^2}$$

$$+ c_{16}^E \frac{\partial^2 u_1}{\partial x_1^2} + \left(c_{12}^E + c_{66}^E\right) \frac{\partial^2 u_1}{\partial x_1 \partial x_2} + \left(c_{14}^E + c_{56}^E\right) \frac{\partial^2 u_1}{\partial x_1 \partial x_3}$$

$$+ \left(c_{25}^E + c_{46}^E\right) \frac{\partial^2 u_2}{\partial x_2 \partial x_3} + c_{26}^E \frac{\partial^2 u_2}{\partial x_2^2} + c_{45}^E \frac{\partial^2 u_2}{\partial x_3^2} + c_{15}^E \frac{\partial^2 u_2}{\partial x_1^2}$$

$$+ \left(c_{56}^E + c_{14}^E\right) \frac{\partial^2 u_3}{\partial x_1 \partial x_2} + \left(c_{13}^E + c_{55}^E\right) \frac{\partial^2 u_3}{\partial x_1 \partial x_3} + \left(c_{36}^E + c_{45}^E\right) \frac{\partial^2 u_3}{\partial x_2 \partial x_3}$$

$$+ c_{46}^E \frac{\partial^2 u_3}{\partial x_2^2} + c_{35}^E \frac{\partial^2 u_3}{\partial x_3^2} + e_{x1} \frac{\partial^2 \phi}{\partial x_1^2} + \left(e_{x6} + e_{y1}\right) \frac{\partial^2 \phi}{\partial x_1 \partial x_2} + \left(e_{x5} + e_{z1}\right)$$

$$\times \frac{\partial^2 \phi}{\partial x_1 \partial x_3} + \left(e_{y5} + e_{z6}\right) \frac{\partial^2 \phi}{\partial x_2 \partial x_3} + e_{y6} \frac{\partial^2 \phi}{\partial x_2^2} + e_{z5} \frac{\partial^2 \phi}{\partial x_3^2} = \rho_m \ddot{u}_1 \quad (7.13)$$

$$c_{16}^E \frac{\partial^2 u_1}{\partial x_1^2} + \left(c_{12}^E + c_{66}^E\right) \frac{\partial^2 u_1}{\partial x_1 \partial x_2} + \left(c_{14}^E + c_{56}^E\right) \frac{\partial^2 u_1}{\partial x_1 \partial x_3} + \left(c_{25}^E + c_{46}^E\right) \frac{\partial^2 u_1}{\partial x_2 \partial x_3}$$

$$+ c_{26}^E \frac{\partial^2 u_1}{\partial x_2^2} + c_{45}^E \frac{\partial^2 u_1}{\partial x_3^2} + c_{66}^E \frac{\partial^2 u_2}{\partial x_1^2} + 2c_{26}^E \frac{\partial^2 u_2}{\partial x_1 \partial x_2} + 2c_{46}^E \frac{\partial^2 u_2}{\partial x_1 \partial x_3}$$

$$+ 2c_{24}^E \frac{\partial^2 u_2}{\partial x_2 \partial x_3} + c_{22}^E \frac{\partial^2 u_2}{\partial x_2^2} + c_{44}^E \frac{\partial^2 u_2}{\partial x_3^2} + c_{56}^E \frac{\partial^2 u_3}{\partial x_1^2} + \left(c_{25}^E + c_{46}^E\right) \frac{\partial^2 u_3}{\partial x_1 \partial x_2}$$

$$+ \left(c_{36}^E + c_{45}^E\right) \frac{\partial^2 u_3}{\partial x_1 \partial x_3} + \left(c_{23}^E + c_{44}^E\right) \frac{\partial^2 u_3}{\partial x_2 \partial x_3} + c_{24}^E \frac{\partial^2 u_3}{\partial x_2^2} + c_{34}^E \frac{\partial^2 u_3}{\partial x_3^2}$$

$$+ e_{x6} \frac{\partial^2 \phi}{\partial x_1^2} + \left(e_{x2} + e_{y6}\right) \frac{\partial^2 \phi}{\partial x_1 \partial x_2} + \left(e_{x4} + e_{z6}\right) \frac{\partial^2 \phi}{\partial x_1 \partial x_3}$$

$$+ \left(e_{y4} + e_{z2}\right) \frac{\partial^2 \phi}{\partial x_2 \partial x_3} + e_{y2} \frac{\partial^2 \phi}{\partial x_2^2} + e_{z4} \frac{\partial^2 \phi}{\partial x_3^2} = \rho_m \ddot{u}_2 \quad (7.14)$$

$$c_{15}^E \frac{\partial^2 u_1}{\partial x_1^2} + \left(c_{56}^E + c_{14}^E\right)\frac{\partial^2 u_1}{\partial x_1 \partial x_2} + \left(c_{13}^E + c_{55}^E\right)\frac{\partial^2 u_1}{\partial x_1 \partial x_3} + \left(c_{36}^E + c_{45}^E\right)\frac{\partial^2 u_1}{\partial x_2 \partial x_3}$$

$$+ c_{46}^E \frac{\partial^2 u_1}{\partial x_2^2} + c_{35}^E \frac{\partial^2 u_1}{\partial x_3^2} + c_{56}^E \frac{\partial^2 u_2}{\partial x_1^2} + \left(c_{25}^E + c_{46}^E\right)\frac{\partial^2 u_2}{\partial x_1 \partial x_2}$$

$$+ \left(c_{36}^E + c_{45}^E\right)\frac{\partial^2 u_2}{\partial x_1 \partial x_3} + \left(c_{23}^E + c_{44}^E\right)\frac{\partial^2 u_2}{\partial x_2 \partial x_3} + c_{24}^E \frac{\partial^2 u_2}{\partial x_2^2} + c_{34}^E \frac{\partial^2 u_2}{\partial x_3^2}$$

$$+ c_{55}^E \frac{\partial^2 u_3}{\partial x_1^2} + 2c_{45}^E \frac{\partial^2 u_3}{\partial x_1 \partial x_2} + 2c_{35}^E \frac{\partial^2 u_3}{\partial x_1 \partial x_3} + 2c_{34}^E \frac{\partial^2 u_3}{\partial x_2 \partial x_3} + c_{44}^E \frac{\partial^2 u_3}{\partial x_2^2}$$

$$+ c_{33}^E \frac{\partial^2 u_3}{\partial x_3^2} + e_{x5}\frac{\partial^2 \phi}{\partial x_1^2} + \left(e_{x4} + e_{y5}\right)\frac{\partial^2 \phi}{\partial x_1 \partial x_2} + \left(e_{x3} + e_{z5}\right)\frac{\partial^2 \phi}{\partial x_1 \partial x_3}$$

$$+ \left(e_{y3} + e_{z4}\right)\frac{\partial^2 \phi}{\partial x_2 \partial x_3} + e_{y4}\frac{\partial^2 \phi}{\partial x_2^2} + e_{z3}\frac{\partial^2 \phi}{\partial x_3^2} = \rho_m \ddot{u}_3 \qquad (7.15)$$

$$e_{x1}\frac{\partial^2 u_1}{\partial x_1^2} + \left(e_{x6} + e_{y1}\right)\frac{\partial^2 u_1}{\partial x_1 \partial x_2} + \left(e_{x5} + e_{z1}\right)\frac{\partial^2 u_1}{\partial x_1 \partial x_3} + \left(e_{y5} + e_{z6}\right)\frac{\partial^2 u_1}{\partial x_2 \partial x_3}$$

$$+ e_{y6}\frac{\partial^2 u_1}{\partial x_2^2} + e_{z5}\frac{\partial^2 u_1}{\partial x_3^2} + e_{x6}\frac{\partial^2 u_2}{\partial x_1^2} + \left(e_{x2} + e_{y6}\right)\frac{\partial^2 u_2}{\partial x_1 \partial x_2}$$

$$+ \left(e_{x4} + e_{z6}\right)\frac{\partial^2 u_2}{\partial x_1 \partial x_3} + \left(e_{y4} + e_{z2}\right)\frac{\partial^2 u_2}{\partial x_2 \partial x_3} + e_{y2}\frac{\partial^2 u_2}{\partial x_2^2} + e_{z4}\frac{\partial^2 u_2}{\partial x_3^2}$$

$$+ e_{x5}\frac{\partial^2 u_3}{\partial x_1^2} + \left(e_{x4} + e_{y5}\right)\frac{\partial^2 u_3}{\partial x_1 \partial x_2} + \left(e_{x3} + e_{z5}\right)\frac{\partial^2 u_3}{\partial x_1 \partial x_3}$$

$$+ \left(e_{y3} + e_{z4}\right)\frac{\partial^2 u_3}{\partial x_2 \partial x_3} + e_{y4}\frac{\partial^2 u_3}{\partial x_2^2} + e_{z3}\frac{\partial^2 u_3}{\partial x_3^2} - \varepsilon_{xx}^S \frac{\partial^2 \phi}{\partial x_1^2} - 2\varepsilon_{xy}^S \frac{\partial^2 \phi}{\partial x_1 \partial x_2}$$

$$- 2\varepsilon_{xz}^S \frac{\partial^2 \phi}{\partial x_1 \partial x_3} - 2\varepsilon_{yz}^S \frac{\partial^2 \phi}{\partial x_1 \partial x_3} - \varepsilon_{yy}^S \frac{\partial^2 \phi}{\partial x_2^2} - \varepsilon_{zz}^S \frac{\partial^2 \phi}{\partial x_3^2} = 0 \qquad (7.16)$$

Although these equations are cumbersome for the most general crystal system, the triclinic, the materials used in practice possess material tensors with a great many zero elements. The latter causes a considerable simplification in the above equations.

As an example, consider plane-wave propagation along the $x_3 = z$ axis of LiNbO$_3$, Table 9 and Sect. 6.3.4. In this case, since the wave is constant in the $x = x_1$ and $y = x_2$ directions, it follows that $\partial/\partial x_1 = 0$ and $\partial/\partial x_2 = 0$, and Eqs. (7.13) to (7.16) can be simplified to

$$c_{55}^E \frac{\partial^2 u_x}{\partial z^2} = \rho_m \frac{\partial^2 u_x}{\partial t^2} \tag{7.17}$$

$$c_{44}^E \frac{\partial^2 u_y}{\partial z^2} = \rho_m \frac{\partial^2 u_y}{\partial t^2} \tag{7.18}$$

and

$$c_{33}^E \frac{\partial^2 u_z}{\partial z^2} + e_{z3} \frac{\partial^2 \phi}{\partial z^2} = \rho_m \frac{\partial^2 u_z}{\partial t^2} \tag{7.19}$$

$$e_{z3} \frac{\partial^2 u_z}{\partial z^2} - \epsilon_{zz}^S \frac{\partial^2 \phi}{\partial z^2} = 0 \tag{7.20}$$

with $c_{44}^E = c_{55}^E$.

Equations (7.17) and (7.18) represent two independent shear waves, propagating in the z direction with velocity $(c_{44}^E/\rho_m)^{1/2}$, and polarizations in the x and y directions, respectively. It is seen that these two degenerate modes, in this particular case, are analogous to the two shear-wave Eqs. (2.91) and (2.92) for wave propagation in an isotropic material. Since none is related to quasi-static potential ϕ, both modes are purely mechanical in nature. The third wave, given by Eqs. (7.19) and (7.20), is the longitudinal wave, which is piezoelectric since it is coupled to the potential ϕ. Upon elimination of ϕ, it follows that

$$c_{33}^E \left[1 + \frac{e_{z3}^2}{\epsilon_{zz}^S c_{33}^E} \right] \frac{\partial^2 u_z}{\partial z^2} = \rho_m \frac{\partial^2 u_z}{\partial t^2} \tag{7.21}$$

which is seen to be analogous to the one-dimensional wave Eq. (5.23) for the longitudinal piezoelectric wave with the fields given by Eq. (5.29), as discussed in Chap. 5. It follows that the piezoelectrically stiffened elastic constant is given by

$$\bar{c}_{33} = c_{33}^E \left[1 + \frac{e_{z3}^2}{\epsilon_{zz}^S c_{33}^E} \right] \tag{7.22}$$

and electromechanical coupling factor by

$$K^2 = \frac{e_{z3}^2}{\epsilon_{zz}^S c_{33}^E} \tag{7.23}$$

In another example, consider the propagation along the $x = X$ axis of ZnO belonging to the hexagonal class (6mm). In this case, $\partial/\partial x_2 = \partial/\partial y = 0$, $\partial/\partial x_3 = \partial/\partial z = 0$, and Eqs. (7.13) to (7.16) reduce to

$$c_{11}^E \frac{\partial^2 u_x}{\partial x^2} = \rho_m \ddot{u}_x \tag{7.24}$$

$$c_{66}^E \frac{\partial^2 u_y}{\partial x^2} = \rho_m \ddot{u}_y \tag{7.25}$$

$$c_{55}^E \frac{\partial^2 u_z}{\partial x^2} + e_{x5} \frac{\partial^2 \phi}{\partial x^2} = \rho_m \ddot{u}_z \tag{7.26}$$

$$e_{x5} \frac{\partial^2 u_z}{\partial x^2} - \varepsilon_{xx}^S \frac{\partial^2 \phi}{\partial x^2} = 0 \tag{7.27}$$

Equation (7.24) represents a longitudinal wave, Eq. (7.25) a shear wave, both purely mechanical in nature, and Eqs. (7.26) and (7.27) a shear wave which is piezoelectric in nature due to its coupling to the quasi-static potential. From the latter two equations, upon elimination of potential, it follows that

$$c_{55}^E \left[1 + \frac{e_{x5}^2}{c_{55}^E \varepsilon_{xx}^S} \right] \frac{\partial^2 u_z}{\partial x^2} = \rho_m \ddot{u}_z \tag{7.28}$$

with

$$K^2 = \frac{e_{x5}^2}{c_{55}^E \varepsilon_{xx}^S} \tag{7.29}$$

being the electromechanical coupling factor for bulk shear wave propagation in the x direction. Since the potential is only a function of propagation direction x and time, the only component of the electric field, $E_x = -\partial\phi/\partial x$ is in the direction of propagation.

As the final example, consider wave propagation along the y axis of crystals, such as LiNbO$_3$, belonging to trigonal class (3m). In this case, Eqs. (7.13) to (7.16) can be reduced to

$$c_{66}^E \frac{\partial^2 u_x}{\partial y^2} = \rho_m \frac{\partial^2 u_x}{\partial t^2} \tag{7.30}$$

$$c_{22}^E \frac{\partial^2 u_y}{\partial y^2} + c_{24}^E \frac{\partial^2 u_z}{\partial y^2} + e_{y2} \frac{\partial^2 \phi}{\partial y^2} = \rho_m \frac{\partial^2 u_y}{\partial t^2} \tag{7.31}$$

$$c_{24}^E \frac{\partial^2 u_y}{\partial y^2} + c_{44}^E \frac{\partial^2 u_z}{\partial y^2} + e_{y4} \frac{\partial^2 \phi}{\partial y^2} = \rho_m \frac{\partial^2 u_z}{\partial t^2} \tag{7.32}$$

$$e_{y2} \frac{\partial^2 u_y}{\partial y^2} + e_{y4} \frac{\partial^2 u_z}{\partial y^2} - \varepsilon_{yy}^S \frac{\partial^2 \phi}{\partial y^2} = 0 \tag{7.33}$$

Again Eq. (7.30) represents a purely mechanical shear wave propagating with velocity $(c_{66}^E/\rho_m)^{1/2}$ in the y direction. The last three equations represent two *coupled waves*, a shear and a longitudinal wave, both piezoelectric in nature. In order to find an equation determining the relationship between frequency and wave number, and necessary for finding the phase and group velocities of the two waves, Eq. (7.33) is substituted into Eqs. (7.31) and (7.32). This results in

$$c_{66}^E \frac{\partial^2 u_x}{\partial y^2} = \rho_m \frac{\partial^2 u_x}{\partial t^2}$$

$$\left[c_{22}^E + \frac{e_{y2}^2}{\varepsilon_{yy}^S} \right] \frac{\partial^2 u_y}{\partial y^2} + \left[c_{24}^E + \frac{e_{y2}e_{y4}}{\varepsilon_{yy}^S} \right] \frac{\partial^2 u_z}{\partial y^2} = \rho_m \frac{\partial^2 u_y}{\partial t^2}$$

$$\left[c_{24}^E + \frac{e_{y2}e_{y4}}{\varepsilon_{yy}^S} \right] \frac{\partial^2 u_y}{\partial y^2} + \left[c_{44}^E + \frac{e_{y4}^2}{\varepsilon_{yy}^S} \right] \frac{\partial^2 u_z}{\partial y^2} = \rho_m \frac{\partial^2 u_z}{\partial t^2} \qquad (7.34)$$

For a plane-wave solution of the form

$$u_j = U_j \exp[\, j(\omega t - ky)], \qquad j = x, y, z$$

$$\phi = \Phi \exp[\, j(\omega t - ky)] \qquad (7.35)$$

where U_j and Φ are constants, this can be written in matrix form as

$$\begin{bmatrix} c_{66}^E k^2 & 0 & 0 \\ 0 & \left[c_{22}^E + e_{y2}^2/\varepsilon_{yy}^S \right] k^2 & \left[c_{24}^E + e_{y2}e_{y4}/\varepsilon_{yy}^S \right] k^2 \\ 0 & \left[c_{24}^E + e_{y2}e_{y4}/\varepsilon_{yy}^S \right] k^2 & \left[c_{44}^E + e_{y4}^2/\varepsilon_{yy}^S \right] k^2 \end{bmatrix} \begin{bmatrix} U_x \\ U_y \\ U_z \end{bmatrix} = \rho_m \omega^2 \begin{bmatrix} U_x \\ U_y \\ U_z \end{bmatrix}$$

$$(7.36)$$

or

$$k^2 [\Gamma][U] = \rho_m \omega^2 [U] \qquad (7.37)$$

where $[\Gamma]$ is called the *Christoffel matrix* when $e_{iJ} = 0$ and the *stiffened Christoffel matrix* when $e_{iJ} \neq 0$. The Christoffel matrix is symmetric and is a function of material constants and wave propagation direction. In order for the solution of Eq. (7.37) to exist,

$$\Delta(\omega, k) \equiv \det | k^2 [\Gamma] - \rho_m \omega^2 [I] | = 0 \qquad (7.38)$$

where $[I]$ is the 3×3 identity matrix. Relations of this type are termed *dispersion relations*, Δ determining the conditions for existence of various wave modes. In full, Eq.

(7.38) can be written as

$$
\Delta(\omega, k) \equiv \left[c_{66}^E k^2 - \rho_m \omega^2 \right] \left\{ \left[c_{22}^E + \frac{e_{y2}^2}{\varepsilon_{yy}^S} \right] k^2 - \rho_m \omega^2 \left[\left(c_{44}^E + \frac{e_{y4}^2}{\varepsilon_{yy}^S} \right) k^2 - \rho_m \omega^2 \right] \right.
$$

$$
\left. - \left[c_{24}^E + \frac{e_{y2} e_{y4}}{\varepsilon_{yy}^S} \right]^2 k^4 \right\} = 0 \tag{7.39}
$$

yielding the phase velocity for the x-polarized shear wave, Eq. (7.30), as $\omega = \pm (c_{66}^E / \rho_m)^{1/2} k$ and a quadratic equation in ω^2 defining the dispersion relation for the two coupled piezoelectric modes. Setting $\bar{v}_A^2 = \omega^2 / k^2$, where $\bar{v}_A$ is the piezoelectrically stiffened acoustic velocity[†] and using stiffened elastic constants, this can be written as

$$
\left[\bar{c}_{22} - \rho_m \bar{v}_A^2 \right] \left[\bar{c}_{44} - \rho_m \bar{v}_A^2 \right] - \left[\bar{c}_{24} \right]^2 = 0 \tag{7.40}
$$

which gives two values for $\bar{v}_A^2$

$$
\left\{ \begin{matrix} \bar{v}_{A1}^2 \\ \bar{v}_{A2}^2 \end{matrix} \right\} = \frac{\bar{c}_{22} + \bar{c}_{24} \pm \left[\left(\bar{c}_{22} - \bar{c}_{44} \right)^2 + 4 \left(\bar{c}_{24} \right)^2 \right]^{1/2}}{2 \rho_m} \tag{7.41}
$$

for wave propagation in the y direction. Both of these waves have y and z components of the displacement and therefore are neither purely transverse nor purely longitudinal. They are termed *quasi-longitudinal* [the plus sign in Eq. (7.41)] or *quasi-transverse* [minus sign in Eq. (7.41)], according to which wave has a larger y or z component of the displacement, respectively. This is shown in Fig. 7.1(a). It can be shown[1] that the particle displacements stay orthogonal among the three waves, which is true in general when none of the three waves is purely longitudinal or purely transverse as shown in Fig. 7.1(b).

It is seen from Eq. (7.39) that even without piezoelectricity, $e_{ij} = 0$, the two modes are coupled. Thus *mode coupling* is an inherent characteristic of anisotropic materials, not necessarily dependent on piezoelectricity, the latter being a property of anisotropic materials without a center of symmetry.

In Chap. 3, Eq. (3.72) it has been shown that in isotropic materials $v_L / v_S > \sqrt{2}$. In anisotropic materials also, the (quasi) longitudinal wave is faster than (quasi) transverse waves but with one notable exception. In the case of TeO_2 (paratellurite), Table 7 in Chap. 6, for wave propagation along the X direction, all three modes are pure but $v_L = 3050$ m/s, $v_{S1} = 3317$ m/s for Y-polarized and $v_{S2} = 2100$ m/s for Z-polarized shear waves. Thus in this case $v_L < v_{S1}$.

By using various material and field matrices developed in the previous chapters, it is possible to represent the four wave Eqs. (7.12) by a single matrix equation.[2] The latter is of considerable importance in numerical studies of various wave problems.

[†]In what follows, in order to avoid confusion with indices, two subscripts will be used for acoustic, that is, phase velocity: v_A and v_P. It is understood that $v_A = v_P$.

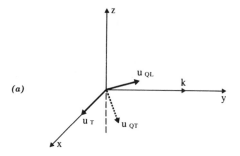

(a)

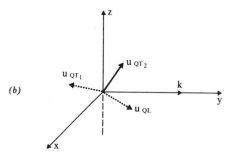

(b)

Figure 7.1. (a) The relationship between the x-polarized pure transverse mode with displacement u_T, the quasi-longitudinal mode with displacement u_{QL} in the zy plane, and the quasi-transverse mode with displacement u_{QT} in the $-zy$ plane for wave propagation along the $Y = y$ axis of crystals belonging to trigonal class ($3m$). (b) A general case when none of the wave is purely longitudinal or purely transverse for wave propagating in the y direction.

Of special interest are directions in crystals that allow the existence of pure longitudinal or pure transverse modes. These are termed the *pure mode directions*, and the corresponding crystal cuts are used extensively for the design of piezoelectric devices. The pure mode directions can be identified from the properties of the Christoffel matrix, which, of course, reflects the symmetry properties of crystals but may also reveal pure modes in the so-called nonsymmetry directions.

7.1 ELECTROMECHANICAL COUPLING FACTOR AND PURE MODE DIRECTIONS

For the wave propagation in a direction given by a unit vector $\vec{n} = (n_x, n_y, n_z)$, which is perpendicular to the wave front defined by $\vec{n} \cdot \vec{r} = $ constant, the solution of Eqs. (7.11) can be sought, for plane waves of arbitrary temporal and spatial variations, in the form

$$u_j = U_i f\left(t - \frac{\vec{n} \cdot \vec{r}}{\bar{v}_A}\right)$$

$$\phi = \Phi f\left(t - \frac{\vec{n} \cdot \vec{r}}{\bar{v}_A}\right) \tag{7.42}$$

where $\vec{n} \cdot \vec{r} = n_i x_i = n_x x + n_y y + n_z z$, and U_i and Φ are constants. From Eq. (7.4) it follows that the electric field is always longitudinal, that is, in the direction of propagation,

$$E_k = -\frac{\partial \phi}{\partial x_k} = \frac{n_k}{\bar{v}_A} \Phi f' \tag{7.43}$$

where f' is a derivative of f, and various terms in Eqs. (7.11), which are repeated here for convenience,

$$c_{ijkl}^E \frac{\partial^2 u_l}{\partial x_k \partial x_j} + e_{ijk} \frac{\partial^2 \phi}{\partial x_k \partial x_j} = \rho_m \ddot{u}_i$$

$$e_{ikl} \frac{\partial^2 u_l}{\partial x_k \partial x_i} - \varepsilon_{ik}^S \frac{\partial^2 \phi}{\partial x_k \partial x_i} = 0 \tag{7.11}$$

are found using Eqs. (7.42) as

$$\frac{\partial^2 u_l}{\partial x_k \partial x_j} = \frac{n_k n_j}{\bar{v}_A^2} U_l f''$$

$$\frac{\partial^2 \phi}{\partial x_k \partial x_j} = \frac{n_k n_j}{\bar{v}_A^2} \Phi f'' \tag{7.44}$$

and

$$\ddot{u}_i = U_i f''$$

Using Eqs. (7.44) and (7.11) it follows that

$$c_{ijkl}^E n_k n_j U_l + e_{ijk} n_k n_j \Phi = \bar{v}_A^2 \rho_m U_i$$

$$e_{ikl} n_k n_i U_l - \varepsilon_{ik}^S n_i n_k \Phi = 0 \tag{7.45}$$

or upon elimination of potential Φ,

$$\left\{ c_{ijkl}^E n_k n_j + \frac{(e_{ijk} n_k n_j)(e_{ikl} n_k n_i)}{\varepsilon_{ik}^S n_i n_k} \right\} U_l = \rho_m \bar{v}_A^2 U_i \tag{7.46}$$

It is to be noted that this procedure is equivalent to the procedure used in Eqs. (7.30) to (7.36) to derive the Christoffel matrix. In order to avoid confusion in indices, Eq. (7.46) can be written as

$$\left\{ c_{ijkl}^E n_k n_j + \frac{(e_{ija} n_j n_a)(e_{bkl} n_k n_b)}{\varepsilon_{pq}^S n_p n_q} \right\} U_l = \rho_m \bar{v}_A^2 U_i \tag{7.47}$$

or

$$n_k n_j \left\{ c_{ijkl}^E + \frac{(e_{ija} n_a)(e_{bkl} n_b)}{\varepsilon_{pq}^S n_p n_q} \right\} U_l = \rho_m \bar{v}_A^2 U_i \qquad (7.48)$$

which allows for the general definition of the *unstiffened Christoffel tensor*[3]

$$\Gamma_{il}^E = n_k n_j c_{ijkl}^E \qquad (7.49)$$

and the *stiffened Christoffel tensor*

$$\bar{\Gamma}_{il} = n_k n_j \bar{c}_{ijkl} \qquad (7.50)$$

where, from Eq. (7.48), it is seen that

$$\bar{c}_{ijkl} = c_{ijkl}^E + \frac{(e_{ija} n_a)(e_{bkl} n_b)}{\varepsilon_{pq}^S n_p n_q} \qquad (7.51)$$

The symmetry of the Christoffel tensor, which has been tabulated for all crystal classes,[1] guarantees the orthogonality of the three displacements, Fig. 7.1. In short subscript notation, using $ij \rightarrow I$ and $kl \rightarrow J$, Eq. (7.51) can be written as

$$\bar{c}_{IJ} = c_{IJ}^E + \frac{[n_b e_{bJ}][e_{Ia} n_a]}{\varepsilon_{pq}^S n_p n_q} \qquad (7.52)$$

and in analogy with Eq. (7.28) the *electromechanical coupling factor* for the wave propagation in the $\vec{n}$ direction is defined as

$$K_{IJ}^2 \equiv \frac{[e_{Ia} n_a][n_b e_{bJ}]}{\left(n_p \varepsilon_{pq}^S n_q \right) c_{IJ}^E} \qquad (7.53)$$

Taking $\vec{n}$ as a row vector,

$$[n] = \begin{bmatrix} n_x & n_y & n_z \end{bmatrix} \qquad (7.54)$$

and keeping in mind the definition of the e and ε matrices, Eqs. (6.11) and (6.5), Eq. (7.53) can be written in the matrix form as

$$K_{IJ}^2 = \frac{[n][e][\check{e}][\tilde{n}]}{[n][\varepsilon][\tilde{n}]} = \frac{[n][e][\widetilde{n}][e]}{[n][\varepsilon][\tilde{n}]} \qquad (7.55)$$

yielding a 6×6 matrix for K_{IJ}^2.

The importance of pure mode directions as related to the coupling factor has been pointed out earlier. Two sets of pure mode directions exist, the first occurring in the so-called symmetry directions in solids, and the second occurring in nonsymmetry directions. Both of these sets can be determined from the properties of components of the Christoffel tensor, Eq. (7.49), written here in matrix form as

$$[\Gamma] = \begin{bmatrix} \Gamma_{xx} & \Gamma_{xy} & \Gamma_{xz} \\ \Gamma_{xy} & \Gamma_{yy} & \Gamma_{yz} \\ \Gamma_{xz} & \Gamma_{yz} & \Gamma_{zz} \end{bmatrix} \tag{7.56}$$

For the first set, the following rules apply[1,4]:

Propagation along a twofold axis: All modes are pure. For instance, if the z axis is a twofold axis, then $\Gamma_{xz} = \Gamma_{yz} = 0$.

Propagation along a threefold, fourfold, or sixfold axis: All modes are pure; shear modes are degenerate. Since a shear wave may have any polarization in the plane perpendicular to the axis, by analogy with optics these three axes are sometimes called *acoustic axes*. In this case, for instance, if the z axis is a ($p > 2$)-order axis, $\Gamma_{xy} = \Gamma_{xz} = \Gamma_{yz} = 0$ and $\Gamma_{xx} = \Gamma_{yy}$.

Propagation in a mirror plane: A pure shear mode, polarized normal to the plane. A mirror plane of the crystal, for instance perpendicular to the z axis, is defined by a generator matrix, Sect. 6.4, of the form

$$\begin{bmatrix} 1 & 0 & 0 \\ 0 & 1 & 0 \\ 0 & 0 & -1 \end{bmatrix}. \tag{7.57}$$

In this case, $\Gamma_{xz} = \Gamma_{yz} = 0$.

Propagation normal to a twofold, fourfold, or sixfold axis: A pure shear mode, polarized parallel to the axis. For instance, if a fourfold axis is parallel to the z axis, $\Gamma_{xz} = \Gamma_{yz} = 0$, and the pure shear mode is polarized along the z axis.

Similarly, the second set of modes, propagating in nonsymmetry directions,[4,5] can be deduced by considering wave propagation in various directions in crystals, Sect. 6.4, and by finding coordinate angles for which components of the transformed Christoffel matrix become equal to zero.

The polarization of acoustic waves is strongly affected by anisotropy of materials. In an isotropic material, a shear wave, like a light wave, may have any type of polarization. In anisotropic material, elliptically polarized shear waves are possible only when the direction of propagation coincides with an acoustic axis. For any other direction, the shear waves are quasi-linearly polarized, as shown in

Fig. 7.1(b). Therefore there is an analogy with the optics of crystals, where, in transparent crystals, a light wave may be elliptically polarized along the optic axis while being linearly polarized in any other direction. In isotropic materials and in anisotropic materials along the pure-mode axis, the polarization of longitudinal waves is collinear with the direction of propagation. In any other direction, as seen from Fig. 7.1(b), the polarization is quasi-linear.

7.2   GROUP VELOCITY

The concept of *group velocity*, that is, the velocity of a wave packet, a group of waves of closely spaced frequencies, is usually identified with the velocity of the modulation envelope of a wave. In what follows, the concept of group velocity is developed for lossless materials only.

For one-dimensional wave propagation, group velocity is defined as

$$v_G = \frac{\partial \omega}{\partial k} \tag{7.58}$$

and in this case the direction and magnitude of the group and the phase velocities coincide,

$$v_P = v_A = \frac{\omega}{k} = v_G \tag{7.59}$$

For three-dimensional wave propagation, Eq. (7.58) is written as

$$\vec{v}_G = \hat{x}\frac{\partial \omega}{\partial k_x} + \hat{y}\frac{\partial \omega}{\partial k_y} + \hat{z}\frac{\partial \omega}{\partial k_z} \tag{7.60}$$

or in component form

$$(\vec{v}_G)_j = \frac{\partial \omega}{\partial k_j}, \qquad k_j = n_j k \tag{7.61}$$

with $j = x, y, z$. Using Eq. (7.59), this can be written as

$$(\vec{v}_G)_j = \frac{\partial v_A}{\partial n_j} \tag{7.62}$$

It is seen from Eqs. (7.61) and (7.62) that in order to find various components, the dispersion relation, Eq. (7.38), which relates the angular frequency and the wave vector k_j, must be known. For the general case treated here, the dispersion relation is obtained from Eq. (7.47) by a procedure similar to the one used for the derivation of Eq. (7.38), as

$$\Delta(\omega, \vec{k}) \equiv \det|k^2 n_k n_j \bar{c}_{ijkl} - k^2 \rho_m \bar{v}_A^2 \delta_{il}| = 0 \tag{7.63}$$

or, on using Eqs. (7.59) and (7.61), as

$$\Delta(\omega, \vec{k}) \equiv \det|k_k k_j \bar{c}_{ijkl} - \rho_m \omega^2 \delta_{il}| = 0. \tag{7.64}$$

Thus either Eqs. (7.61) and (7.64) or Eqs. (7.62) and (7.63) can be used for the determination of group velocity. Another expression can be derived by multiplying Eq. (7.48) by U_i and differentiating, that is,

$$n_k n_j \bar{c}_{ijkl} U_l U_i = \rho_m \bar{v}_A^2 U_i^2 \tag{7.65}$$

and

$$2n_k \bar{c}_{ijkl} U_l U_i = 2\rho_m \bar{v}_A \frac{\partial \bar{v}_A}{\partial n_j} U_i^2 \tag{7.66}$$

Using Eq. (7.62), it follows that

$$(\vec{v}_G)_j = \frac{\partial \bar{v}_A}{\partial n_j} = \frac{\bar{c}_{ijkl} n_k U_l U_i}{\rho_m v_A U_i^2}. \tag{7.67}$$

Since $U_i^2 = U_1^2 + U_2^2 + U_3^2$, Eq. (7.65) represents a single equation, and its physical meaning is that in a plane wave the kinetic energy is equal to the stiffened strain energy. This conclusion has also been reached in Chap. 5, Eq. (5.141), for one-dimensional wave propagation.

It is well known that in linear systems the group velocity is independent of the wave magnitude as seen from Eq. (7.60). This fact must be kept in mind when dealing with Eq. (7.67), which has been derived to show the relationship between the group and energy velocities. The latter is defined in the following section.

7.3   POYNTING VECTOR IN ANISOTROPIC MATERIALS

The energy conservation theorem for one-dimensional wave propagation in piezoelectric materials along a pure-mode axis has been derived in Sect. 5.8. Furthermore, the energy conservation theorem for wave propagation in isotropic materials has been considered in Sect. 4.1. A characteristic of isotropic materials is that the direction of the Poynting vector coincides with the direction of wave propagation, that is, the phase velocity direction, normal to the phase fronts. However, in general, as we shall see in this section, in anisotropic materials the two directions do not coincide.

From the derivations in Sect. 4.1 and 5.8, it is evident that the magnitude of the Poynting vector is equal to the amount of acoustic energy crossing the unit surface in unit time, its directions being that of the energy transport. If the total stored energy density of the wave is denoted by u_T, then a velocity $\vec{v}_\epsilon$, termed the *energy velocity*, is defined as

$$\vec{\Pi} = u_T \vec{v}_\epsilon \qquad (7.68)$$

where the direction of $\vec{v}_\epsilon$ is, in general, different from the direction of phase velocity $\vec{v}_p = (\omega/k)\vec{n}$, where $\vec{n} = \vec{k}/k$, that is, $\vec{v}_\epsilon \cdot \vec{v}_p = v_p v_\epsilon \cos\theta$, Fig. 7.2. The angle θ is called the *power flow angle*, and the phenomenon when θ is nonzero is sometimes referred to as *acoustic beam steering*, this being another inherent characteristic of anisotropic materials.

To simplify the mathematics, the acoustical Poynting theorem for anisotropic (piezoelectric or nonpiezoelectric) materials will be derived using symbolic notation.[2] First, the equation

$$\nabla \cdot \vec{\vec{T}} = \rho_m \frac{\partial^2 \vec{u}}{\partial t^2} - \vec{F}^b \qquad (2.20)$$

is multiplied scalarly by $\vec{v}$, and the equation

$$\frac{\partial \vec{\vec{S}}}{\partial t} = \nabla_s \vec{v} \qquad (7.69)$$

by $\vec{\vec{T}}$. Then, by addition, it follows that

$$\vec{v} \cdot (\nabla \cdot \vec{\vec{T}}) + \vec{\vec{T}} : \nabla_s \vec{v} = \vec{\vec{T}} : \frac{\partial \vec{\vec{S}}}{\partial t} + \rho_m \vec{v} \cdot \frac{\partial \vec{v}}{\partial t} - \vec{v} \cdot \vec{F}^b \qquad (7.70)$$

In analogy with the well-known vector identity

$$\nabla \cdot (\phi \vec{A}) = \phi \nabla \cdot \vec{A} + \vec{A} \cdot \nabla \phi \qquad (7.71)$$

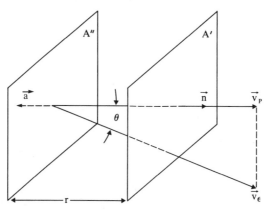

Figure 7.2. Schematic used for derivation of Eq. (7.96) and for illustration of Eq. (7.68).

it can be shown[6] that

$$\nabla \cdot (\vec{\phi} \cdot \overleftrightarrow{A}) = \vec{\phi} \cdot (\nabla \cdot \overleftrightarrow{A}) + \overleftrightarrow{A} : \nabla_s \vec{\phi} \tag{7.72}$$

is applicable to tensors, where $\overleftrightarrow{A}$ is a second-rank symmetric tensor. Equation (7.69) can now be written as

$$\nabla \cdot (\vec{v} \cdot \overleftrightarrow{T}) = \overleftrightarrow{T} : \frac{\partial \overleftrightarrow{S}}{\partial t} + \rho_m \vec{v} \cdot \frac{\partial \vec{v}}{\partial t} - \vec{v} \cdot \vec{F}^b \tag{7.73}$$

Furthermore, Eqs. (6.14) are modified to include acoustic losses (Sect. 4.1) and are written as

$$\overleftrightarrow{T} = \overleftrightarrow{c}^E : \overleftrightarrow{S} + \overleftrightarrow{\eta} : \frac{\partial \overleftrightarrow{S}}{\partial t} - \overleftrightarrow{e} \cdot \vec{E} \tag{7.74}$$

and

$$\vec{D} = \overleftrightarrow{e} : \overleftrightarrow{S} + \overleftrightarrow{\varepsilon}^S \cdot \vec{E} \tag{7.75}$$

Combining Eq. (7.73) with the definition of the acoustic Poynting vector

$$\vec{\Pi} = -\vec{v} \cdot \overleftrightarrow{T} \tag{4.29}$$

it follows that

$$\nabla \cdot \vec{\Pi} = -\rho_m \vec{v} \cdot \frac{\partial \vec{v}}{\partial t} - \overleftrightarrow{T} : \frac{\partial \overleftrightarrow{S}}{\partial t} + \vec{v} \cdot \vec{F}^b \tag{7.76}$$

Since the piezoelectric wave has associated electric fields, it may carry electric power flow. The latter must be taken into account when computing the total energy. The *electric power flow density* (*electric Poynting vector*) $\vec{\Pi}_E$ in quasi-static approximation can be derived from the identity

$$\nabla \cdot (\vec{E} \times \vec{H}) = \vec{H} \cdot \nabla \vec{E} - \vec{E} \cdot \nabla \times \vec{H} \tag{7.77}$$

which is written in terms of quasi-static potential, and Eqs. (5.17) and (5.18) can be written as

$$\nabla \cdot (\vec{E} \times \vec{H}) = \nabla \phi \cdot (\nabla \times \vec{H}) \tag{7.78}$$

Using Eqs. (7.78) and (7.71), it follows that

$$\nabla \cdot (\vec{E} \times \vec{H}) = \nabla \cdot (\phi \nabla \times \vec{H}) \tag{7.79}$$

where, in general,

$$\nabla \times \vec{H} = \frac{\partial \vec{D}}{\partial t} + \vec{J}_s + \vec{J}_c \tag{7.80}$$

and where $\vec{J}_s$ and $\vec{J}_c$ are the source and conduction current densities, respectively, and

$$\nabla \cdot \vec{D} = \rho_e \tag{7.81}$$

with ρ_e being the free charge source density. In source-free, $\vec{J}_s = 0$, and lossless, $\vec{J}_c = 0$, piezoelectrics, it follows from Eqs. (7.79) and (7.80) that

$$\nabla \cdot (\vec{E} \times \vec{H}) \equiv \nabla \cdot \vec{\Pi}_E = \nabla \cdot \left(\phi \frac{\partial \vec{D}}{\partial t} \right) \tag{7.82}$$

or

$$\vec{\Pi}_E = \phi \frac{\partial \vec{D}}{\partial t} \tag{7.83}$$

Thus the divergence of the total power flow density is obtained by addition of Eqs. (7.76) and (7.78) and is given by

$$\nabla \cdot \vec{\Pi}_T = \nabla \cdot (\vec{\Pi} + \vec{\Pi}_E)$$

$$= -\rho_m \vec{v} \cdot \frac{\partial \vec{v}}{\partial t} - \vec{\vec{T}} : \frac{\partial \vec{\vec{S}}}{\partial t} + \nabla \phi \cdot \frac{\partial \vec{D}}{\partial t} + \vec{v} \cdot \vec{F}^b + \phi \frac{\partial}{\partial t} (\nabla \cdot \vec{D}) \tag{7.84}$$

or, upon integration over a volume V bounded by surface A, as

$$\iint_A \vec{\Pi}_T \cdot \vec{a} \, dA = - \iiint_V \left(\rho_m \vec{v} \cdot \frac{\partial \vec{v}}{\partial t} + \vec{\vec{T}} : \frac{\partial \vec{\vec{S}}}{\partial t} + \vec{E} \cdot \frac{\partial \vec{D}}{\partial t} \right) dV$$

$$+ \iiint_V \left(\vec{v} \cdot \vec{F}^b + \phi \frac{\partial \rho_e}{\partial t} \right) dV \tag{7.85}$$

where Eqs. (5.18) and (7.81) have been used and $\vec{a}$ is the surface normal. The third term in the right-hand side of Eq. (7.85) represents the rate of the stored electric field energy, and the fifth the source term due to free charge density, Eq. (7.81).

As in Sect. 5.8, consider Eq. (7.85) for propagation far away from the source, $\vec{F}^b = 0$, $\rho_e = 0$, and in the lossless material. It follows that

$$\iint_A \vec{\Pi}_T \cdot \vec{a} \, dA = - \iiint_V \left(\rho_m \vec{v} \cdot \frac{\partial \vec{v}}{\partial t} + \vec{\vec{T}} : \frac{\partial \vec{\vec{S}}}{\partial t} + \vec{E} \cdot \frac{\partial \vec{D}}{\partial t} \right) dV \tag{7.86}$$

and from Eq. (7.81) that $\nabla \cdot \vec{D} = 0$. For plane-wave propagation of the form given by Eq. (7.42), the latter reduces to

$$\vec{n} \cdot \vec{D} = 0 \tag{7.87}$$

and the equation $\nabla \times \vec{E} = 0$ reduces to

$$\vec{n} \times \vec{E} = 0 \tag{7.88}$$

Thus $\vec{E}$ is collinear with $\vec{n}$, that is,

$$\vec{E} = C\vec{n} \tag{7.89}$$

where C is a constant to be determined and $\vec{D}$ is orthogonal to $\vec{k}$. Thus the rate of the electric energy density in the plane piezoelectric wave, Eq. (7.86), is given by

$$\frac{\partial u_E}{\partial t} = \vec{E} \cdot \frac{\partial \vec{D}}{\partial t} = 0 \tag{7.90}$$

However, the electric power flow, Eq. (7.83), is nonzero and is orthogonal to the direction of propagation. The constant C in Eq. (7.89) is found[7] by using Eqs. (7.89), (7.87), and (7.75) and is given by

$$C = - \frac{\vec{n} \cdot \overset{\leftrightarrow}{e} : \overset{\leftrightarrow}{S}}{\vec{n} \cdot \overset{\leftrightarrow}{\varepsilon}^S \cdot \vec{n}} \tag{7.91}$$

The expression for the electric field, Eqs. (7.89) and (7.91), is substituted in Eq. (7.74), and the resulting expression for $\overset{\leftrightarrow}{T}$ with $\overset{\leftrightarrow}{\eta} = 0$ is then substituted in Eq. (7.86), which, in view of Eq. (7.90), is now written as

$$\iint\limits_{A} \vec{\Pi}_T \cdot \vec{a} \, dA = - \frac{\partial}{\partial t} \iiint\limits_{V} u_T \, dV \tag{7.92}$$

where the total energy density is given by

$$u_T = \frac{\rho_m \vec{v} \cdot \vec{v}}{2} + \frac{\overset{\leftrightarrow}{S} : \overset{\leftrightarrow}{c} : \overset{\leftrightarrow}{S}}{2} \tag{7.93}$$

and where

$$\overset{\leftrightarrow}{c} = \overset{\leftrightarrow}{c}^E + \frac{\left(\overset{\leftrightarrow}{e} \cdot \vec{n} \right)\left(\vec{n} \cdot \overset{\leftrightarrow}{e} \right)}{\vec{n} \cdot \overset{\leftrightarrow}{\varepsilon}^S \cdot \vec{n}} \tag{7.94}$$

Equation (7.94) is the symbolic representation of the stiffened elastic constant previously defined by Eq. (7.51).

Now consider plane-wave propagation in the geometry shown in Fig. 7.2, where $\vec{a}$, the surface normal, is opposite to the wave propagation direction $\vec{n}v_P = \vec{v}_P$. For a wave packet, surface A' can be chosen to coincide with the surface of constant phase at the beginning of the wave packet, and surface A'' to coincide with the surface of constant phase at the end of the wave packet. In this case it follows from Eqs. (7.68) and (7.92) that

$$-\vec{v}_\epsilon \cdot \vec{n} \iint_A u_T \, dA = -\frac{\partial}{\partial t} \iiint_V u_T \, dA \, dr$$

where r is the coordinate measured in the direction of propagation, Fig. 7.2, or

$$\vec{v}_\epsilon \cdot \vec{n} u_T A = \frac{\partial}{\partial t}(u_T A r) = u_T A \frac{\partial r}{\partial t} \tag{7.95}$$

However, from the concept of constant phase, $r - v_P t = \text{constant}$, it follows that $\partial r / \partial t = v_P$, or

$$\vec{v}_\epsilon \cdot \vec{n} = v_P = v_A \tag{7.96}$$

Thus, as shown in Fig. 7.2, the magnitude of the phase velocity is equal to the projection of energy velocity to the direction of propagation. The relationship between the energy and the group velocity will be considered in the next two sections, first for anisotropic nonpiezoelectrics and then for piezoelectric materials.

7.3.1 Anisotropic Nonpiezoelectric Materials

In this case, the expression for the total energy density,[4] Eq. (7.93), is given by

$$u_T = \frac{\rho_m v_i^2}{2} + \frac{1}{2} c_{ijkl} S_{ij} S_{kl} \tag{7.97}$$

which on using Eq. (3.15) and symmetry conditions Eq. (3.35) reduces to

$$u_T = \frac{\rho_m}{2} \left(\frac{\partial u_i}{\partial t}\right)^2 + \frac{1}{2} c_{ijkl} \frac{\partial u_i}{\partial x_j} \frac{\partial u_l}{\partial x_k} \tag{7.98}$$

Combining Eqs. (7.98) and (7.42) leads to

$$u_T = \left(\rho_m U_i^2 - \frac{c_{ijkl} n_j n_k U_i U_l}{v_A^2}\right) \left(\frac{f'^2}{2}\right) \tag{7.99}$$

The expression in the first set of parentheses is seen to be identical to Eq. (7.48) when multiplied by U_i and with $e_{ija} = 0$. Thus $u_T = 0$, meaning that the kinetic and strain energy densities are equal, and Eq. (7.92) reduces to

$$\iint\limits_A \vec{\Pi} \cdot \vec{a} \, dA = 0 \qquad (7.100)$$

ensuring wave energy conservation. In order to find the energy velocity, the total energy density, which is equal to twice the strain (or kinetic) energy density, is first found from Eqs. (7.42) and (7.98) as

$$u_T = 2\frac{\rho_m}{2}\left(\frac{\partial u_i}{\partial t}\right)^2 = \rho_m U_i^2 f'^2 \qquad (7.101)$$

and the Poynting vector from Eqs. (4.29) and (7.42) as

$$\Pi_i = -T_{ij}\frac{\partial u_j}{\partial t} = -c_{ijkl}S_{kl}\frac{\partial u_j}{\partial t}$$

$$= -c_{ijkl}\frac{\partial u_l}{\partial x_k}\frac{\partial u_j}{\partial t} = c_{ijkl}U_l U_j\frac{n_k}{v_A}f'^2 \qquad (7.102)$$

Upon division of Eq. (7.102) by Eq. (7.101), the energy velocity is found as

$$(\vec{v}_\epsilon)_i = \frac{\Pi_i}{u_T} = \frac{c_{ijkl}U_j U_l n_k}{\rho_m U_i^2 v_A} \qquad (7.103)$$

which for unit displacement $U_i^2 = 1$, with c_{ijkl} and v_A in place of $\bar{c}_{ijkl}$ and $\bar{v}_A$, respectively, and in view of symmetry conditions imposed on i and j, is identical with Eq. (7.67). Thus in anisotropic nonpiezoelectrics the group and energy velocities are equal. Since there are three different types of waves (two quasi-shear and one quasi-longitudinal), three different group (energy) velocities can be found. (In isotropic materials the group and phase velocities of the longitudinal wave are equal. The same is true for shear waves.) In anisotropic nonpiezoelectrics along the pure mode directions, the group and phase velocities of a longitudinal wave are equal. With the exception of the trigonal crystal system, the same is true for shear waves. In materials belonging to the trigonal crystal system, for wave propagation along a threefold acoustic axis, the energy velocities of the two shear waves can exhibit rather complicated behavior, allowing for a *cone of propagating directions*.[8,9] By analogy with optics, this phenomenon is termed *internal conical refraction*.

7.3.2 Piezoelectric Materials

By using the same procedure as in Sect. 7.3.1, it can be shown that the wave energy in the plane wave is conserved in this case also, the kinetic energy density being equal to the stiffened strain energy density, $\frac{1}{2}\bar{c}_{ijkl}(\partial u_i/\partial x_j)$ $(\partial u_l/\partial x_k)$, Eq. (7.93).

The Poynting vector in this case is obtained from Eqs. (4.29) and (7.83) as

$$(\vec{\Pi}_T)_i = -T_{ij}\frac{\partial u_j}{\partial t} + \phi\frac{\partial D_j}{\partial t} \tag{7.104}$$

or, taking Eqs. (7.75), (7.89), and (7.91) into account, as

$$(\vec{\Pi}_T)_i = -\bar{c}_{ijkl}\frac{\partial u_l}{\partial x_k}\frac{\partial u_j}{\partial t} + \phi\frac{\partial D_i}{\partial t} \tag{7.105}$$

Using Eq. (7.42) in the first term on the right-hand side of Eq. (7.105), it is found that

$$(\vec{\Pi}_T)_i = \frac{\bar{c}_{ijkl}U_lU_jn_kf'^2}{\bar{v}_A} + \phi\frac{\partial D_i}{\partial t} \tag{7.106}$$

and upon division with the total energy density u_T^E, which is in terms of symbols identical to Eq. (7.101), it follows that

$$(\vec{v}_\epsilon^E)_i = \frac{\bar{c}_{ijkl}U_jU_ln_k}{\rho_mU_i^2\bar{v}_A} + \frac{\phi}{u_T}\frac{\partial D_i}{\partial t} \tag{7.107}$$

where the superscript E has been added to indicate a piezoelectric material. When compared with Eq. (7.67), this can be written as

$$(\vec{v}_\epsilon^E)_i = (\vec{v}_G)_i + \frac{\phi}{u_T}\frac{\partial D_i}{\partial t} \tag{7.108}$$

Thus in piezoelectric materials the group and energy velocities are not the same. It is seen that the effect of piezoelectricity on the energy velocity, Eqs. (7.108) and (7.103), is to stiffen the elastic constants, see, for instance, Eq. (5.30), and to change the energy velocity in accordance with the second term on the right-hand side of Eq. (7.108). The latter is the contribution[2] of the electric Poynting vector, which causes a further increase of the power flow angle, from θ to θ_E, as shown in Fig. 7.3. Equation (7.96) remains in force, and with reference to Fig. 7.3 is now written as

$$\vec{v}_\epsilon^E \cdot \vec{n} = \bar{v}_A \tag{7.109}$$

where $\bar{v}_A$ is the piezoelectrically stiffened acoustic velocity given, for instance, for quasi-longitudinal or quasi-transverse waves by Eq. (7.41). For the quasi-longitudinal wave propagating along the y axis of LiNbO$_3$, which is a strong piezoelectric, considered in Eqs. (7.30) to (7.41), the electric Poynting vector increases the power flow angle from $\theta = 8.8°$ to $\theta_E = 9.2°$, which is a dif-

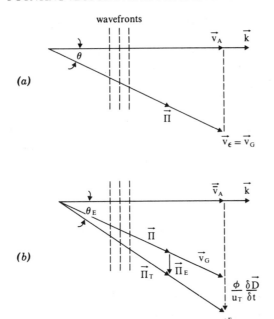

Figure 7.3. (a) Relationship between the acoustic Poynting vector Π, power flow angle θ, energy velocity v_ϵ, and phase velocity v_A in anisotropic nonpiezoelectrics. (b) Total Poynting vector Π_T, electric Poynting vector Π_E, power flow angle θ_E, stiffened phase velocity $\bar{v}_A$, and energy velocity v_ϵ^E in piezoelectrics.

ference of less than 5%. This has been experimentally confirmed.[7] In weak piezoelectrics such as quartz the effect is negligible, and the group and energy velocities are equal, to a very good approximation ($\pm 0.5\%$).

In our discussions of piezoelectricity, two types of stiffened elastic constants have been defined, namely,

$$\bar{c}_{ijkl} = c_{ijkl}^E + \frac{e_{ija}n_a e_{bkl}n_b}{\varepsilon_{pq}^S n_p n_q} \tag{7.51}$$

which are direction dependent and valid for $\vec{n} \cdot \vec{D} = 0$, and the stiffened elastic constant, Eq. (5.30), obtainable from Eqs. (7.74) and (7.75) in more general form as

$$c_{ijkl}^D = c_{ijkl}^E + e_{ija}(\vec{\varepsilon}^S)_{ab}^{-1} e_{bkl} \tag{7.110}$$

which are direction independent and are valid for $\vec{D} = 0$. For the one-dimensional case treated in Sect. 5.8, there is no difference between $\bar{c}$ and c^D, since in this case $\vec{n} \cdot \vec{D} = 0$ and $\vec{D} = 0$. Equations (7.110) are relevant to static measurements, while Eqs. (7.51) are to be used for wave propagation problems.[2]

7.4 CHARACTERISTIC SURFACES

From the preceding sections it is seen that acoustic propagation in anisotropic materials is a rather complicated topic requiring, in many cases, involved analytical and numerical computations. A better understanding of numerical results is achieved by representing the data in a graphical form.

In analogy with crystal optics, where three surfaces are used to illustrate the laws of wave propagation, similar surfaces are used in acoustics to demonstrate some salient features of acoustic wave propagation. These surfaces, termed *characteristic surfaces*, are defined as follows:

The *normal surface* (*velocity surface*) is defined as the locus of the phase velocities, centered at some point O_N versus the wave vector direction, $\vec{n} = \vec{k}/k$. The radius vector of this surface is therefore given by

$$\vec{r}_N = \vec{n}v_P \tag{7.111}$$

Since for each direction of the wave vector, there are, in general, three different phase velocities, the surface has three sheets, corresponding to (quasi) longitudinal and two (quasi) shear modes. The surface is determined from Eq. (7.63) as

$$\det\left[n_k n_j c_{ijkl} - \rho_m v_P^2 \delta_{il}\right] = 0 \tag{7.112}$$

For instance, for isotropic materials, the normal surface consists of two concentric spheres, the radius of the larger corresponding to the longitudinal wave, and the radius of the smaller corresponding to two shear waves of equal velocities.

The *slowness surface*, also termed the *reciprocal velocity surface*, *index surface*, *refraction surface*, or *inverse surface*, is the locus of $1/v_P$ versus wave vector direction, centered at some point O_S. Its radius vector is therefore

$$\vec{r}_S = \frac{\vec{n}}{v_P} = \frac{k}{\omega} \tag{7.113}$$

and the surface can be computed from Eq. (7.112). Again this surface has three sheets corresponding to the longitudinal and two transverse waves and consists, for isotropic materials, of two concentric spheres.

The *ray surface*, also termed the *energy surface* or *wave surface*, is the locus of v_ϵ centered at some point O_ϵ versus its direction. Its radius vector is given by

$$\vec{r}_\epsilon = \vec{v}_\epsilon \tag{7.114}$$

representing the energy transport in unit time in all directions. As such, this surface has the most concrete physical meaning, differing only in scale from the actual wavefront propagating in all directions from a point source. Thus the ray surface is also a surface of constant phase.

From Eqs. (7.111) to (7.114) it is seen that the following relationships exist:

$$\vec{r}_S \cdot \vec{r}_N = 1 \tag{7.115}$$

and taking into account Eq. (7.109),

$$\vec{r}_S \cdot \vec{r}_\epsilon = 1 \tag{7.116}$$

and

$$\vec{r}_\epsilon \cdot \vec{r}_N = \vec{v}_\epsilon \cdot \vec{v}_P = v_\epsilon v_P \cos \theta = v_P^2 \tag{7.117}$$

From Eqs. (7.115) to (7.112), it is seen that the energy velocity must always be orthogonal to the slowness surface and that the wave vector must always be orthogonal to the ray surface.

The characteristic surfaces are generally complicated, and hence their properties are usually represented by sections of various planes such as YX, YZ, ZX, as shown in Fig. 7.4 for slowness surface sheets or other symmetry planes. The sections of the slowness surface sheets for various crystal systems have been systematically studied and tabulated.[4, 10] They are of great help in problems

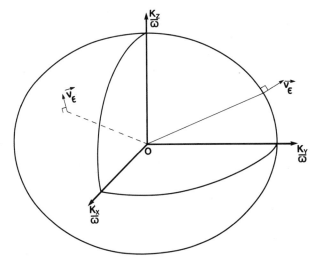

Figure 7.4. Three cross section (XY, YZ, XZ) of an idealized slowness surface. The energy velocity $\vec{v}_\epsilon$ must always be normal to the slowness surface.

considering reflection and refraction of acoustic waves. Using an acousto-optic technique known as the Schaefer-Bergmann technique,[11-13] the slowness surface sections can be visualized as shown in Fig. 7.5, where four sheets are seen. The resulting photograph is called an *elastogram*. The section shown is the *XZ* plane in LiNbO$_3$, and the fourth (outer) sheet is due to the SAW mode. This should be compared with the numerical results shown in Fig. 7.6.

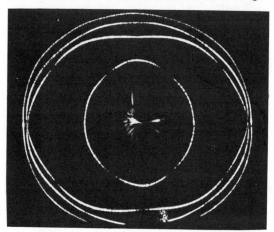

Figure 7.5. Acousto-optic visualization of the *XZ* section of the slowness surface for LiNbO$_3$. (*Courtesy of G. I. Stegeman, Optical Sciences Center, University of Arizona, Tucson.*)

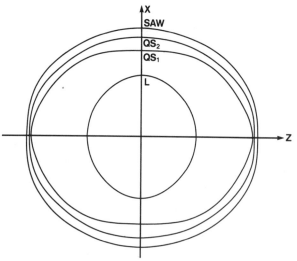

Figure 7.6. Numerical data obtained for an *XZ* section of the slowness surface for LiNbO$_3$. SAW is the Rayleigh wave, QS_1 and QS_2 are the two shear waves, and L is the longitudinal wave, (*Courtesy of G. I. Stegeman, Optical Sciences Center, University of Arizona, Tucson.*)

REFERENCES

1. B. A. Auld, *Acoustic Fields and Waves in Solids*, Vol. I, Wiley, New York, 1973, Chap. 7.
2. Ref. 1, Chap. 8.
3. E. W. Christoffel, *Ann. Math. (Pura Appl. Ser.)*, **8**(2), 193 (1877).
4. E. Dieulesaint and D. Royer, *Elastic Waves in Solids*, Wiley, New York, 1980, Chap. 5.
5. Ref. 1, App. 3.
6. Ref. 1, Chap. 5.
7. J. F. Havlice, W. L. Bond, and L. B. Wigton, "Elastic Poynting Vector in a Piezoelectric Medium," *IEEE Trans. Sonics Ultrason.*, **SU-17**(4), 246–249 (1970).
8. J. M. Liu and R. E. Green, Jr., "Optical Probing of Ultrasonic Diffraction in Single Crystal Sodium Chloride," *J. Acoust. Soc. Am.*, **53**(2), 468–478 (1973).
9. B. A. Auld, *Acoustic Fields and Waves in Solids*, Vol. II, Wiley, New York, 1973, Chap. 9.
10. Ref. 1, Chap. 7 and App. 3.
11. C. Schaefer and L. Bergmann, *Naturwiss.*, **22**, 685(1934); **23**, 799 (1935).
12. L. E. Hargrove and K. Achyuthan in *Physical Acoustics* (W. P. Mason, Ed.), Vol. 2B, Academic, New York, 1965.
13. G. W. Willard, *J. Acoust. Soc. Am.*, **23**, 83(1951).

EXERCISES

1. Consider plane wave propagation along the $x_3 = z$ axis of LiTaO$_3$. Find the velocities of various modes and the numerical value of the electromechanical coupling factor for the piezoelectric mode.

2. Consider plane-wave propagation along the poling axis of PZT-5A. Find the velocities of various modes and the numerical value of the electromechanical coupling factor for the piezoelectric mode.

3. Consider wave propagation along the y axis of LiTaO$_3$. Find the numerical values of the stiffened acoustic velocities of the two coupled piezoelectric modes. Can you determine the polarization of the two modes?

4. Consider propagation in the YZ plane of LiNbO$_3$.
   (a) Using $\vec{n} = (0, \cos \phi, \sin \phi)$, find the matrix for K_{IJ}^2, Eq. (7.55).
   (b) Show that only two modes are coupled.
   (c) Show that for $\phi = 36°$ or for $\phi = 163°$, where the angle ϕ is measured from the Y axis, either the electromechanical coupling coefficient for quasi-transversal or the one for the quasi-longitudinal wave is zero.

Chapter 8

Surface Acoustic Wave Transducers

The properties of surface acoustic waves[1] on isotropic materials have been studied in Chap. 4. The anisotropy of materials affects the surface acoustic waves in much the same way as the bulk waves; the phenomena of beam steering,[2] focusing,[3,4] polarization change,[5] and mode coupling[6] do take place here as well. The anisotropy does not change the character of SAW fundamentally except that it allows, through piezoelectricity if it is present, a conversion of electromagnetic into acoustic energy and vice versa.

In device design, the properties of SAW on piezoelectric materials are of primary interest.[1] As in the case of bulk piezoelectric waves, the surface acoustic piezoelectric wave has electric field components coupled to displacement components. The coupling can be used in a variety of ways for excitation and detection of SAW. This is usually achieved by deposition of a metallic electrode structure on a piezoelectric substrate. The most common electrode structure, known as the uniform interdigital transducer (IDT), is shown in Fig. 8.1. It consists of a series of uniform metal electrodes in the form of two interspaced combs deposited onto the surface of the piezoelectric substrate. By applying an electric potential of the appropriate frequency between the two metal combs, acoustic waves are excited and made to propagate both on the surface and in the bulk of the substrate, as shown in Fig. 8.1(a) where both SAW and bulk acoustic wave (BAW) are illustrated. The detection process is achieved in a similar fashion. Acoustic waves incident in the transducer will establish electric potentials and RF currents in the electrodes, which in turn can be connected to an external electrical load.

The studies of wave generation and detection with SAW transducers have evolved in two different directions. The problem is analytically intractable,[7] and various approximations and numerical techniques must be used. The second approach is to use the equivalent circuits[8] developed for bulk waves in Chap. 5. The SAW energy is well confined within a Rayleigh wavelength from the top of the substrate, and the wave propagation in this region can be approximated by a homogeneous bulk wave, thus allowing the use of the equivalent circuits shown in Figs. 5.5 and 5.6.

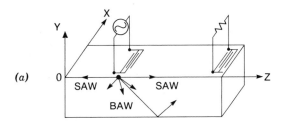

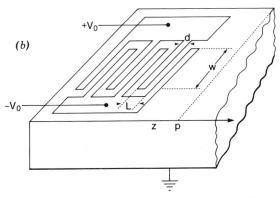

Figure 8.1. (*a*) A SAW device with an input and an output transducer. (*b*) A uniform transducer defined by half-spatial period L, electrode width d, and aperture width W. The observation line P used for derivation of the impulse response is at distance $z + L$ from the first source.

Many theories have been developed and several formulas derived to explain and calculate various parameters of SAW transducers on piezoelectric substrates. These theories, within the framework of their initial assumptions and thus accuracy, lead to or give more-or-less the same results. Chronologically the transducer electrostatic problem[9] and the transducer excitation and detection problem[10] were considered first. The concept of SAW coupling coefficient was then developed[11] and the transducer capacitance found,[12] and both of these were used in developing the equivalent circuit model.[13] The concept of "surface impedance" was also developed,[14] in addition to the normal mode theory.[15] The δ-function model was then derived,[16] and the concept of an effective dielectric constant was used to find the transducer impedance.[17, 18] Subsequently, the "surface impedance" concept was applied to the SAW transducer problem,[19] and there was further development in the application of normal mode theory to various transducer structures.[20] Field theory,[21] based on the exponential coupling of modes, was also developed. Later there were further developments in the application of normal mode theory[22] and the field theory.[7]

The radiation problem, consisting of a treatment of the SAW transducer as the radiator of both surface and bulk waves, was also considered, both analytically[23, 24] and numerically.[25-27]

Many of the listed theories are beyond the scope of this textbook, and interested readers are referred to the original papers.

In what follows, an approximate analytical technique[28] is used to derive some basic properties of IDT. The SAW propagation in anisotropic materials is described by

$$
c_{ijkl}^{E} \frac{\partial^2 u_l}{\partial x_k\, \partial x_j} + e_{ijk} \frac{\partial^2 \phi}{\partial x_k\, \partial x_j} = \rho_m \ddot{u}_1
$$

$$
e_{ikl} \frac{\partial^2 u_l}{\partial x_k\, \partial x_i} - \varepsilon_{ik}^{S} \frac{\partial^2 \phi}{\partial x_k\, \partial x_i} = 0
$$

(7.11)

subject to boundary conditions on the surface of the substrate. In the transducer region, if σ_e is the volume charge density in the immediate vicinity of electrodes, then at the electrodes the boundary condition is

$$
\frac{\partial D_i}{\partial x_i} = \sigma_e
$$

(8.1)

In seeking an analytical approximation, it is assumed that the length of the fingers in the $x_1 = x$ direction is much larger than the Rayleigh wavelength, and therefore $\partial/\partial x_1 = \partial/\partial x \approx 0$. In order to conform with the models previously derived for bulk waves, it is also assumed that the SAW has only a nonzero displacement component in the $x_3 = z$ direction. Furthermore, in evaluating Eq. (8.1) the piezoelectric constants in Eq. (7.8) are assumed to be small, that is,

$$
D_i \simeq -\varepsilon_{ik}^{S} \frac{\partial \phi}{\partial x_k}
$$

(8.2)

Consider, under these assumptions, the problem of excitation of SAW propagating along the Z axis on YZ-cut crystal such as LiNbO$_3$, Fig. 6.4. Using material constants belonging to the trigonal system, which includes quartz (Sect. 6.3), Eqs. (7.11), (8.1), and (8.2) can be reduced to

$$
c_{33}^{E} \frac{\partial^2 u_z}{\partial z^2} - \rho_m \frac{\partial^2 u_z}{\partial t^2} + c_{44}^{E} \frac{\partial^2 u_z}{\partial y^2} = -e_{x5} \frac{\partial^2 \phi}{\partial y^2} - e_{z3} \frac{\partial^2 \phi}{\partial z^2}
$$

$$
e_{z3} \frac{\partial^2 u_z}{\partial z^2} + e_{x5} \frac{\partial^2 u_z}{\partial y^2} = \varepsilon_{yy}^{S} \frac{\partial^2 \phi}{\partial y^2} + \varepsilon_{zz}^{S} \frac{\partial^2 \phi}{\partial z^2}
$$

$$
\varepsilon_{yy}^{S} \frac{\partial^2 \phi}{\partial y^2} + \varepsilon_{zz}^{S} \frac{\partial^2 \phi}{\partial z^2} = -\sigma_e
$$

(8.3)

Combining Eqs. (8.3) it follows that

$$\frac{\partial^2 u_z}{\partial z^2} - \frac{1}{v^2}\frac{\partial^2 u_z}{\partial t^2} = \mu\left(-\frac{\partial E_z}{\partial z} + \frac{\eta\sigma_e}{\varepsilon_{yy}^S}\right) \tag{8.4}$$

where v, μ, and η are constants. In this approximate approach the precise value of these constants cannot be obtained. For instance, when all surface boundary conditions are satisfied, $v = v_R$, where v_R is the Rayleigh wave velocity. This problem is considered in Sect. 8.2. In Eq. (8.4), σ_e can be written as

$$\sigma_e = \sigma_e(t, z) = \sigma_{e1}(z)\sigma_{e2}(t)$$

where in the region just under the electrodes $\sigma_{e1}(z)$ is proportional to $E_y(t, z)$. Therefore, the wave equation (8.4) can be written as

$$\frac{\partial^2 u_z}{\partial z^2} - \frac{1}{v_R^2}\frac{\partial^2 u_z}{\partial t^2} = B\sigma(t, z) \tag{8.5}$$

with

$$\sigma(t, z) = \sigma_1(z)\sigma_2(t) \tag{8.6}$$

and

$$\sigma_1(z) = \begin{cases} A_1\dfrac{dE_z(z)}{dz}, & \text{``in-line'' model} \\[2mm] A_2 E_y(z), & \text{``crossed-field'' model} \end{cases} \tag{8.7}$$

and where B, A_1, and A_2 are constants.[†] In studies of IDTs it is customary to discriminate between the equivalent circuit model based on the assumption that the longitudinal electric field E_z is predominant ("in-line" model) and the equivalent circuit model based on the assumption that the transverse electric field E_y is predominant ("crossed-field" model). In seeking the solution of Eq. (8.5) under impulse excitation, it has been shown[28] that the response obtained from the in-line model is pertinent to the electroded (source) region of the transducer. In this region the SAW grows exponentially in amplitude,[7,21] and the generation process can be understood by considering the E_y field to be associated with the energy storage supplied by an outside generator and the E_z field with a mechanism whereby the stored electric energy is transferred into the spatially growing SAW under the transducer. Since our interest is in the response outside the electroded region of the transducer, the derivation of the impulse response will be based on the crossed-field model.

[†]All three constants are proportional to the excitation strength of the corresponding body forces; see, for instance, Eq. (3.2).

The solution of Eq. (8.5) for impulse excitation, $\sigma_2(t) = \delta(t)$ in Eq. (8.6) for the crossed-field model, is given by[28]

$$u_z(t, z) = A' \int_{-\infty}^{+\infty} \chi\left[t - \frac{|z - \zeta|}{v_R}\right] E_y(\zeta)\, d\zeta \tag{8.8}$$

for progressive waves propagating in the $+z$ direction and where the integration is performed over the length (z direction in Fig. 8.1) of the transducer. Furthermore, χ is the familiar step function, equal to unity for $t \geqslant 0$ and zero otherwise. For purposes of impulse-response modeling, it is convenient to define a partial potential given by

$$\frac{d\phi_y}{dz} = E_y\left[\frac{z}{L}\right] \tag{8.9}$$

where L is defined in Fig. 8.1, allowing the solution of Eq. (8.8) to be written as

$$u_z(t, z) = A\phi_y\left[\frac{z - tv_R}{L}\right] \tag{8.10}$$

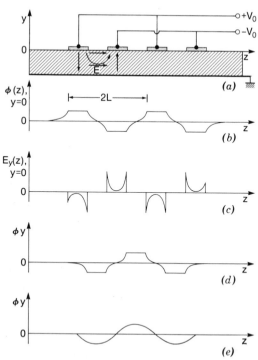

Figure 8.2. (*a*) An IDT with a sketch of the electric field configuration between two electrodes. (*b*) The potential. (*c*) Transverse electric field. (*d*) Partial potential. (*e*) Sine-function modeling of the partial potential.

It is seen that in this case the response carries the image of the partial potential ϕ_y. In Fig. 8.2, both the true potential $\phi(z)$ and the partial potential $\phi_y(z)$ are shown. The partial potential can be modeled in various ways; for instance, both rectangles and sine functions have been used. The sine function modeling is shown in Fig. 8.2(e). In this case each electrode is represented by a half spatial period of the sine function. The spatial period $2L$ is equal to the resonant wavelength of an electrode pair,

$$\beta_0 = \frac{\omega_0}{v_R} = \frac{\pi}{L} \tag{8.11}$$

Therefore, with sine-function modeling the impulse response for an electrode can be written as

$$u_z(t, z) = A\left[\chi(z - tv_R) - \chi(z - tv_R - L)\right]\sin \beta_0(z - tv_R) \tag{8.12}$$

However, in practice it is customary to model the electric field rather than the particle displacement. The electric field, Fig. 8.3(a), is usually modeled either

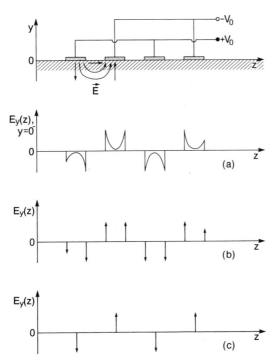

Figure 8.3. (a) The transverse electric field. (b) Double δ-function modeling and (c) Single δ-function modeling of the transverse electric field.

as a double δ-function model, Fig. 8.3(*b*), or a single δ-function model, Fig. 8.3(*c*). The difference between the double δ-function model and the single δ-function model is that in the latter case the spectral response of a single electrode is neglected. The spectral bandwidth of a single electrode is much larger than that of the transducer, and therefore the single δ-function model is one of the approximations used in design practice.

In the single δ-function model, the δ function is positioned in the middle of the electrode, and the electric field is written as

$$E_y = E_y^0 \delta \left[\frac{z - tv_R}{L} \right] \tag{8.13}$$

where E_y^0 is the value of the electric field at the edge of the electrode. Using Eqs. (8.9) and (8.10), it follows that

$$u_z(t, z) = LA\phi_y^0 \chi \left[\frac{z - tv_R}{L} \right] \tag{8.14}$$

where ϕ_y^0 is the potential at the electrode edge.

8.1 CONDUCTANCE AND IMPULSE RESPONSE

In the transducer model that will be used for finding the radiation conductance and impulse response, it will be assumed that the metallization ratio is 0.5, that is, that the width d of the electrodes is equal to $L/2$. The electric field will be assumed to decrease with the thickness of substrate as $E_y^0 \exp(y/2L)$, $y < 0$, where $2L$ is the Rayleigh wavelength equal to v_R/f_0. Consider the volume consisting of a half-section of length L, aperture w, and the thickness of the substrate, as shown in Fig. 8.1(*b*).

The power of a SAW generated under the transducer and propagating in the $+z$ (or $-z$) direction is not conserved but increases with distance at the expense of the stored electric energy, which is in turn supplied by the generator. By applying Poynting's theorem over the volume, under assumptions of no losses, zero magnetic energy, and equal static and kinetic energy, it can be shown that

$$\Delta P_a = - \frac{\Delta W_E}{\Delta t} \tag{8.15}$$

where ΔP_a is the increase of the acoustic power in both waves propagating in the $\pm z$ direction after each wave has traveled the length L in a time Δt, and $\Delta W_E / \Delta t$ is the rate of decrease of the peak stored electric energy. The depletion of the stored energy at the resonant frequency $f_0 = 1/2\Delta t$ as the single wave passes the length L is given by

$$\frac{\Delta W_E'}{W_E'} = -K^2 \tag{8.16}$$

where K^2 is the familiar piezoelectric coupling factor defined in Eq. (5.30). For both waves this is given by[†]

$$\frac{\Delta W_E}{W_E} = -2K^2 \tag{8.17}$$

Combining Eqs. (8.17) and (8.15) it follows that

$$\Delta P_a = 2K^2 \frac{W_E}{\Delta t} = 4f_0 K^2 W_E = 4f_0 K^2 \left[\frac{wC_s'(2V_0)^2}{2} \right] \tag{8.18}$$

where $2V_0$ is the peak voltage between electrodes as shown in Fig. 8.1(b), C_s' is the capacitance per single L-long section per unit aperture w, and $W_E = wC_s'(2V_0)^2/2$ is the peak electric energy stored in C_s'. The conductance of an L-long section due to the presence of both waves is given by

$$G = \frac{2\Delta P_a}{(2V_0)^2} = 4f_0 K^2 C_s' w \tag{8.19}$$

To simplify the mathematics when calculating the transducer conductance it will be assumed that the amplitude of the waves increases linearly, rather than exponentially, with distance in the direction of propagation. The power of the waves will be increasing quadratically with distance, that is, with the number of sections. Thus the conductance $G_0^{(M)}$ of M, L-long sections for both waves will be

$$G_0^{(M)} = 4f_0 K^2 C_s' M^2 w \tag{8.20}$$

Sometimes it is more convenient to work with $2L$-long, rather than L-long, sections. The conductance $G_0^{(N)}$ of N, $2L$-long sections is found by setting $M = 2N$ and $C_s' = C_s/2$ in Eq. (8.20), where C_s is the capacitance of a single $2L$-long section per unit aperture width w and is given by

$$G_0^{(N)} = 8f_0 K^2 C_s N^2 w \tag{8.21}$$

In order to find the off-resonance behavior of the conductance, consider M, δ sources as shown in Fig. 8.3(c). Assuming noninteracting sources and neglecting diffraction, losses, and beam steering, the total impulse response at some line P parallel to the electrode length, Fig. 8.1(b), and outside the transducer region in the $+z$ direction is given by

$$h(t, z) = \sum_{m=1}^{M} A_0(-1)^m \delta\left(\frac{z_m - tv_R}{L}\right) = \frac{A_0 L}{v_R} \sum_{m=1}^{M} (-1)^m \delta\left(t - \frac{z_m}{v_R}\right) \tag{8.22}$$

[†]More accurate methods for finding K^2 are referenced in Sect. 8.2.

and $\tilde{H}(\omega, z)$, the inverse Fourier transform, of $h(t, z)$ can be defined as

$$\tilde{H}(\omega, z) = \frac{\tilde{E}_y(\omega, z)}{E_y^0} \tag{8.23}$$

Since E_y varies along the thickness of the substrate at $E_y^0 \exp(y/2L)$, it follows after integration along the thickness that $V_0 = \phi_y^0 = 2E_y^0 L$. Furthermore, in Eq. (8.22) $z + L$ is the distance from the closest source to the observation line P, v_R is the Rayleigh velocity, and $z_m = z + Lm$, $m = 1, 2, \ldots, M$ determines the location of the sources.

The transfer function, obtained by the Fourier transform of Eq. (8.22), is given by

$$\tilde{H}(\omega, z) = \frac{A_0 L}{v_R} \exp\left(-\frac{j\omega z}{v_R}\right) \sum_{m=1}^{M} (-1)^m \exp\left(-\frac{j\omega Lm}{v_R}\right) \tag{8.24}$$

In the above equation z/v_R represents, approximately, the time delay for wave propagation between the closest source and the observation line P. Since $2L = v_R/f_0$, the sum in Eq. (8.24) can be written as

$$\sum_{m=1}^{M} \exp\left[-m\pi\left(\frac{f - f_0}{f_0}\right)\right] \simeq M\frac{\sin X_m}{X_m} \exp\left[j\left(\frac{M\pi}{2}\right) - j\left(\frac{M\pi f}{2f_0}\right)\right] \tag{8.25}$$

where $\sin[\pi(f - f_0)/2f_0] \simeq \pi(f - f_0)/2f_0$ has been used and where $X_M = M\pi(f - f_0)/2f_0$.

To determine the constant A_0 in Eq. (8.24), consider the power at the resonance $f = f_0$ of the transducer of M, L-long sections, excited with peak voltage of $2V_0(t) = \delta(t)$ V. First the SAW is approximated with a homogeneous bulk wave of cross-sectional area $2Lw$, and the piezoelectric effect is neglected. The acoustic power of the wave is found from

$$P_a = 2Lw\left(\frac{-v_z^* T_3}{2}\right) = \frac{|F|^2}{2\bar{Z}_R} \tag{8.26}$$

where F is the force and $\bar{Z}_R = 2Lw\rho_m v_R$ is the impedance of the wave. Using Eqs. (8.14) and (8.26) with impulse excitation $4E_y^0 L = \delta(t)$ V, it follows that $P_a = |H(\omega, z)|^2/2\bar{Z}_R$. For a single wave, the electric power delivered to load, $G_0^{(M)}/2$ at $f = f_0$, is found from Eq. (8.20) and is equated to P_a,

$$\frac{|\tilde{H}(\omega_0, z)|^2}{2\bar{Z}_R} = \frac{G_0^{(M)}[2V_0(\omega)]^2}{4} = \frac{G_0^{(M)}}{4} \tag{8.27}$$

resulting in

$$A_0 = \frac{v_R}{L} K\left(\bar{Z}_R 2C_s' w f_0\right)^{1/2} \tag{8.28}$$

Using Eqs. (8.22), (8.24), and (8.28), the transfer function of a uniform transducer consisting of M, L-long sections is given by

$$\tilde{H}(\omega, z) = \xi_M K\left(\overline{Z}_R 2C_s' w f_0\right)^{1/2} M \frac{\sin X_M}{X_M} \exp\left(\frac{-jM\pi f}{2f_0}\right) \exp\left(-\frac{j\omega z}{v_R}\right)$$

(8.29)

where $\xi_M = \exp(jM\pi/2)$ and the impulse response is

$$h(t, z) = 2K\left(\overline{Z}_R 2C_s' w f_0^3\right)^{1/2} \sum_{m=1}^{M} (-1)^m \delta\left(\frac{z_m - tv_R}{L}\right)$$

(8.30)

With $M = 2N$ and $2C_s' = C_s$, the corresponding expressions for a transducer consisting of N, $2L$-long sections are readily found.

Using Eq. (8.26) off resonance, the off-resonance radiation conductance is found as

$$G(\omega) \simeq G_0^{(M)} \frac{\sin^2 X_M}{X_M^2}$$

(8.31)

and by Hilbert transform, the off-resonance radiation susceptance is obtained as

$$B(\omega) \simeq G_0^{(M)} \frac{\sin 2X_M - 2X_M}{2X_M^2} + \omega MC_s' w$$

(8.32)

where the static capacitance term $\omega MC_s' w$ has been added. The behavior of the radiation conductance and susceptance is similar to behavior of radiation resistance and reactance of meander-type transducers, Fig. 3.2(b). The electric field response is found from Eqs. (8.23) and (8.29) as

$$E_y(t, z) = \frac{V_0 K}{L}\left(\overline{Z}_R 2C_s' w f_0^3\right)^{1/2} \sum_{m=1}^{M} (-1)^m \delta\left(\frac{z_m - tv_R}{L}\right)$$

(8.33)

In Eqs. (8.30) and (8.33), the $f_0^{3/2}$ frequency dependence is the most notable characteristic that has been experimentally confirmed.[29] If the transducer is phase apodized, that is, if the separation between adjacent fingers varies with distance, the resonant frequency will vary as well, and $f_0^{3/2}$ will no longer be a constant. The latter will cause amplitude distortion in the frequency band of interest. Another important point is that the δ-function model for the transfer function, Eq. (8.28), results in the phase that is a linear function of frequency.

In the equations developed, the Rayleigh wave velocity v_R on anisotropic substrates and the capacitance per unit aperture C_s' have not been determined.

In the following section, the determination of these two constants is considered, as well as a more precise treatment of the coupling factor.

8.2 SAW VELOCITY, THE COUPLING FACTOR, AND TRANSDUCER CAPACITANCE

For a particular direction of propagation, the SAW velocity on anisotropic substrates can be obtained with a numerical procedure[11, 30, 31] from Eq. (7.11) under prescribed boundary conditions. Two cases are of interest, a free surface of the substrate and a surface coated with a thin layer of a perfect conductor assumed to be of zero mass. In the latter case the only function of the conducting layer is to short out the electric field of the SAW at the surface of the substrate. The two cases correspond to the metallized and unmetallized regions of the IDT, Fig. 8.2(a). The velocity of the free surface is usually denoted by $v_R = v_\infty$ (metallized layer at infinity), and the velocity of the metallized sample by v_0 (metallized layer at the surface). In the transducer region for a transducer of metallization ratio 0.5, the SAW velocity of the transducer v_T is obtained[32] as $v_T = 2v_0 v_\infty/(v_0 + v_\infty)$. The velocity v_T is used for the computation of various parameters of the transducer. For instance, for a Y-cut plate of LiNbO$_3$ and for propagation in the Z direction,[31] $v_\infty = v_R = 3488$ m/s and $v_0 = 3404$ m/s.

In the modeling of the transducer, the piezoelectric coupling factor[11, 30] has been defined in Eq. (8.16). However, a more precise definition[32] of the coupling factor for SAW, K_S^2, can be obtained by analyzing the IDT as an infinite periodic structure. In this case it can be shown that

$$K_S^2 = -\frac{\Delta W_E'}{W_E'} \simeq \frac{2(v_\infty - v_0)}{v_\infty} = \frac{2\Delta v}{v_\infty} \qquad (8.34)$$

In practice,[30] the values of $2\Delta v/v_\infty$ and $v_R = v_\infty$ are usually tabulated as shown in Table 11. Thus K_S should be substituted for K in Eqs. (8.21) and (8.28) to (8.33) in order to obtain more accurate results.[†] Equation (8.34) is analogous to Eq. (5.27) for bulk waves.

An approximate expression for the static capacitance C_s', defined in the preceding section, can be found[12] by considering the transducer as an infinite periodic structure. The piezoelectric terms are assumed small, and Eqs. (7.6) and (7.11) are written as

$$D_i \simeq \varepsilon_{ik} E_k$$

$$\varepsilon_{ik} \frac{\partial^2 \phi}{\partial x_i \, \partial x_k} \simeq 0 \qquad (8.35)$$

[†]For more precise analytical expressions for the coupling factor, see the original papers related to the theories listed at the beginning of the chapter.

TABLE 11   SAW Velocity and Coupling Constant for Some Materials

Material	Orientation	$v_R = v_\infty$, m/s	$K_S^2 = 2\Delta v/v_R$
$LiNbO_3$	Y plate, Z propagation	3488	0.048
$Bi_{12}GeO_{20}$	001 plate, 110 propagation	1681	0.0136
$LiTaO_3$	Z plate, Y propagation	3329	0.0118
Quartz	ST plate, X propagation	3158	0.00116

The transducer consists of an infinite number of spatial half-periods L of metallization ratio a, $a < 1$, when the width of each electrode is $d = aL$. By expanding the potential ϕ in terms of spatial harmonics, A_{2n+1}, $n = 0, 1, 2, \ldots$, the amplitude of the spatial harmonics and the capacitance C_s' per single L-long section per unit aperture w are found as

$$A_{2n+1} \simeq \frac{2\pi V_0}{L} \frac{P_n(2k^2 - 1)}{K'(k)} \tag{8.36}$$

$$C_s' = 0.5\varepsilon_0 (1 + \varepsilon_r) \frac{K(k)}{K'(k)} \tag{8.37}$$

$$\varepsilon_r = \frac{\left(\varepsilon_{yy}^T \varepsilon_{zz}^T - \varepsilon_{yz}^T\right)^{1/2}}{\varepsilon_0}$$

where P_n is the Legendre function of the first kind, $k = \cos[\pi(1 - a)/2]$, and $K(k)$ and $K'(k)$ are the complete elliptic integrals of the first kind. The analysis[12] of Eq. (8.36) indicates that for $a = 0.5$ the amplitudes of the third and seventh spatial harmonics are zero, while the first and fifth harmonics are nonzero, the remaining spatial harmonics being negligible. The majority of SAW devices operate on the first harmonic, $a \approx 0.5$. With $a \approx 0.2$, SAW devices operating on the third harmonic are sometimes used as well, although in this case the transducer coupling losses increase by about 6 dB. For transducers with a small number of electrodes, the value of C_s' can be determined by numerical methods.[33]

In Eq. (8.37), a useful approximation for the ratio of the two elliptic functions is given by[34]

$$\frac{K(k)}{K'(k)} \simeq 1.47a^2 + 0.244a + 0.535 \tag{8.37a}$$

where $a = d/L$.

8.3   EQUIVALENT CIRCUIT MODELS

It was pointed out early in this chapter that in modeling of the SAW IDT, the in-line model can be used for studying the response in the transducer region and the crossed-field model for the response outside the transducer region. In the past, both models of equivalent circuits[13] have been used for the response outside the transducer region. In what follows the in-line circuit model will be developed for SAW. The in-line circuit model is particularly applicable to multilayer configurations of the thickness-driven piezoelectric transducers[35] (Fig. 5.22 and Exercise 9 in Chap. 5).

The in-line circuit model, characterized with an electric field in line with the displacement, uses the equivalent circuits shown in Figs. 5.5 and 5.6. Each unmetallized region is represented by the circuit in Fig. 5.5(b), and each metallized region by the same circuit with $n = 0$. This is shown in Fig. 8.4. To find the overall characteristics of the transducer, the circuit in Fig. 5.5(b) is redrawn as shown in Fig. 8.5 in order to express the quantities on the left-hand side of the network, F_m, v_m, V_m, I_m, as functions of the quantities on the right-hand side, $F_{m+1}, v_{m+1}, V_{m+1}, I_{m+1}$. This can be written in matrix form as

$$\begin{bmatrix} F_m \\ v_m \\ V_m \\ I_m \end{bmatrix} = [T_m] \begin{bmatrix} F_{m+1} \\ v_{m+1} \\ V_{m+1} \\ I_{m+1} \end{bmatrix} \tag{8.38}$$

where $T_m = T_m^0$ for metallized (electroded) sections and $T_m = T_m^\infty$ for unmetallized sections. Using the definitions

$$\kappa^2 \equiv \frac{n^2}{\omega_0 C_0 \overline{Z}_0}, \qquad \gamma \equiv \beta_a l$$

$$s \equiv \frac{\kappa^2 \sin \gamma}{\gamma}, \qquad c \equiv \frac{\kappa^2 (1 - \cos \gamma)}{\gamma} \tag{8.39}$$

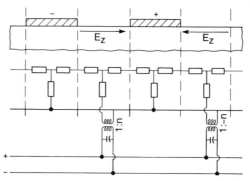

Figure 8.4.   The modeling of an IDT using the Mason in-line circuit model.

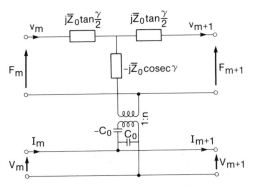

Figure 8.5. The Mason in-line circuit model.

where $\omega_0 = \pi v_L^D/l$ and $\beta_a l = \pi f/f_0$, it can be shown[35] that

$$
[T_m^\infty] = \frac{1}{1-s}
\begin{bmatrix}
\cos \gamma - s & j\bar{Z}_0(\sin \gamma - 2c) & n(1 - \cos \gamma) & 0 \\
j \sin \gamma/\bar{Z}_0 & \cos \gamma - s & -jn \sin \gamma/\bar{Z}_0 & 0 \\
0 & 0 & 1 - s & 0 \\
-jn \sin \gamma/\bar{Z}_0 & n(1 - \cos \gamma) & j\omega C_0 & 1 - s
\end{bmatrix}
$$

$$(8.40)$$

where the T_m^0 matrix is obtained from the above expression by setting $n = c = s = C_0 = 0$. The transfer matrix for the entire transducer having $M - 1$ unmetallized and M electroded sections is obtained by multiplying the transfer matrices for all sections,

$$
[T] = \left(\prod_{m=1}^{M-1} [T_m^\infty] \right) \left(\prod_{m=1}^{M} [T_m^0] \right)
$$

$$(8.41)$$

The adjacent electrodes are of alternate polarity, and therefore the value of n in the expressions for adjacent T_m^∞ matrices must alternate in sign. Since the transducer is terminated at both ends by mechanical impedance $\bar{Z}_0$, it follows that

$$V_{M+1} = V_1, \qquad F_1 = -\bar{Z}_0 v_1$$

$$F_{M+1} = \bar{Z}_0 v_{M+1}, \qquad I_{M+1} = 0 \qquad (8.42)$$

The last condition follows from the fact that after the last transducer section there is no current flow. The input impedance is found as $Z_{\text{in}} = V_1/I_1$.

Before the described model can be used for SAW IDT, the capacitance C_0, the transformer ratio n [or the parameter κ, Eq. (8.39)], the mechanical impedance $\bar{Z}_0$, and γ must be specified. Usually it is taken[13] that $C_0/w \simeq C_s'$ and $\kappa \simeq K_S$, where C_s' and K_S are given by Eqs. (8.37) and (8.34). Furthermore, $\bar{Z}_0 \simeq \rho_m v_\infty A$ and $\gamma = \omega l/v_\infty$ for unelectroded and $\bar{Z}_0^e \simeq \rho_m v_0 A$ and $\gamma^e = \omega l_e/v_0$ for electroded sections with $A = w\tau \simeq w2L$, where $2L$ is the resonant (Rayleigh) wavelength of the transducer and $l = (1 - a)L$ and $l_e = aL$.

8.3.1   The Crossed-Field Model

The crossed-field model[35] for bulk waves consists of a bulk wave with a single displacement component u_z propagating perpendicular to the electric field E_y in an electroded piezoelectric bar, Fig. 8.6. The following assumptions are made.

The electrical boundary conditions are $E_x = E_z = 0$, $E_y = V_3 e^{j\omega t}/\tau$, where τ is the thickness of the bar, and $\partial E_y/\partial z = 0$, which is an assumption of the uniform field. The mechanical boundary conditions are $v = v_1$, $F = F_1$ at $z = 0$, and $v = -v_2$, $F = -F_2$ at $z = l$. It is assumed that the only existing strain and stress components are S_3 and T_3. Under these conditions for class (2) piezoelectric crystals, Sect. 6.2.2, the constitutive relations reduce to

$$S_3 = s_{33}^E T_3 + d_{y3} E_y$$

$$D_y = d_{y3} T_3 + \varepsilon_{yy}^T E_y \tag{8.43}$$

with equation of motion

$$\frac{\partial T_3}{\partial z} = \rho_m \ddot{u}_z \tag{8.44}$$

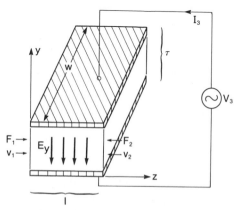

Figure 8.6. Bulk transducer configuration used for derivation of the Mason crossed-field circuit model.

Since $S_3 = \partial u_z / \partial z$, the last two equations, combined with $\partial E_y / \partial z = 0$, result in the wave equation

$$\frac{\partial^2 u_z}{\partial z^2} = s_{33}^E \rho_m \ddot{u}_z \tag{8.45}$$

with solution

$$u_z = [B_1 \sin \beta_a z + B_2 \cos \beta_a z] e^{j\omega t} \tag{8.46}$$

where $\beta_a = \omega / v_a$, $v_a = (s_{33}^E \rho_m)^{-1/2}$. Upon application of mechanical and electrical boundary conditions it follows that

$$F_1 = \frac{\overline{Z}_0}{j \tan \beta_a l} v_1 + \frac{\overline{Z}_0}{j \sin \beta_a l} v_2 + n V_3$$

$$F_2 = \frac{\overline{Z}_0}{j \sin \beta_a l} v_1 + \frac{\overline{Z}_0}{j \tan \beta_a l} v_2 + n V_3$$

$$I_3 = -n v_1 - n v_2 + j \omega C_0 V_3 \tag{8.47}$$

where

$$\overline{Z}_0 = A \rho_m v_a, \qquad A = w\tau \tag{8.48}$$

$$n = \frac{w d_{y3}}{l s_{33}^E} \tag{8.49}$$

and

$$C_0 = \frac{wl}{\tau} \left(\varepsilon_{yy}^T - \frac{d_{y3}^2}{s_{33}^E} \right) \tag{8.50}$$

Equation (8.47) can be expressed in matrix form as

$$\begin{bmatrix} F_1 \\ F_2 \\ V_3 \end{bmatrix} = -j \begin{bmatrix} \dfrac{\overline{Z}_0}{\tan \beta_a l} + \dfrac{n^2}{\omega C_0} & \dfrac{\overline{Z}_0}{\sin \beta_a l} + \dfrac{n^2}{\omega C_0} & \dfrac{n}{\omega C_0} \\[2mm] \dfrac{\overline{Z}_0}{\sin \beta_a l} + \dfrac{n^2}{\omega C_0} & \dfrac{\overline{Z}_0}{\tan \beta_a l} + \dfrac{n^2}{\omega C_0} & \dfrac{n}{\omega C_0} \\[2mm] \dfrac{n}{\omega C_0} & \dfrac{n}{\omega C_0} & \dfrac{1}{\omega C_0} \end{bmatrix} \begin{bmatrix} v_1 \\ v_2 \\ I_3 \end{bmatrix} \tag{8.51}$$

resulting in the equivalent circuit shown in Fig. 8.7, called the Mason equivalent circuit for the crossed-field model.[36] With the negative capacitance shorted, the equivalent circuit in Fig. 8.5 can be used to derive the T_m^0 matrix, which is given by

$$
[T_m^0] = \begin{bmatrix}
\cos\gamma & j\overline{Z}_0\sin\gamma & n(1-\cos\gamma) & 0 \\
j\dfrac{\sin\gamma}{\overline{Z}_0} & \cos\gamma & -jn\dfrac{\sin\gamma}{\overline{Z}_0} & 0 \\
0 & 0 & 1 & 0 \\
-jn\dfrac{\sin\gamma}{\overline{Z}_0} & n(1-\cos\gamma) & j\omega C_0(1+s) & 1
\end{bmatrix}
\tag{8.52}
$$

where s and γ are defined by Eqs. (8.39) and $\overline{Z}_0$, n, and C_0 by Eqs. (8.48), (8.49), and (8.50), respectively. The coupling constant κ is again determined using Eq. (8.39). The T_m^∞ matrix is obtained from Eq. (8.52) by setting $n = C_0 = 0$. The transfer matrix for the entire transducer is again determined by Eq. (8.41), where the value of n in the expressions for adjacent T_m^0 matrices must alternate in sign. The conditions given by Eq. (8.42) apply here as well. When applied to SAW devices, Fig. 8.8, again $C_0/w \simeq C_s'$ and $\kappa \simeq K_S$. The expressions for the mechanical impedances and propagation constants for the electroded and unelectroded regions remain the same. In the SAW device consisting of identical transmitting and receiving transducers, Fig. 8.1(a), the insertion loss of the device can be found by reciprocity.[16]

In the analysis of SAW IDT, other equivalent circuits are also being used as alternatives to the Mason equivalent circuit shown in Fig. 8.7.[8,37,38] Further advancements in this area are the scattering matrix representation with a signal-flow graph technique[39] and the capacitive circuit model based on the electrostatic field equations.[40]

In some SAW devices the presence of the thin metallic film used for fabrication of IDT must be also accounted for.[41] Equivalent circuits[42,43] have been developed to take this effect into account.

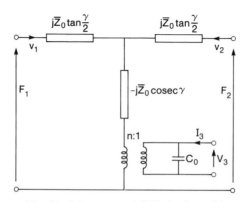

8.7. The Mason crossed-field circuit model.

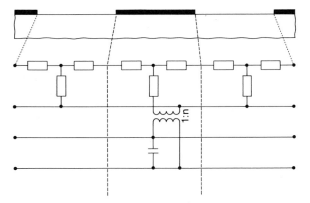

Figure 8.8.   The modeling of an IDT using the Mason crossed-field circuit model.

8.4   INPUT ADMITTANCE AND MATCHING

From Eqs. (10.30) and (10.31) the input conductance and susceptance of an IDT consisting of N, $2L$-long sections is given by

$$G(\omega) \simeq G_0^{(N)} \frac{\sin^2 X_N}{X_N^2} \qquad (8.53)$$

and

$$B(\omega) \simeq G_0^{(N)} \frac{(\sin 2X_N - 2X_N)}{2X_N^2} + \omega C_T \qquad (8.54)$$

where

$$G_0^{(N)} = 8f_0 K_S^2 C_s N^2 w \qquad (8.21)$$

$$X_N = \frac{N\pi(f - f_0)}{f_0} \qquad (8.55)$$

$$K_S^2 = \frac{2\Delta v}{v_\infty} \qquad (8.34)$$

and

$$C_T = NC_s w \qquad (8.56)$$

where C_T is the total static capacitance of the transducer, $C_s = 2C_s'$, and $N = M/2$. Thus the input admittance of a SAW IDT consisting of N, $2L$-long sections can be represented by the equivalent circuit shown in Fig. 8.9, where $B_a(\omega) = B(\omega) - \omega C_T$. The presence of the capacitance C_T at the input makes the conversion of electric to acoustic signals inefficient. A series inductance L_0 is usually employed to resonate the capacitance C_T at the resonant frequency. Since the matching procedure inevitably affects the bandwidth of the transducer, two Q factors are defined. The acoustic Q factor Q_a is obtained from Eq. (8.53) as the reciprocal of the relative bandwidth [see Eq. (3.35), for instance] and is given by

$$Q_a \simeq N \tag{8.57}$$

and the electric Q factor Q_e as

$$Q_e = \frac{\omega_0 C_T}{G_0^{(N)}} = \frac{\pi}{4K_S^2 N} \tag{8.58}$$

The total bandwidth of the transducer will be controlled by the larger of the two Q factors. Since Q_e is inversely proportional to N and Q_a is directly proportional to N, in order to maximize the bandwidth it follows that we must have $Q_e \simeq Q_a$, or

$$N_0^2 = \frac{\pi}{4K_S^2} \tag{8.59}$$

In the design of SAW devices, high conversion efficiency and good impedance matching are generally the objectives. The basic parameters to be determined in the design of a uniform IDT are the periodicity of the transducer $2L$, the electrode width d, the number of $2L$-long sections N, and the transducer aperture w.

The electrode width is usually chosen to be one-fourth of the transducer periodicity, $d = L/2$, and such transducers are referred to as $\lambda/4$ transducers. If it is desired to reduce the in-band reflections, $d = L/4$, and such transducers are referred to as double (split) electrode transducers or $\lambda/8$ transducers.[44, 45] The latter choice may cause difficulties associated with the production of thin lines[46] at high frequencies. Other types of transducers such as the "dogleg" structure[47] are also used to change the impedance level at the input.

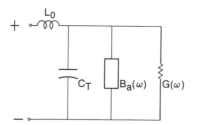

Figure 8.9. The equivalent circuit of the input admittance of an IDT with tuning inductance L_0.

Consider a problem of determining the characteristics of an IDT operating at $f_0 = 100$ MHz with SAW propagating along the Z axis of a Y-cut LiNbO$_3$ substrate. In this case, $v_\infty = 3488$ m/s and $v_0 = 3404$ m/s. The effective SAW velocity in the transducer region is $v_T = 2v_0 v_\infty/(v_0 + v_\infty) = 3445$ m/s. The transducer periodicity is given by

$$2L = \frac{v_T}{f_0} = 34.45 \ \mu m$$

and the electrode linewidth as

$$d = aL = \frac{L}{2} = 8.6 \ \mu m$$

The number of $2L$-long sections N_0 for optimum conversion bandwidth is obtained from Eq. (8.59) as

$$N_0 = \frac{\sqrt{\pi}}{2K_S} \simeq 4.5$$

where the value for K_S has been obtained from Table 11. If it is desired to match the transducer input impedance at resonant frequency to $R_0 = 50 \ \Omega$, the series inductance L_0 is used to tune out the static capacitance $L_0 = 1/\omega_0^2 C_T$, and the transducer aperture w is determined from

$$R_0 = j\omega_0 L_0 + \frac{1}{G_0^{(N)} + j\omega_0 C_T}$$

Since $G_0^{(N)} \ll \omega_0 C_T$, it follows that

$$\frac{w}{2L} = \frac{2K_S^2}{\pi^2 v_T R_0 C_s}$$

From Eqs. (8.37) and (8.37a) for $a = 0.5$, it is found that $K(k)/K'(k) \simeq 1$ and $C_s = 2C_s' \simeq 455$ pF/m. It follows that $w/2L \simeq 124$, $C_T = N_0 C_s w \simeq 9.2$ pF, and $L_0 \simeq 0.3 \ \mu H$.

Larger bandwidths can be obtained by using either dispersive (chirp) SAW transducers[47] or by series connection of several uniform SAW transducers,[48] each operating at a different center frequency (array transducers). Another commonly used method for obtaining wider bandwidths is to decrease the electrical Q factor Q_e which also causes an increase in the transducer conversion loss. The latter procedure can be illustrated by considering the equivalent circuit shown in Fig. 8.9. In the vicinity of resonant frequency, the radiation susceptance $B_a(\omega)$ and the radiation conductance $G_0^{(N)}(\omega)$ are much smaller than the static susceptance ωC_T. The input impedance of the transducer can be approximated by

$$Z_{in} \simeq R_a(\omega) + \frac{1}{j\omega C_T}$$

where

$$R_a(\omega) = \frac{G_0^{(N)}}{\omega^2 C_T^2}.$$

The transducer conversion efficiency can be defined as the ratio of the power available from the generator, $V_g^2/4R_g$, and the power delivered to $R_a(\omega)$. For a transducer tuned with inductance L_0 (Fig. 8.9) and driven by the generator of voltage V_g and generator impedance R_g, the conversion loss, taking into account the bidirectionality, is found as

$$L_c = \frac{(R_g + R_a)^2 + (\omega L_0 - 1/\omega C_T)^2}{2R_g R_a}$$

$$= \frac{(R_g + R_a)^2}{2R_g R_a}\left[1 + Q_L^2\left(\frac{\omega}{\omega_0} - \frac{\omega_0}{\omega}\right)^2\right]$$

where Q_L is the loaded Q factor given by

$$Q_L = \frac{1}{\omega_0 C_T(R_g + R_a)}$$

For $R_g = R_a$ and at the resonant frequency, the conversion loss, $10\log|L_c|$, is 3 dB and $Q_L = Q_e/2$. Therefore, in this case the electric (radiation) Q factor Q_e will dominate the bandshape of the transducer. In the case $R_g \gg R_a$, the conversion loss will increase to $10\log(R_g/2R_a)$, and the bandshape will be dominated by the loaded Q, $Q_L \simeq 1/\omega_0 C_T R_g$.

More than an octave bandwidth can be obtained[49] by using a $\lambda/4$ transmission line of properly chosen characteristic impedance. This procedure also causes an increase in the transducer conversion loss. The optimum relative bandwidth, Eq. (8.59), is equal to $1/N_0$. Thus piezoelectric material with large K_S will provide simultaneously large bandwidth and low insertion loss.

8.5 SIGNAL-PROCESSING FUNCTIONS

The impulse response and the transfer function of an IDT consisting of M, L-long sections, Eqs. (8.30) and (8.29), can be used to deduce the overall transfer function and impulse response of the device consisting of two identical transducers as shown in Fig. 8.1(a). The transfer function (frequency response) of the device is the product of the transfer functions of its transducers. Alternatively, the impulse response of the device is the convolution of its two transducer impulse responses. In Eq. (8.29) the expression $M\pi f/2f_0$ is equal to $NL\omega/v_R$, where $l = 2LN$ is the total length of the transducer. It follows that the phase center of a uniform transducer is at the cross-sectional plane of symmetry normal to the z axis, and the overall transfer function of the device shown in Fig. 8.1(a) will be

$$H_T(\omega, z) = \pm 4K_S^2 f_0 C_s w \overline{Z}_R N^2 \frac{\sin^2 X_N}{X_N^2} \exp\left(\frac{-jb}{v_R}\right) \tag{8.60}$$

where $b = 2NL + z$ is the separation between the phase centers of the transducers and b/v_R is the total input-to-output delay. The measured $20\log|H_T|$ for a device centered at $f_0 = 105$ MHz is shown in Fig. 8.10(a). Various

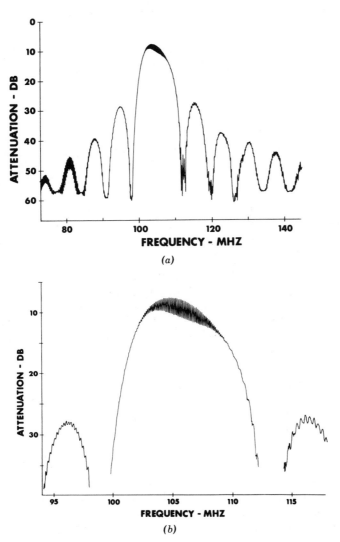

Figure 8.10. (a) The measured $20\log|H_T|$, Eq. (8.60), for a device centered at $f_0 \simeq 105$ MHz. (b) The in-band triple-transit echo.

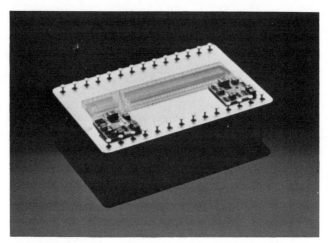

Figure 8.11. A SAW bandpass filter with input and output amplifier. (*Courtesy of Anderson Laboratories, Inc., Bloomfield, CT.*)

spurious effects such as bulk-mode generation, triple-transit echo,[†] and losses cause departure from the ideal $(\sin^2 x)/x^2$ response. The in-band triple transit echo,[50] shown in Fig. 8.10(*b*), is an important parameter in the design of precision SAW filters.[45] Since each transducer is a bidirectional device, there will be a 3-dB loss at the input and, by reciprocity, another 3-dB loss at the output of the device shown in Fig. 8.1(*a*). The difference between 10-dB attenuation in Fig. 8.10 and 6-dB loss due to bidirectionality can be accounted for by propagation, diffraction, and beam-steering losses[51] and bulk-wave generation losses.[27] The losses due to bidirectionality can be eliminated by employing unidirectional SAW transducers.[52] A SAW bandpass filter with input and output amplifiers is shown in Fig. 8.11.

In the introduction, the transfer function, impulse response, and expressions for the electric field and displacement were derived in terms of the δ-function model, a model used for the design of various SAW devices.[16, 53, 54] Other models can also be used. For instance, for the sine model, instead of Eq. (8.13), the equation

$$E_y = E_y^0 \sin[\beta_0(z - tv_R)][\chi(z - tv_R) - \chi(z - tv_R - L)] \quad (8.61)$$

is used with $\beta_0 = \pi/L$ for each electrode. Therefore each electrode is represented by one-half spatial period of the sine function occupying an interval L/v_R in time, or a linewidth L in space, Figs. 8.12(*b*) and (*c*), respectively. Other representations in sine-function modeling are also possible.[29]

[†]The reflection problem in electroded arrays was first considered by S. G. Joshi and R. M. White.[10]

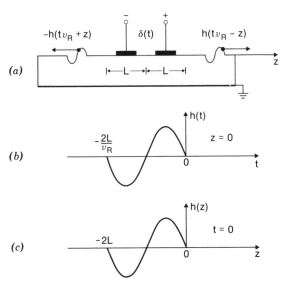

(a)

$-h(t v_R + z)$ $\delta(t)$ $h(t v_R - z)$

$|{\leftarrow} L {\rightarrow}|{\leftarrow} L {\rightarrow}|$

(b)

$h(t)$ $z = 0$

$-\dfrac{2L}{v_R}$

(c)

$h(z)$ $t = 0$

$-2L$

Figure 8.12. (*a*) A two-electrode transducer under impulse excitation, and the corresponding impulse response (*b*) at $z = 0$ and (*c*) at $t = 0$ using sine-function modeling.

For instance, the impulse response outside the transducer region of a transducer consisting of N, $2L$-long sections, Fig. 8.13(*a*), for the wave propagating in the $+z$ direction can be written as[†]

$$h(t, z) = -2K\Pi_H(t, z)\left(\overline{Z}_R C_s w f_0^3\right)^{1/2}\sin(\omega_0 t - \beta_0 z), \qquad z > 0 \qquad (8.62)$$

(a)

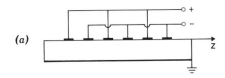

(b)

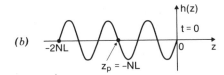

$h(z)$ $t = 0$

$-2NL$ $z_p = -NL$

(c)

$h(t)$ $z = 0$

$-NT$

Figure 8.13. (*a*) A transducer consisting of N, $2L$-long sections. (*b*) The impulse response at $t = 0$ with phase center at $z_p = -NL$. (*c*) The impulse response at $z = 0$.

261

where

$$\Pi_H(t, z) = \chi\left(t - \frac{z}{v_R}\right) - \chi\left(t - \frac{z}{v_R} - NT\right) \tag{8.63}$$

$\omega_0 = v_R\beta_0$, $T = 1/f_0$, and $l = 2LN = Nv_RT$ is the total length of the transducer. The corresponding transfer function is given by

$$\tilde{H}(\omega, z) = \xi_N K(\bar{Z}_R C_s w f_0)^{1/2} N \frac{\sin X_N}{X_N} \exp\left[-j\omega\left(\frac{N}{2f_0} + \frac{z}{v_R}\right)\right] \tag{8.64}$$

From the last equation, the phase center z_p of the array is found by setting the phase equal to zero, resulting in $z_p = -NL$, as shown in Fig. 8.13(b).

The electric field, Eq. (10.33), can now be written as

$$E(t, z) = \frac{-V_0 K}{L}(\bar{Z}_R C_s w f_0^3)^{1/2} \Pi_H(t, z)\sin(\omega_0 t - \beta_0 z) \tag{8.65}$$

where one can either set $t = 0$ and model the fields in space $-2NL \leqslant z \leqslant 0$ (Fig. 8.13(b)) or set $z = 0$ and model the fields in time $-NT \leqslant t \leqslant 0$ (Fig. 8.13(c)). Another important point is that $\Pi_H(t, z)$ and $(\sin X_N)/X_N$ in Eqs. (8.62) and (8.64) represent a Fourier transform pair. This fact applies also to many other classes of transducer configurations and is extensively used in synthesis of SAW filters.[29, 16]

The IDT considered so far has alternate electrodes driven by electrical signals 180° out of phase. It is also possible to phase the electrodes so that the spatial phase distribution corresponds to a prescribed sequence of phase shifts. For instance, the conventional IDT shown in Fig. 8.14(a) produces a two-cycle impulse response with no discontinuity in phase. A transducer with a phase reversal, Fig. 8.14(b), produces a phase discontinuity of π. Thus it is possible to perform a phase coding in the following way:

Logic state	Phase
0	0
1	π

$$\tag{8.66}$$

which forms the basis for biphase-coded SAW devices.[55] The corresponding potentials and fields computed for a finite SAW device[33] are shown in Figs. 8.15(a) and (b).

As an example, consider the problem of designing a SAW device with phase coding given by the sequence

$$1 \quad 0 \quad 0 \quad 1 \quad 0 \quad 1 \quad 1$$

[†] The symbol Π_H denotes, within the context of this chapter, the difference between two step (Heaviside) functions and should be distinguished from the Π used previously for the Poynting vector.

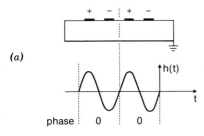

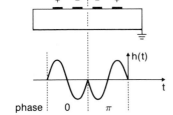

(a)

(b)

Figure 8.14. The illustration of the concept of phase reversal. (*a*) A uniform transducer. (*b*) A uniform transducer with phase reversal.

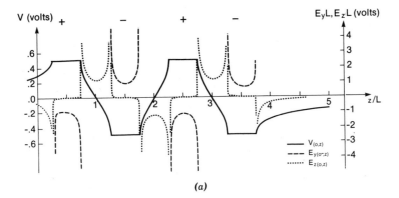

(a)

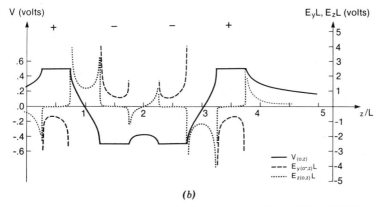

(b)

Figure 8.15. (*a*) The computed fields of the device shown in Fig. 8.14(*a*). (*b*) The computed fields of the device shown in Fig. 8.14(*b*).

as shown in Fig. 8.16(a), where T is the code duration and τ is the chip duration, $T = 7\tau$. The required impulse response, taking, for instance, a single cycle per chip duration (other choices are also possible), is easily modeled as shown in Fig. 8.16(b), and the SAW device capable of delivering the impulse response is shown in Fig. 8.16(c).

The amplitude and phase response of the transducer can be shaped[29, 45, 56] by using the various weighting (apodization) techniques shown in Fig. 8.17. Amplitude apodization is a technique where the overlap between electrodes of opposite polarity (the active region) changes along the transducer length. This technique can be used for shaping the magnitude of the transfer function[16] while keeping the phase essentially linear. Amplitude apodization can also be used for phase shaping.[57] The synthesized amplitude and impulse response, using δ-function modeling, of a SAW bandpass filter are shown in Figs. 8.18(a) and 8.19(a), respectively, and the corresponding experimental results in Figs. 8.18(b) and 8.19(b).

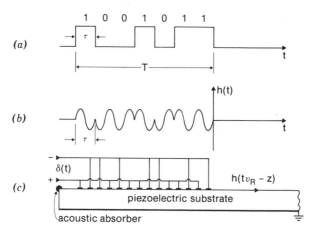

Figure 8.16. Illustration of the phase coding technique. (a) The required code sequence. (b) The corresponding impulse response at $z = 0$. (c) The SAW device capable of delivering the impulse response shown in (b).

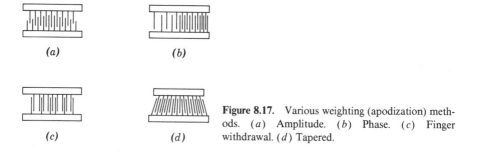

Figure 8.17. Various weighting (apodization) methods. (a) Amplitude. (b) Phase. (c) Finger withdrawal. (d) Tapered.

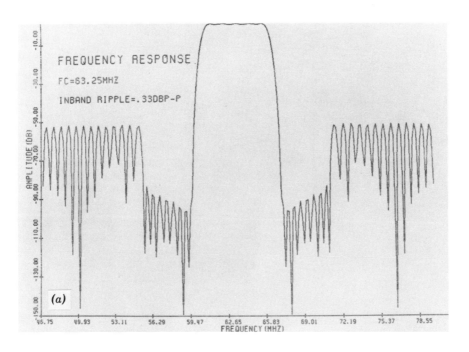

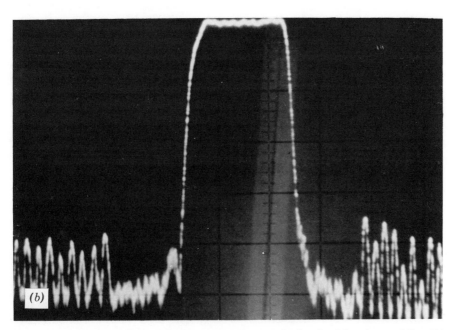

Figure 8.18. (*a*) The synthesized SAW bandpass filter amplitude using δ-function modeling. (*b*) The measured amplitude response.

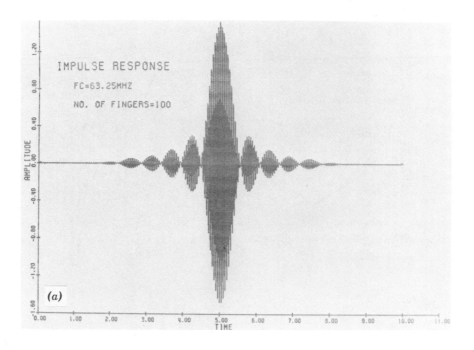

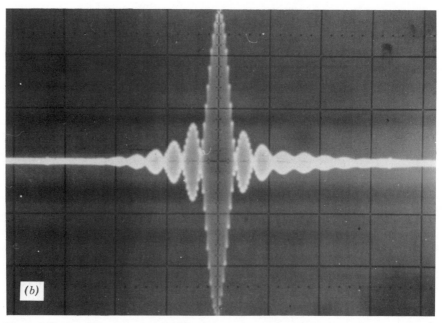

Figure 8.19. (*a*) The synthesized SAW bandpass filter impulse response having he amplitude characteristic shown in Fig. 8.18(*a*). (*b*) The measured impulse response.

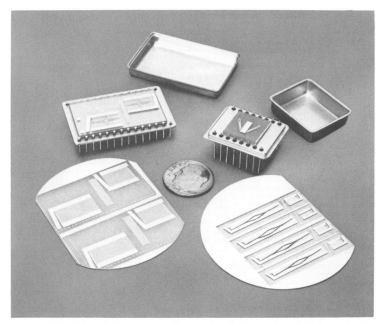

Figure 8.20. Several SAW devices. In particular the device at the top left uses a multistrip coupler[61] to suppress the bulk mode generation. (*Courtesy of Anderson Laboratories, Inc., Bloomfield, CT.*)

Phase weighting is used to shape the phase of the transfer function while keeping the magnitude essentially constant. The most important application of this technique is in pulse-compression (chirp) filters.[29,45,47] The withdrawal weighting technique,[58,59] Fig. 8.17(*c*), consists in selective withdrawal of electrodes to equate the number of electrodes with the prescribed weighting function. Tapered weighting[60] is another technique useful in producing flat bandpass filters. Several weighting techniques can be combined in a single transducer to achieve the desired response. Several SAW devices are shown in Fig. 8.20. In particular, the device at the top left uses a multistrip coupler[61] to suppress the bulk mode generation.

The SAW transducers can also be used for efficient generation of shallow bulk waves,[62,63] also called surface-skimming bulk waves, and bulk waves.[27] The bulk power density, as a function of the angle θ radiated into the substrate from an eight-electrode transducer, is shown in Fig. 8.21. The radiation patterns have been computed at radius $R = 2$ mm from the phase center of the transducer, made on Y-cut lithium niobate, for propagation of SAW along the Z-axis with $L = 17.44$ μm, $a = 0.5$, and $f_0 = 100$ MHz. For the same transducer, the interchange of power between the SAW and the bulk longitudinal (L) and shear (S) modes, termed the *mode coupling*, is shown as a function of frequency in Fig. 8.22. It is seen that the three modes are always present and that the shear wave generation is predominant.

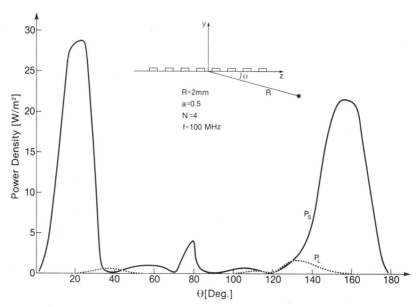

Figure 8.21. The bulk wave power density as a function of angle θ radiated from $N = 4$ uniform transducer on YZ lithium niobate. P_S and P_L are the shear and the longitudinal power densities, respectively. These were computed at $R = 2$ mm from the phase center of the transducer. The parameters are $L = 17.44$ μm, $a = 0.5$, $f_0 = 100$ MHz.

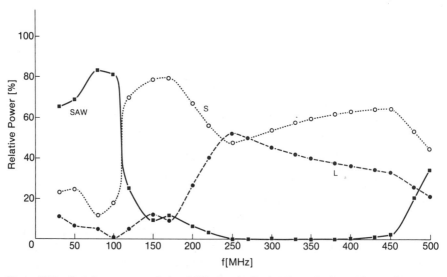

Figure 8.22. Relative powers of the SAW, longitudinal (L), and shear (S) modes versus frequency for the transducer shown in the inset of Fig. 8.21.

If the propagation surface of a SAW device is ideally smooth, the propagation losses, as explained in Chap. 4, are due to the air-loading and viscous mechanisms. The propagation losses α [Np/m] are given by[3]

$$\alpha(f) = Af^2 + Bf$$

where f is in GHz. The values of the constants A and B for several piezoelectric materials are given in Table 12.

Another important class of SAW devices are the reflection grating filters[41] and resonators,[64] which consist of an array of etched grooves on the surface of a piezoelectric plate. The SAW resonators made with this technology possess unloaded Q factors ranging from 5×10^3 at a frequency of about 1 GHz up to about 10^5 at a frequency of 0.1 GHz.[65-67] A 100-MHz SAW resonator is shown in Fig. 8.23.

Still another important group of SAW devices are the SAW layered devices, usually made of sputtered semiconductor piezoelectrics. They provide a bridge between SAW technology and the semiconductor microcircuit technology. Presently the most important approach is the ZnO on Si method[68,69] with a few attempts to use GaAs.[70] The SAW properties of piezoelectric semiconductors[71-73] and silicon[74] are known.

The pertinent SAW velocities for ZnO half-space are[73] $v_\infty = 2681$ m/s and $v_0 = 2668$ m/s, resulting in the coupling factor $K_S^2 = 0.0097$. The value of K_S can be increased considerably when layered structures such as ZnO on oxidized silicon (SiO_2) on Si are used.[68,69] Such structures allow the propagation not only of Rayleigh waves but also of Sezawa waves.[20] The configuration, consisting of Si-SiO_2 onto which IDT transducers are deposited and then ZnO is sputtered, gives the maximum value[68] of $K_S^2 = 0.028$ for $hk_R = 2.9$, where h is the thickness of the ZnO layer and k_R is the SAW wave number. The ZnO films can be deposited by RF magnetron sputtering.[75] Layered devices are finding increasing application in convolvers and correlators.[76,77]

TABLE 12   Constants of the Viscous and Air Loading Losses for Some Materials

Material	Orientation	A, Np/GHz^2	B, Np/GHz
LiNbO$_3$	Y plate, Z propagation	29	6.27
Bi$_{12}$GeO$_{20}$	001 plate, 110 propagation	99	13
Quartz	ST plate, X propagation	106	7
GaAs	110 plate, X propagation	156	13

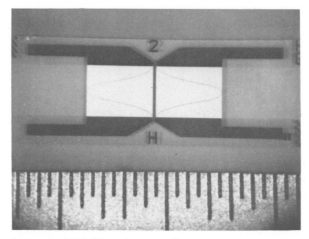

Figure 8.23. A two-port 100-MHz surface acoustic wave resonator on quartz. The apodization patterns of the input and output transducers (bright central areas) are clearly visible, although the individual 8-μm wide electrodes are not. The dark areas along the edges are bonding pads. Light diffracted from the arrays of 300-nm deep grooves makes them visible at the ends of the device. Scale: 0.5 mm/division. (*Courtesy of W. R. Shreve, Hewlett-Packard Co., Palo Alto, California.*)

Another layered device is the SAW amplifier[78] first demonstrated in 1967 with CdS. The SAW amplifier employs the traveling-wave interaction principle already described in Sect. 5.9. A monolithic amplifier[79] for continuous wave (CW) operation at room temperature, consisting of a thin film of InSb, has exhibited 23-dB gain at a frequency of 340 MHz.

Magnetostatic wave devices[80, 81] have recently[82] emerged as very promising for applications in the 1 to 10-GHz range. The magnetostatic surface waves have typical acoustic velocities of 3×10^5 m/s. They are 100 times faster than SAW on piezoelectric substrates and allow sufficient spatial resolution for fabrication of meander-line transducers in the 10-GHz range and beyond.

REFERENCES

1. R. M. White, "Surface Elastic Waves," *Proc. IEEE*, **58**(8), 1238–1276 (1970).

2. M. S. Kharusi and G. W. Farnell, "Diffraction and Beam Steering for Surface-Wave Comb Structures on Anisotropic Substrates," *IEEE Trans. Sonics Ultrason.*, **SU-18**, 35–42 (1971).

3. T. L. Szabo and A. J. Slobodnik, Jr., "The Effect of Diffraction on the Design of Acoustic Surface Wave Devices," *IEEE Trans. Sonics Ultrason.*, **SU-20**, 240–251 (1973).

4. I. M. Mason, "Anisotropy, Diffraction Scaling, Surface Wave Lenses, and Focusing," *J. Acoust. Soc. Am.*, **53**, 1123–1128 (1973).

5. G. W. Farnell, "Types and Properties of Surface Waves," in *Acoustic Surface Waves* (A. A. Oliner, Ed.), Springer-Verlag, New York, 1978.

6. B. A. Auld, *Acoustic Fields and Waves in Solids*, Vol. II, Wiley, 1973, Chap. 10.

7. M. J. Zuliani and V. M. Ristic, "Field Theory for Interdigital Transducers," *J. Appl. Phys.*, **49**(6), 3018–3024 (1978).

8. W. R. Smith, "Basics of the SAW Interdigital Transducer," *Wave Electron.*, **2**, 25–63 (1976).

9. G. A. Coquin, and H. F. Tiersten, "Analysis of the Excitation and Detection of Piezoelectric Surface Waves in Quartz by Means of Surface Electrodes," *J. Acoust. Soc. Am.*, **41**, 921–939 (1966).

10. S. G. Joshi and R. M. White, "Excitation and Detection of Surface Elastic Waves in Piezoelectric Crystals," *J. Acoust. Soc. Am.*, **46**, 17–27 (1968).

11. J. J. Campbell and W. R. Jones, "A Method for Estimating Optimal Crystal Cuts and Propagation Directions for Excitation of Piezoelectric Surface Waves," *IEEE Trans. Sonics Ultrason.*, **SU-15**, 209–218 (1968).

12. H. Engan, "Excitation of Elastic Surface Waves by Spatial Harmonics of Interdigital Transducers," *IEEE Trans. Electron. Devices*, **ED-16**(12), 1014–1017 (1969).

13. W. R. Smith et al., "Analysis of Interdigital Surface Wave Transducers by Use of an Equivalent Circuit Model," *IEEE Trans. Microwave Theory Tech.*, **MTT-17**, 856–864 (1969).

14. K. A. Ingebrigtsen, "Surface Waves in Piezoelectrics," *J. Appl. Phys.*, **40**, 2681–2686 (1969).

15. B. A. Auld and G. S. Kino, "Normal Mode Theory for Acoustic Waves and its Application to Interdigital Transducers," *IEEE Trans. Electron. Dev.*, **ED-18**, 898–908 (1971).

16. R. H. Tancrell and M. G. Holland, "Acoustic Surface Wave Filters," *Proc. IEEE*, **59**, 393–409 (1971).

17. P. R. Emtage, "Self-Consistent Theory of Interdigital Transducers," *J. Acoust. Soc. Am.*, **51**, 1142–1155 (1972).

18. A. K. Ganguly and M. O. Vassell, "Frequency Response of Acoustic Surface Wave Filters. I," *J. Appl. Phys.*, **44**, 1072–1085 (1973).

19. K. Blotekjaer, K. A. Ingebrigtsen, and H. Skeie, "Acoustic Surface Waves in Piezoelectric Materials with Periodic Metal Strips on the Surface," *IEEE Trans. Electron. Dev.*, **ED-20**, 1139–1146 (1973).

20. B. A. Auld, *Acoustic Fields and Waves in Solids*, Vol. II, Wiley, New York, 1973, Chaps. 10, 12, and 13.

21. W. S. Goruk, P. J. Vella, G. I. Stegeman, and V. M. Ristic, "An Exponential Coupling Theory for Interdigital Transducers," *1974 Ultrason. Symp. Proc.*, 402–405 (1974).

22. R. S. Wagers, "Evaluation of Finger Withdrawal Transducer Admittances by Normal Mode Analysis," *IEEE Trans. Sonics Ultrason.*, **SU-25**, 85–92 (1978).

23. E. Dieulesaint and D. Royer, *Elastic Waves in Solids*, Wiley, New York, 1980, Chap. 7. (Original published in French by Masson, Paris, 1974.)

24. C. F. Vasile, "Radiation from a Line Charge Source Located Near an Air–Piezoelectric Interface," *Proc. IEEE*, **64**, 636–639 (1976).

25. R. F. Milsom, N. H. C. Reilly, and M. Redwood, "Analysis of Generation and Detection of Surface and Bulk Acoustic Waves by Interdigital Transducers," *IEEE Trans. Sonics Ultrason.*, **24**, 147–166 (1977).

26. K. Yashiro and N. Goto, "Analysis of Generation of Acoustic Waves on the Surface of a Semi-Infinite Piezoelectric Solid," *IEEE Trans. Sonics Ultrason.*, **25**, 146–153 (1978).

27. V. M. Ristic and A. M. Hussein, "Radiation Patterns of Bulk Modes from a Finite Interdigital Transducer on $Y–Z$ LiNbO$_3$," *IEEE Trans. Sonics Ultrason.*, **SU-27**(1), 15–21 (1980).

28. V. M. Ristic, "Modeling of the Impulse Response in SAW Transducers," *IEEE Trans. Sonics Ultrason.*, **SU-26**(4), 266–271 (1979).

29. C. S. Hartmann, D. T. Bell, and R. C. Rosenfeld, "Impulse Model Design of Acoustic Surface-Wave Filters," *IEEE Trans. Microwave Theory Techniq.*, **MTT-21**, 162–175 (1973).

30. A. J. Slobodnik, Jr., "Surface Acoustic Waves and SAW Materials," *Proc. IEEE*, **64**(5), 581–595 (1976).

31. A. J. Slobodnik, Jr., E. D. Conway, and R. T. Delmonico, *Microwave Acoustics Handbook*, Vol. 1A, AFCRL, TR-73-0597, 1973, unpubl.

32. V. M. Ristic and M. J. Zuliani, "Redefinition of the Coupling Coefficient for Surface-Acoustic-Wave Interdigital Transducer," *J. Appl. Phys.*, **49**(11), 5672 (1978).

33. V. M. Ristic and A. M. Hussein, "Surface Charge and Field Distribution in a Finite SAW Transducer," *IEEE Trans. Microwave Theory Techniq.*, **MTT-27**(11), 897–901 (1979).

34. G. W. Farnell, I. A. Cermak, P. Silvester, and S. K. Wong, "Capacitance and Field Distributions for Interdigital Surface-Wave Transducers," *IEEE Trans. Sonics Ultrason.*, **SU-17**, 188–195 (1970).

35. E. K. Sitting, "Transmission Parameters of Thickness-Driven Piezoelectric Transducers Arranged in Multilayer Configurations," *IEEE Trans. Sonics Ultrason.*, **SU-14**, 167–174 (1967).

36. W. P. Mason, *Electromechanical Transducers and Wave Filters*, 2nd ed., Van Nostrand, New York, 1948.

37. R. Krimholtz, "Equivalent Circuits for Transducers Having Arbitrary Asymmetrical Piezoelectric Excitation," *IEEE Trans. Sonics Ultrason.*, 427–436 (1972).

38. R. Krimholtz, D. A. Leedom, and G. L. Matthaei, "New Equivalent Circuits for Elementary Piezoelectric Transducers," *Electron. Lett.*, **6**, 388–389 (1970).

39. R. C. Rosenfeld, S. H. Arneson, and T. F. O'Shea, "Analysis of Asymmetric Three-Transducer Configuration," *Proc. 1976 Ultrason. Symp.*, 682–685 (1976).

40. S. Datta and B. J. Hunsinger, "Analysis of Energy Storage Effects on SAW Propagation in Periodic Arrays," *IEEE Trans. Sonics Ultrason.*, **SU-27**, 333–341 (1980).

41. R. C. Williamson, "Reflection Grating Filters," in *Surface Wave Filters* (H. Matthews, Ed.), Wiley, New York, 1977.

42. V. M. Ristic, "The Equivalent Circuit of a Step Discontinuity in Surface Acoustic Wave Metallic Gratings," *J. Appl. Phys.*, **50**(11), 6751–6753 (1979).

43. R. C. M. Li and J. Melngailis, "The Influence of Stored Energy at Step Discontinuities on the Behavior of Surface-Wave Gratings," *IEEE Trans. Sonics Ultrason.*, **SU-22**, 189–198 (1975).

44. T. W. Bristol et al., "Application of Double Electrodes in Acoustic Surface Wave Device Design," *Proc. 1972 Ultrason. Symp.*, 343–345 (1972).

45. R. H. Tancrell, "Principles of Surface Wave Filter Design," in *Surface Wave Filters* (H. Matthews, Ed.), Wiley, New York, 1977.

46. H. I. Smith, "Surface Wave Device Fabrication," in *Surface Wave Filters* (H. Matthews, Ed.), Wiley, New York, 1977.

47. H. M. Gerard, "Surface Wave Interdigital Electrode Chirp Filters," in *Surface Wave Filters* (H. Matthews, Ed.), Wiley, New York, 1977.

48. L. T. Nguyen and C. S. Tsai, "Efficient Wideband Guided Wave Acousto-optic Bragg Diffraction Using Phased Surface Acoustic Wave Array in $LiNbO_3$ Waveguides," *Appl. Opt.*, **16**(5), 1297–1304 (1977).

49. T. M. Reeder and W. R. Sperry, "Broad-Band Coupling to High-Q Resonant Loads," *IEEE Trans. Microwave Theory Techniq.*, **MTT-20**(7), 453–458 (1972).

50. D. H. Hurlburt and E. L. Adler, "Group Delay Ripple Resulting from Multiple Reflections in SAW Devices," *Proc. IEEE*, **65**, 167 (1977).

51. A. J. Slobodnik, "A Review of Material Trade-offs in the Design of Acoustic Surface Wave Devices at VHF and Microwave Frequencies," *IEEE Trans. Sonics Ultrason.*, **SU-20**(4), 315–323 (1973).

52. C. S. Hartmann, W. S. Jones, and H. Vollers, "Wideband Unidirectional Surface Wave Transducers," *IEEE Trans. Sonics Ultrason.*, **SU-19**, 378–381 (1972).

53. C. Atzeni and L. Masotti, in *Acoustic Surface Wave and Acousto-optic Devices* (T. Kallard, Ed.), Vol. 4, Optosonic Press, New York, 1971, pp. 69–80.

54. E. Dieulesaint and D. Royer, *Elastic Waves in Solids*, Wiley, New York, 1980, Chap. 7.

55. D. T. Bell, Jr., and L. T. Claiborne, "Phase Coded Generators and Correlators," in *Surface Wave Filters* (H. Matthews, Ed.), Wiley, New York, 1977, Chap. 7.

56. R. F. Milsom, M. Redwood, and N. H. C. Reilly, "The Interdigital Transducer," in *Surface Wave Filters* (H. Matthews, Ed.), Wiley, New York, 1977, Chap. 2.

57. J. P. Reilly, C. K. Campbell, and M. K. Suthers, "The Design of SAW Bandpass Filters Exhibiting Arbitrary Phase and Amplitude Response Characteristics," *IEEE Trans. Sonics Ultrason.*, **SU-24**, 301–305 (1977).

58. C. S. Hartmann, "Weighting Interdigital Surface Wave Transducers by Selective Withdrawal of Electrodes," *Proc. 1973 Ultrason. Symp.*, 423–426 (1973).

59. K. R. Laker, E. Cohen, T. L. Szabo, and J. A. Pustaver, Jr., "Computer-Aided Design of Withdrawal-Weighted SAW Bandpass Filters," *IEEE Trans. Circuits Syst.*, **CAS-25**, 241–251 (1978).

60. A. P. van den Heuvel, "Use of Rotated Electrodes for Amplitude Weighting in Interdigital Surface-Wave Transducers," *Appl. Phys. Lett.*, **21**(6), 280–282 (1972).

61. F. G. Marshall, C. O. Newton, and E. G. S. Paige, "Theory and Design of the Surface Acoustic Wave Multistrip Coupler," *IEEE Trans. Sonics Ultrason.*, **SU-20**, 124–133 (1973).

62. G. W. Farnell and E. Akcakaya, "Field Distributions for Surface-Skimming Bulk-Wave Transducers," *IEEE Trans. Sonics Ultrason.*, **SU-26**, 202–205 (1979).

63. M. F. Lewis, "Surface Skimming Bulk Waves," *Proc. 1977 Ultrason. Symp.*, 744–752 (1977).

64. W. J. Tanski, "Surface Acoustic Wave Resonators on Quartz," *IEEE Trans. Sonics Ultrason.*, **SU-26**, 93–104 (1979).

65. P. S. Cross, "Properties of Reflected Arrays for Surface Acoustic Resonators," *IEEE Trans. Sonics Ultrason.*, **SU-23**, 255–262 (1976).

66. W. R. Shreve, "Surface Wave Two-Port Resonator Equivalent Circuit," *Proc. 1975 Ultrason. Symp.*, 295–298 (1975).

67. P. S. Cross, P. Rissman, and W. R. Shreve, "Microwave SAW Resonator Fabricated with Direct-Writing Electron-Beam Lithography," *Proc. 1980 Ultrason. Symp.*, 158–163 (1980).

68. L. P. Solie, "Piezoelectric Waves in Layered Substrates," *J. Appl. Phys.*, **44**, 619 (1973).

69. G. S. Kino and R. S. Wagers, "Theory of Interdigital Couplers on Non-Piezoelectric Substrates," *J. Appl. Phys.*, **44**, 1480–1488 (1973).

70. T. Grudkowski and C. F. Quate, "Acoustic-Readout of Charge Storage on GaAs," *Appl. Phys. Lett.*, **25**, 99–101 (1974).

71. C. C. Tseng and R. M. White, "Propagation of Piezoelectric and Elastic Surface Waves on the Basal Plane of Hexagonal Piezoelectric Crystals," *J. Appl. Phys.*, **38**, 4274–4280 (1967).

72. C. C. Tseng, "Elastic Surface Waves on Free Surface and Metallized Surface of CdS, ZnO, and PZT-4," *J. Appl. Phys.*, **38**, 4281–4284 (1967).

73. A. J. Slobodnik, R. T. Delmonico, and E. D. Conway, *Microwave Acoustic Handbook*, Vol. 2, ARCRL, TR-74-0536, 1974 (unpubl.).

74. R. G. Pratt and T. C. Lim, "Acoustic Surface Waves on Silicon," *Appl. Phys. Lett.*, **15**, 403–405 (1969).

75. T. Shiosaki, "High-Speed Fabrication of High-Quality Sputtered ZnO Thin Films for Bulk and Surface Wave Applications," *Proc. 1978 Ultrason. Symp.*, 100–110 (1978).

76. B. T. Khuri-Yakub and G. S. Kino, "A Monolithic Zinc-Oxide-on-Silicon Convolver," *Appl. Phys. Lett.*, **25**, 188–190 (1974).

77. F. C. Lo, R. L. Gunshor, and R. F. Pierret, "Monolithic (ZnO) Sezawa-Mode *pn*-Diode-Array Memory Correlator," *Appl. Phys. Lett.*, **34**, 725–726 (1979).

78. R. M. White, "Surface Elastic Wave Propagation and Amplification," *IEEE Trans. Electron Devices*, **ED-14**, 181–189 (1967).

79. L. A. Coldren and G. S. Kino, "CW Monolithic Acoustic Surface Wave Amplifier Incorporated in $\Delta v/v$ Waveguide," *Appl. Phys. Lett.*, **23**, 117–118 (1973).

80. R. W. Voltmer, R. M. White, and C. W. Turner, "Magnetostrictive Generation of Surface Elastic Waves," *Appl. Phys. Lett.*, **15**, 153–154 (1969).

81. W. R. Robbins, "Approximate Theory of Magnetoelastic Surface Wave Transducers on YIG," *IEEE Trans. Sonics Ultrason.*, **SU-26**, 230–234 (1979).

82. *Proc. 1980 Ultrason. Symp.*, Vols. 1 and 2, 1980. Several papers deal with various aspects of magnetostatic devices.

EXERCISES

1. Using the double δ-function model, derive expressions for the impulse response and the transfer function of a uniform transducer with M electrodes. Compare the results with Eqs. (8.29) and (8.30), and identify the transfer function of a single electrode. Compare the bandwidths of a single electrode and the transducer.

2. Compute the conductance of a uniform transducer having $M = 10$, L-long sections made on Y-cut LiNbO$_3$ crystal for wave propagation in the Z direction. Use $a = 0.5$, $f_0 = 100$ MHz, and $w = 1$ cm. Find the peak electric energy stored in C'_s if the transducer is operating with $V_0 = 50$ mV.

3. Using Eqs. (8.9), (8.10), and (8.33), derive an expression for the displacement u_z of a SAW generated by a uniform transducer having M electrodes.

4. (a) Find the transducer optimum relative bandwidth for the three materials listed in Table 11.
(b) From Fig. 8.10(a) or (b), find the numerical value of the relative bandwidth and the number N of $2L$-long sections.

5. Consider a uniform transducer consisting of M electroded sections. Often the crossed-field circuit model for the transducer is derived by assuming $l = L$ in $\beta_a l = \pi f/f_0$, in which case Eq. (8.41) is written as

$$[T] \simeq \prod_{m=1}^{M} [T_m^0]$$

(a) Show that the T_1^0 matrix of the crossed-field circuit model at synchronism, $f = f_0$ ($\gamma = \pi$), reduces to

$$[T_1^0] = \begin{bmatrix} -1 & 0 & 2n & 0 \\ 0 & -1 & 0 & 0 \\ 0 & 0 & 1 & 0 \\ 0 & 2n & j\omega_0 C_0 & 1 \end{bmatrix}$$

and that the transfer matrix of the transducer can be written as

$$[T] \simeq \begin{bmatrix} (-1)^M & 0 & 2Mn & 0 \\ 0 & (-1)^M & 0 & 0 \\ 0 & 0 & 1 & 0 \\ 0 & (-1)^{M+1}2Mn & j\omega_0 MC_0 & 1 \end{bmatrix}.$$

(b) Using the above equation and Eqs. (8.38) and (8.42), show that the electrical input admittance of the transducer at resonance is given by

$$Y_{in} = \frac{I_1}{V_1} = \frac{2}{\pi} M^2 \kappa^2 \omega_0 C_0 + j\omega_0 MC_0$$

Compare this result with Eq. (8.20).

6. Repeat Exercise 5, using the in-line circuit model. In this case,

$$[T] \simeq \prod_{m=1}^{M} [T_m^\infty]$$

and at resonance, $\gamma = \pi$,

$$[T_1^\infty] = \begin{bmatrix} -1 & -j4Z_0\kappa^2/\pi & 2n & 0 \\ 0 & -1 & 0 & 0 \\ 0 & 0 & 1 & 0 \\ 0 & 2n & j\omega_0 C_0 & 1 \end{bmatrix}$$

and

$$[T] = \begin{bmatrix} (-1)^M & (-1)^M j4MZ_0\kappa^2/\pi & 2Mn & 0 \\ 0 & (-1)^M & 0 & 0 \\ 0 & 0 & 1 & 0 \\ 0 & (-1)^{M+1}2Mn & j\omega_0 MC_0 & 1 \end{bmatrix}.$$

Show that the input admittance can be written as

$$Y_{in} = \frac{2M^2\kappa^2\omega_0 C_0}{\pi(1 + 4M^2\kappa^4/\pi^2)} + j\omega_0 MC_0\left(1 - \frac{4M^2\kappa^4}{\pi^2}\right)$$

which in the limit $(2M\kappa^2/\pi)^2 \ll 1$ becomes identical to the corresponding expression derived using the crossed-field circuit model. Express the last inequality in terms of the electrical Q factor, Eq. (8.58).

7. Derive an expression for the impulse response corresponding to the transfer function given by Eq. (8.60).

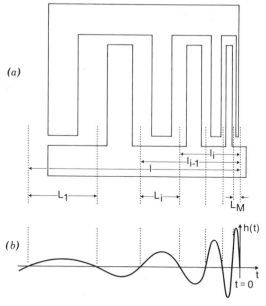

Figure 8.24.

8. Consider the impulse response at $z = 0$ of a phase-apodized transducer in Fig. 8.24(a). Show that the impulse, Fig. 8.24(b), can be written as

$$h(t) \simeq 2K_S \left(2\bar{Z}_R C_s' w\right)^{1/2} \sum_{i=1}^{M} (-1)^i \Pi_i(t) f_i^{3/2} \sin \omega_i \left(t + \frac{l_i}{v_R}\right)$$

where

$$\Pi_i(t) = \chi\left(t + \frac{l_{i-1}}{v_R}\right) - \chi\left(t + \frac{l_i}{v_R}\right)$$

and where $\omega_i = 2\pi f_i = \pi v_R/L_i$, $l_0 = l$, and $l_M = 0$.

9. Consider a dispersive transducer where the apodization function can be assumed continuous,

$$f(t) = \frac{1}{2\pi} \frac{d\phi}{dt} = f_0 + pt, \qquad -T \leqslant t \leqslant 0$$

(a) Derive an expression for the impulse response of the transducer using sine-function modeling.
(b) Find an expression for the electrode positioning law (see Sect. 3.4).
(c) Repeat the exercise using the functions given by Eqs. (3.82) and (3.83).

10. Show that the conclusions reached in Sect. 3.4 for nonuniform sampling of chirp meander-line transducers apply to SAW chirp transducers as well.

Chapter 9

Reflection and Refraction of Bulk Waves and the Coupling Prism

Due to the finite dimensions of acoustic devices, most design problems tend to be dominated by reflection and refraction phenomena at the boundaries. The boundary conditions for an intimate contact between two solids have been discussed in Chap. 1, and the boundary conditions for a solid–vacuum interface in Chap. 4. Other boundary conditions such as those existing between a solid and a liquid are also of interest in practical devices. In order to include this important case in our considerations, in this chapter we will consider the propagation of acoustic waves in liquids. To simplify the presentation, the mechanisms of absorption, diffraction, and scattering of acoustic waves in liquids[1] will not be discussed, however.

9.1 ACOUSTIC WAVES IN LIQUIDS

With the exception of liquid crystals, all nonviscous liquids support the propagation of longitudinal bulk waves only. In this case, the coefficient of rigidity is equal to zero, $\mu = 0$, and the wave equation (2.57) reduces to

$$\rho_m \frac{\partial^2 u_i}{\partial t^2} = \lambda \frac{\partial^2 u_l}{\partial x_i \partial x_l} \tag{9.1}$$

The hydrostatic pressure p, defined as $p = -T$, is obtained from Eq. (2.46) as

$$p = -T = \lambda \Delta \tag{9.2}$$

Equation (2.2) is reduced to

$$T_{ij} = -p\delta_{ij} \quad \text{or} \quad \overset{\leftrightarrow}{T} = -p\overset{\leftrightarrow}{I} \tag{9.3}$$

where $\overset{\leftrightarrow}{I}$ is the second-order identity tensor, and Eq. (2.23) becomes

$$[\nabla_{iJ}] \rightarrow \left[\delta_{ij}\frac{\partial}{\partial r_j}\right] \qquad (9.4)$$

where the operator on the right-hand side is given by a 3×3 matrix. Using the last three equations it follows from Eq. (2.20) that

$$\nabla \cdot \overset{\leftrightarrow}{T} = -\nabla p = \rho_m \frac{\partial \vec{v}}{\partial t} \qquad (9.5)$$

Equation (2.50) for dilatation can be written as

$$\Delta = \frac{\partial u_l}{\partial x_l} \qquad (9.6)$$

and when combined with Eq. (9.1), which is first differentiated with respect to x_i, yields

$$\rho_m \frac{\partial^2 \Delta}{\partial t^2} = \lambda \frac{\partial^2 \Delta}{\partial x_i^2} \qquad (9.7)$$

where $\partial^2/\partial x_i^2$ is the Laplacian operator. Thus in liquids the change of volume per unit volume propagates with velocity

$$v_l = \left(\frac{\lambda}{\rho_m}\right)^{1/2} = \frac{1}{(\kappa\rho_m)^{1/2}} \qquad (9.8)$$

where κ is termed the compressibility of the fluid, $\kappa = 1/K = 1/\lambda$, where K is the coefficient of incompressibility defined in Eq. (2.55), with $K \approx 2.2 \times 10^9$ N/m² for water.

In Chap. 2, the displacement potential ϕ, Eq. (2.66), has been used for solving the wave equation. In acoustics of fluids, the velocity potential φ is often used. The latter is defined as

$$\vec{v} = -\nabla\varphi \qquad \text{or} \qquad v_i = -\frac{\partial\varphi}{\partial x_i} \qquad (9.9)$$

Both potential functions ϕ and φ lead to identical results. In order to obtain a relation between the pressure and the velocity potential, Eq. (9.8) is differentiated with respect to t,

$$\frac{\partial\Delta}{\partial t} = \frac{\partial^2 u_l}{\partial x_l \partial t} = \frac{\partial v_l}{\partial x_l} = \nabla \cdot \vec{v} = -\nabla^2\varphi \qquad (9.10)$$

and since the potential φ also satisfies the wave equation

$$\kappa \rho_m \frac{\partial^2 \varphi}{\partial t^2} = \nabla^2 \varphi \tag{9.11}$$

it follows from Eqs. (9.2), (9.10), and (9.11) that

$$p = \rho_m \frac{\partial \varphi}{\partial t} \tag{9.12}$$

The wave equation (9.7) when combined with Eq. (9.2) can also be expressed in terms of pressure as

$$\nabla^2 p = \frac{1}{v_l^2} \frac{\partial^2 p}{\partial t^2} \tag{9.13}$$

The plane wave solution of this equation is well known and is left as an exercise at the end of the chapter.

If the wave has spherical symmetry, the pressure is a function of the radial distance r (and the time t) only, being independent of angular coordinates, and Eq. (9.13) is written as

$$\frac{1}{r^2} \frac{\partial}{\partial r} \left(r^2 \frac{\partial p}{\partial r} \right) = \frac{1}{v_l^2} \frac{\partial^2 p}{\partial t^2} \tag{9.14}$$

with a general solution for the outgoing wave

$$p = p_0 \frac{f(t - r/v_l)}{r} \tag{9.15}$$

The particle (RF) radial velocity v_r is obtained from Eq. (9.5) as

$$\frac{\partial p}{\partial r} = -\rho_m \frac{\partial v_r}{\partial t} \tag{9.16}$$

For harmonic waves of the form $\exp[j(\omega t - \beta r)]$ it follows that

$$p = p_0 \frac{e^{j(\omega t - \beta r)}}{r} \tag{9.17}$$

and

$$v_r = \frac{p}{\rho_m v_l} \left(1 + \frac{1}{j\beta r} \right) \tag{9.18}$$

where $\beta = \omega/v_l$. The complex Poynting vector is obtained from Eqs. (4.38), (9.17), (9.2), and (9.18), and is given by

$$\vec{\Pi} = \frac{p\vec{v}_r^{\,*}}{2} = \frac{|p|^2}{2\rho_m v_l}\left(1 - \frac{1}{j\beta r}\right)\hat{r} \tag{9.19}$$

Using Eq. (4.37) it can be shown that the imaginary part of the Poynting vector indicates the existence of energy storage into near fields, which in this case are localized, approximately, in the spherical region $\beta r < 1$. As seen from Eq. (4.43), the imaginary part does not contribute to the average power flow. In acoustics of fluids, the real part of the Poynting vector is termed *intensity* and is given by

$$\vec{I} = \text{Re}\{\vec{\Pi}\} = \text{Re}\left\{\frac{-\overleftrightarrow{T}\cdot\vec{v}^*}{2}\right\} = \text{Re}\left\{\frac{p\vec{v}^*}{2}\right\} \tag{9.20}$$

The average power is then obtained by applying Eq. (4.43), where A is a closed surface surrounding the source. In the radiating fields region, $\beta r \gg 1$, it follows from Eqs. (9.19) and (9.20) that $I \simeq \Pi$ and from Eq. (9.18) that $v_r \simeq p/\rho_m v_l$. The latter expression is used as the defining equation for the wave impedance

$$Z_l = \frac{p}{v_r} = \rho_m v_l \tag{9.21}$$

which coincides with the corresponding equation for plane waves.

As an example, consider a spherical source of radiation in a liquid, where the radius of the sphere is a harmonic function of time, $R(t) = a\sin\omega t$. It is assumed that the sphere is small at all frequencies, that is, $\beta a \ll 1$. The velocity of the surface of the sphere is therefore $v_R = dR/dt = v_0\cos\omega t$, where $v_0 = a\omega$. Since the sphere is in an intimate contact with the surrounding liquid, the particle radial velocity of the liquid, close to the sphere surface, is equal to the velocity v_0, that is, $v_r(a) = v_0$. Using Eqs. (9.17) and (9.18) for $\beta a \ll 1$, it follows that $p_0 = j\rho_m v_l\beta v_0 a^2$, and from Eq. (9.20) the intensity is found as

$$I = \frac{Z_l\beta^2 a^4 v_0^2}{2r^2} \tag{9.22}$$

It is seen that the intensity decreases with r^2 (which is termed *spherical divergence*), the average power being constant through any closed surface surrounding the source.

The strength[2] of the spherical source, q_S, is defined as the product of the velocity magnitude and the surface area, $q_S = v_0 4\pi a^2$. Using the latter expression, Eq. (9.17) can be written as

$$p_S = \left(\frac{j\rho_m v_l\beta}{4\pi r}\right)q_S e^{j(\omega t - \beta r)} \tag{9.23}$$

For a hemispherical source[†] of the same radius and velocity magnitude radiating into half-space, the strength would be $q_H = v_0 2\pi a^2$. Using symmetry considerations it is easy to show that the expression for pressure, p_H, in terms of the velocity amplitude for the hemispherical source, is identical to the corresponding equation for the spherical source. It follows that

$$\frac{p_H}{q_H} = 2\frac{p_S}{q_S} \qquad (9.24)$$

Besides the displacement potential defined in Chap. 2 and the velocity potential defined in this section, another potential termed *pressure potential* can also be defined. In the latter case the gradient of the potential is equal to the pressure, thus the name. The potential functions are routinely used in the studies of various reflection and refraction problems.[3] For instance, in the problems involving liquid/solid boundaries, velocity potentials defined in the solid as

$$\vec{v} = \nabla\phi + \nabla \times \vec{\psi} \qquad (9.25)$$

are often used and in the liquid the velocity potential

$$\vec{v} = \nabla\varphi' \qquad (9.26)$$

which differs in sign from the velocity potential, Eq. (9.9), previously defined. In this case Eq. (9.12) is replaced by

$$p = -\rho_m \frac{\partial\varphi'}{\partial t} \qquad (9.27)$$

Potential functions are not unique,[‡] and for historical reasons, the formulation given by Eqs. (9.9) and (9.12) is often used in the theory of acoustic radiation and diffraction, Chap. 10, while the formulation given by Eqs. (9.25) to (9.27) is often applied in problems involving reflection and refraction.

9.2 BOUNDARY CONDITIONS

Consider an interface between two materials defined by the plane $z = 0$. If the two materials are solids, then at the interface we must have the continuity of both the normal and tangential (shear) components of the stress as well as the continuity of both the normal and shear components of displacements (velocity). At the interface between a liquid and a solid, the continuity of the normal components of stress and normal components of displacement (velocity) is required, while the sum of all shear stresses at $z = 0$ must be equal to zero. At

---

[†] It is assumed that the hemispherical source is surrounded by an infinite baffle.
[‡] In the text φ' and φ are used for potential functions in liquids, and ϕ and $\vec{\psi}$ for potential functions in solids.

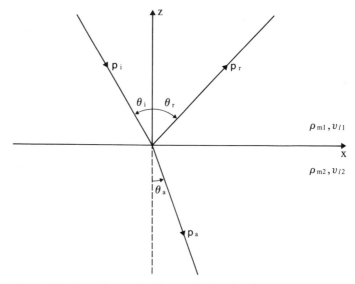

Figure 9.1. Acoustic wave incident on the interface between two liquids.

an interface between two liquids, only the continuity of the normal components of stress and normal components of displacement (velocity) is required. Finally, at an interface between a solid and a vacuum, both the normal and tangential components of stress must be equal to zero.

When an acoustic wave is obliquely incident on a boundary, parts of the incident energy are transmitted through, reflected from, or propagated along the boundary. Unlike electromagnetic waves, where in isotropic materials incident and reflected wave, wave numbers are always equal for oblique incidence, in acoustics, due to mode conversion at the boundary, the incident and reflected wave numbers are equal in exceptional cases only. For instance, the particle velocity of an acoustic wave (either shear or longitudinal) obliquely incident at the interface between two isotropic solids can be decomposed into two components, one perpendicular and the other parallel to the boundary. The corresponding stresses will then act as the generating fields for the reflected and transmitted longitudinal and shear waves, respectively.

As an example, consider the problem of finding the reflection and transmission coefficients of an acoustic wave incident on the interface between two liquids, Fig. 9.1

The expressions for the potentials of the incident, reflected, and refracted waves respectively, can be written as

$$\varphi^i = A^i \exp j\left(\omega t - \beta_x^i x - \beta_z^i z\right)$$

$$\varphi^r = A^r \exp j\left(\omega t - \beta_x^r x - \beta_z^r z\right) \qquad (9.28)$$

$$\varphi^a = A^a \exp j\left(\omega t - \beta_x^a x - \beta_z^a z\right)$$

The pressures are obtained from Eq. (9.12) as

$$p^i = j\omega\rho_{m1}A^i\exp j(\omega t - \beta_x^i x - \beta_z^i z)$$

$$p^r = j\omega\rho_{m1}A^r\exp j(\omega t - \beta_x^r x - \beta_z^r z) \tag{9.29}$$

$$p^a = j\omega\rho_{m2}A^a\exp j(\omega t - \beta_x^a x - \beta_z^a z)$$

and the normal components of velocities from Eq. (9.9) as

$$v_z^i = j\beta_z^i\varphi^i$$

$$v_z^r = j\beta_z^r\varphi^r \tag{9.30}$$

$$v_z^a = j\beta_z^a\varphi^a$$

At the boundary $z = 0$,

$$p^i + p^r = p^a \tag{9.31}$$

$$v_z^i + v_z^r = v_z^a \tag{9.32}$$

From Eq. (9.31) it follows that

$$\beta_x^i = \beta_x^r = \beta_x^a \tag{9.33}$$

and

$$\rho_{m1}A^i + \rho_{m1}A^r = \rho_{m2}A^a \tag{9.34}$$

and from Eq. (9.32) that

$$\beta_z^i A^i + \beta_z^r A^r = \beta_z^a A^a. \tag{9.35}$$

Since $\beta_x^i = \beta^i\sin\theta_i$, $\beta_x^r = \beta^r\sin\theta_r$, and $\beta_x^a = \beta^a\sin\theta_a$, with $\beta^i = \beta^r = \omega/v_{l1}$ and $\beta^a = \omega/v_{l2}$, Eq. (9.33) results in Snell's law,

$$\frac{\sin\theta_i}{v_{l1}} = \frac{\sin\theta_r}{v_{l1}} = \frac{\sin\theta_a}{v_{l2}} \tag{9.36}$$

Solving Eqs. (9.34) and (9.35) for unknown coefficients, it is found that

$$\frac{A^r}{A^i} = \frac{\rho_{m2}v_{l2}/\cos\theta_a - \rho_{m1}v_{l1}/\cos\theta_i}{\rho_{m2}v_{l2}/\cos\theta_a + \rho_{m1}v_{l1}/\cos\theta_i} \tag{9.37}$$

and

$$\frac{A^a}{A^i} = \left(\frac{\rho_{m1}}{\rho_{m2}}\right)\frac{2\rho_{m2}v_{l2}/\cos\theta_a}{\rho_{m2}v_{l2}/\cos\theta_a + \rho_{m1}v_{l1}/\cos\theta_i} \tag{9.38}$$

The pressure reflection coefficient is obtained from Eqs. (9.27) as

$$r = \left(\frac{p^r}{p^i}\right)_{z=0} = \frac{A^r}{A^i} \tag{9.39}$$

and the pressure transmission coefficient as

$$\tau = \left(\frac{p^a}{p^i}\right)_{z=0} = \frac{\rho_{m2}}{\rho_{m1}}\frac{A^a}{A^i} \tag{9.40}$$

Finally, if we denote the impedances for oblique incidence[3] as $Z_1 = \rho_{m1}v_{l1}/\cos\theta_i$ and $Z_2 = \rho_{m2}v_{l2}/\cos\theta_a$, Eqs. (9.39) and (9.40) can be written as

$$r = \frac{Z_2 - Z_1}{Z_2 + Z_1} \quad \text{and} \quad \tau = \frac{2Z_2}{Z_2 + Z_1} \tag{1.23}$$

which are the expressions given in Chap. 1 for normal incidence.

In experimental investigations, the intensity reflection coefficients I^r/I^i and I^a/I^i are often used. Since in a plane wave $p = Z_l v$, it follows that

$$I^i = \frac{|p_i|^2}{2Z_1}, \quad I^r = \frac{|p_r|^2}{2Z_1}, \quad \text{and } I^a = \frac{|p_a|^2}{2Z_2}$$

The law of conservation of energy applies. The energy carried away from the boundary by the reflected and refracted waves must be equal to the energy in the incident wave. With the exception of "critical angles," to be defined in the next paragraph, only the components of the intensity orthogonal to the boundary are of interest, that is,

$$I^i \cos\theta_i = I^r \cos\theta_r + I^a \cos\theta_a \tag{9.41}$$

or

$$1 = r^2 + \tau^2 \frac{Z_1}{Z_2}\frac{\cos\theta_a}{\cos\theta_i}.$$

From Snell's law, Eq. (9.36), which can be written as

$$\theta_a = \arcsin\left(\frac{v_{l2}}{v_{l1}}\sin\theta_i\right) \tag{9.42}$$

it is seen that the reflection and transmission coefficients are real for $v_{l2} < v_{l1}$. In the case $v_{l2} > v_{l1}$, when $(v_{l2}\sin\theta_i)/v_{l1} = 1$ in Eq. (9.42), a total reflection occurs defined by the critical angle

$$\sin\theta_c = \frac{v_{l1}}{v_{l2}} \tag{9.43}$$

In the vicinity of the critical angle, part of the acoustic energy travels along the interface between the two liquids. In this case, Eq. (9.36) can be written as

$$\frac{\sin \theta_i}{v_{l1}} = \frac{\sin 90°}{v_R} \tag{9.44}$$

where v_R is the velocity of the wave propagating along the boundary. The relation $\sin \theta_R = v_{l1}/v_R$ determines the condition for maximum excitation of the wave propagating along the boundary, and the incident angle θ_R is often referred to as the *Rayleigh angle*.

As an example consider the incidence from air [$v_{l1} = 340$ m/s, $\rho_{m1} = 1.2$ kg/m^3] into water [$v_{l2} = 1480$ m/s, $\rho_{m2} = 1000$ kg/m^3]. The critical angle occurs at $\theta_c \simeq 13°$. Thus, it would appear that total internal reflection will occur for $\theta_i > \theta_c$. However, if the water is assumed lossy, it can be shown[3] that total internal reflection does not occur for any angle of incidence. The acoustic wave penetrates the water and propagates as an evanescent wave.

Another interesting case occurs when $A^r = 0$ in Eq. (9.37), resulting in

$$\frac{\rho_{m2}v_{l2}}{\cos \theta_a} = \frac{\rho_{m1}v_{l1}}{\cos \theta_i}$$

Combining the last equation with Eq. (9.36), it follows that the condition for no reflection is satisfied for incident angle

$$\sin^2 \theta_i^B = \frac{(\rho_{m2}/\rho_{m1})^2 - (v_{l1}/v_{l2})^2}{(\rho_{m2}/\rho_{m1})^2 - 1}$$

provided that

$$0 < \sin^2 \theta_i^B < 1.$$

The incident angle so deduced, if it exists, is analogous to the Brewster angle in optics.

9.3   REFLECTION AND REFRACTION AT A LIQUID–SOLID INTERFACE

In this case[3] both types of waves, longitudinal and shear, are generated in the solid, as shown in Fig. 9.2. The velocity potential in the liquid can be written as

$$\vec{v} = \nabla \varphi' \tag{9.26}$$

and in the solid by

$$\vec{v} = \nabla \phi + \nabla \times \vec{\psi} \tag{9.25}$$

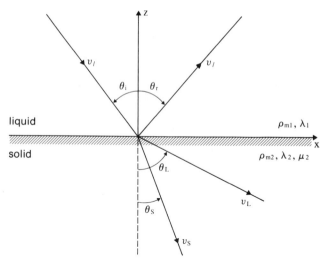

Figure 9.2. Acoustic wave incident from the liquid on the interface between liquid and an isotropic solid.

For plane waves, assuming that all the wave vectors are in the xz plane, the potentials can be expressed as $\varphi' = \varphi'(x, z)\exp j\omega t$, $\phi = \phi(x, z)\exp j\omega t$, and $\vec{\psi} = \hat{y}\psi(x, z)\exp j\omega t$. In the liquid, the potentials of the incident and reflected waves can be expressed as

$$\varphi_i' = A \exp j\left[\omega t - \beta \sin \theta_i x + \beta \cos \theta_i z\right]$$

$$\varphi_r' = A R_L \exp j\left[\omega t - \beta \sin \theta_r x - \beta \cos \theta_r z\right] \tag{9.45}$$

In the solid, the potential of the longitudinal wave can be further specified as

$$\phi = A T_L \exp j\left[\omega t - \beta_L \sin \theta_L x + \beta_L \cos \theta_L z\right] \tag{9.46}$$

and the potential of the shear wave as

$$\psi = A T_S \exp j\left[\omega t - \beta_S \sin \theta_S x + \beta_S \cos \theta_S z\right] \tag{9.47}$$

where R_L, T_L, T_S are the sought potential reflection and transmission coefficients. Furthermore, in the above equations, β, β_L, and β_S are the wave numbers of the longitudinal wave in liquid, and the longitudinal and shear waves in the solid, respectively. In the liquid, the pressure is obtained using Eqs. (9.2) and (9.26) as

$$p = \frac{\lambda_1}{j\omega} \nabla^2 \varphi' \tag{9.48}$$

where $\lambda_1 = \rho_{m1} v_l^2 = \rho_{m1} \omega^2 / \beta^2$, where v_l is the acoustic wave velocity. In the solid, the stresses are obtained from Eqs. (2.46) and (2.48) as

$$j\omega T_{zz} = \lambda_2 \left(\frac{\partial v_x}{\partial x} + \frac{\partial v_z}{\partial z} \right) + 2\mu_2 \frac{\partial v_z}{\partial z} \tag{9.49}$$

and

$$j\omega T_{xz} = \mu_2 \left(\frac{\partial v_x}{\partial z} + \frac{\partial v_z}{\partial x} \right) \tag{9.50}$$

with

$$v_x = \frac{\partial \phi}{\partial x} - \frac{\partial \psi}{\partial z}$$

$$v_z = \frac{\partial \phi}{\partial z} + \frac{\partial \psi}{\partial x} \tag{9.51}$$

and

$$\lambda_2 + 2\mu_2 = \rho_{m2} v_L^2 = \frac{\rho_{m2} \omega^2}{\beta_L^2}$$

$$\mu_2 = \rho_{m2} v_S^2 = \frac{\rho_{m2} \omega^2}{\beta_S^2}$$

where ρ_{m2} is the mass density of the solid, and v_L, β_L and v_S, β_S are the acoustic velocities and wave numbers of the longitudinal and shear waves in the solid, respectively. From Eqs. (9.49) to (9.51), it follows that

$$j\omega T_{zz} = \lambda_2 \nabla^2 \phi + 2\mu_2 \left(\frac{\partial^2 \phi}{\partial z^2} + \frac{\partial^2 \phi}{\partial x \partial z} \right) \tag{9.52}$$

and

$$j\omega T_{xz} = \mu_2 \left(2 \frac{\partial^2 \phi}{\partial x \partial z} + \frac{\partial^2 \psi}{\partial x^2} - \frac{\partial^2 \psi}{\partial z^2} \right) \tag{9.53}$$

The boundary conditions at $z = 0$ require the continuity of the normal velocity,

$$\frac{\partial \varphi'}{\partial z} = \frac{\partial \phi}{\partial z} + \frac{\partial \psi}{\partial x} \tag{9.54}$$

with $\varphi' = \varphi_i' + \varphi_r'$, zero tangential stress, Eq. (9.53),

$$\frac{\partial^2 \psi}{\partial x^2} + 2\frac{\partial^2 \phi}{\partial x \partial z} - \frac{\partial^2 \psi}{\partial z^2} = 0 \tag{9.55}$$

and the continuity of the normal stresses, Eqs. (9.48) and (9.52), that is,

$$\lambda_1 \nabla^2 \varphi' = \lambda_2 \nabla^2 \phi + 2\mu_2 \left(\frac{\partial^2 \phi}{\partial z^2} + \frac{\partial^2 \psi}{\partial x \partial z} \right) \tag{9.56}$$

Using Eqs. (9.45) to (9.47) and (9.54), Snell's law is again obtained in the form

$$\frac{\sin \theta_i}{v_l} = \frac{\sin \theta_r}{v_l} = \frac{\sin \theta_L}{v_L} = \frac{\sin \theta_S}{v_S} \tag{9.57}$$

and the relation

$$\beta \cos \theta_i R_L + \beta_L \cos \theta_L T_L - \beta_S \sin \theta_S T_S = \beta \cos \theta_i \tag{9.58}$$

From Eqs. (9.55), (9.46), and (9.47) it follows that

$$\beta_L^2 \sin 2\theta_L T_L + \beta_S^2 \cos 2\theta_S T_S = 0 \tag{9.59}$$

and from Eqs. (9.45) to (9.47) and (9.56)

$$\rho_{m1} R_L + \rho_{m2} \left[2\frac{\beta_L^2}{\beta_S^2} \sin^2 \theta_L - 1 \right] T_L + \rho_{m2} \sin 2\theta_S T_S = 0 \tag{9.60}$$

Using Eqs. (9.58) to (9.60), it is now possible to solve for the unknown constants. The latter, on using Eq. (9.57), can be written as

$$R_L = \frac{Z_L \cos^2 2\theta_S + Z_S \sin^2 2\theta_S - Z_0}{Z_L \cos^2 2\theta_S + Z_S \sin^2 2\theta_S + Z_0} \tag{9.61}$$

$$T_L = \left(\frac{\rho_{m1}}{\rho_{m2}} \right) \frac{2Z_L \cos 2\theta_S}{Z_L \cos^2 2\theta_S + Z_S \sin^2 2\theta_S + Z_0} \tag{9.62}$$

and

$$T_S = -\left(\frac{\rho_{m1}}{\rho_{m2}} \right) \frac{2Z_S \sin 2\theta_S}{Z_1 \cos^2 2\theta_S + Z_S \sin^2 2\theta_S + Z_0} \tag{9.63}$$

where

$$Z_0 = \frac{\rho_{m1} v_l}{\cos \theta_i}, \quad Z_L = \frac{\rho_{m2} v_L}{\cos \theta_L}, \quad Z_S = \frac{\rho_{m2} v_S}{\cos \theta_S} \tag{9.64}$$

It is easy to show that the pressure reflection coefficient r is equal to the potential reflection coefficient R_L, Eq. (9.61). By considering the impedance

$$Z_t = Z_L \cos^2 2\theta_S + Z_S \sin^2 2\theta_S \qquad (9.65)$$

as the total impedance of the boundary seen by the incident longitudinal wave, the reflection coefficient can be written as $r = (Z_t - Z_0)/(Z_t + Z_0)$.

The critical angles are determined from Eq. (9.57). For the majority of cases, $v_L > v_S > v_l$, resulting in $\theta_L > \theta_S$ as shown in Fig. 9.2. The two critical angles of incidence are determined by the requirement that $\sin \theta_S$ and $\sin \theta_L$ are greater than unity. As the angle of incidence is increased from zero (measured with respect to the normal on the boundary), the first critical angle of incidence, θ_{cL}, is reached when $\theta_L = 90°$, that is, when the transmitted longitudinal wave propagates along the boundary. For $\theta_i > \theta_{cL}$, only the shear wave is transmitted to the solid, while the longitudinal wave is totally reflected. When θ_i is further increased, the second critical angle of incidence, θ_{cS}, is reached when the shear wave propagates along the boundary. For $\theta_i > \theta_{cS}$, the evanescent wave propagates along the boundary, causing excitation of a Rayleigh (SAW) wave at the liquid/solid interface. The maximum excitation of the Rayleigh wave occurs for the incident angle

$$\sin \theta_R = \frac{v_l}{v_R} \qquad (9.66)$$

where v_R is the velocity of the Rayleigh wave at the liquid–solid interface.[4,5]

Equations (9.61) to (9.63), except for a small interval near the Rayleigh angle, show reasonable agreement with experiment[6] and as such are used routinely in engineering practice. The same conclusion applies to other equations such as those describing the oblique incidence at the solid–solid interfaces.[7] However, excitation near the Rayleigh angle is of special interest, as it is commonly used in devices to generate Rayleigh waves on nonpiezoelectric substrates. Schoch[8] was the first to formulate a theory related to Rayleigh angle excitation and derived an expression for the displacement of the reflected beam, now known as Schoch displacement.

9.4 SCHOCH DISPLACEMENT AND ANISOTROPY

At or near the Rayleigh angle, a conversion process occurs where the longitudinal wave excites the Rayleigh wave, which after propagating over a distance of few wavelengths, reradiates its energy as a longitudinal wave into the fluid medium. This results in a displacement of the reflected beam position with respect to the incident beam position, and as such is analogous to the Goos-Hänchen[9-11] effect in optics. This lateral displacement is usually on the order of 10 to 30 Rayleigh wavelengths. The underlying physics is very intricate, and many aspects of the phenomenon have been investigated.[6,12-16]

In practice, the acoustic beams are of finite width and their wavefronts are not planar. For experimental purposes, the diffraction effects are usually neglected if an acoustic beam has a width of 100 wavelengths or more, and the resulting wave is treated to a reasonable degree of approximation as a plane wave. However, the analysis of phenomena such as Schoch displacement requires that acoustic beams of finite spatial limits be used, that is, the beams must be bounded.

Bounded beams[3, 8, 12] can be treated as a superposition of an infinite number of plane waves.[†] Starting with the particle velocity of the incident beam, $v_i = v_i(x, z)$ in the coordinate system shown in Fig. 9.1, this can be expressed in terms of the Fourier integral as

$$v_i(x, z) = \frac{1}{2\pi} \int_{-\infty}^{+\infty} V_i(k_x) \exp[j(k_x x + k_z z)] \, dk_x \qquad (9.67)$$

where $k_x^2 + k_z^2 = k^2$, and

$$V_i(k_x) = \int_{-\infty}^{+\infty} v_i(x, z) \exp[-j(k_x x + k_z z)] \, dx \qquad (9.68)$$

Thus, according to Eq. (9.67), $v_i(x, z)$ is a sum of an infinite number ($m = 1, 2, \ldots$) of plane waves, each wave consisting of an amplitude $V_i(k_m)$ and being incident at an angle $\sin \theta_m = k/k_m$, all waves being of the same frequency. Similarly, the transmitted and reflected bounded beams are obtained as sums of an infinite number of plane waves,

$$v_r(x, z) = \frac{1}{2\pi} \int_{-\infty}^{+\infty} R(k_x) V_i(k_x) \exp[j(k_x x + k_z z)] \, dk_x \qquad (9.69)$$

and

$$v_t(x, z) = \frac{1}{2\pi} \int_{-\infty}^{+\infty} T(k_x) V_i(k_x) \exp[j(k_x x + k_z z)] \, dk_x \qquad (9.70)$$

where $R(k_x)$ and $T(k_x)$ are the velocity reflection and transmission coefficients for plane waves.

Wedge transducers,[17, 18] Fig. 9.3, are used routinely in engineering practice for excitation of SAW on nonpiezoelectric substrates, with $\sin \theta_R = v_l/v_R$, as given by Eq. (9.66). The velocity v_R is the velocity of the Stonely surface waves[19, 20] propagating at the interface between two solids. Wedges are also used for excitation of shear waves in solids for the purposes of NDT. For instance, if the particle velocity of the incident shear wave is parallel to the

---

[†]Expansion into an angular spectrum of plane waves is routinely used in optics. See, for instance, M. Born and E. Wolf, *Principles of Optics*, Pergamon, New York, 1975, Chap. 11.

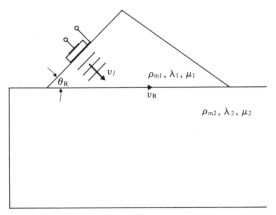

Figure 9.3. Wedge transducer used for excitation of Stonely and Rayleigh waves.

interface, no longitudinal waves will be generated in either solid. In many situations, wedges are used with liquid couplant (gel). In this case, in the fluid, a layer (longitudinal) wave propagates and is continually being reconverted into Rayleigh waves at the two planes of the solids. If the parallelism of the two solid planes is progressively decreased, the rays of the layer wave coalesce into a single ray. The existence of many rays in the layer wave[21] is due to a phase change (Schoch displacement) that the wave undergoes upon reflection from the boundaries. Wedge transducers are analogous to the coupling prisms used in integrated optics.[22,23]

In anisotropic solids,[24] an analytical form of Snell's law is difficult to obtain. Using a graphical procedure[24] on slowness surfaces, defined in Chap. 7, the scattering angles can be determined.

REFERENCES

1. L. D. Landau and E. M. Lifschitz, *Fluid Mechanics*, Addison-Wesley, London, 1959.

2. L. E. Kinsler and A. R. Frey, *Fundamentals of Acoustics*, 2nd ed., Wiley, New York, 1962, Chap. 7.

3. L. M. Brekhovskikh, *Waves in Layered Media*, Academic, New York, 1960, Chap. 1.

4. K. Dransfeld and E. Salzmann, "Excitation, Detection, and Attenuation of High-Frequency Elastic Surface Waves," in *Physical Acoustics* (W. P. Mason and R. N. Thurston, Eds.), Vol. 7, Academic, New York, 1970.

5. N. G. Brower, D. E. Himberger, and W. G. Mayer, "Restrictions on the Existence of Leaky Rayleigh Waves," *IEEE Trans. Sonics Ultrason.*, **SU-26**(4), 306–308 (1979).

6. W. G. Neubauer, "Ultrasonic Reflection of a Bounded Beam at Rayleigh and Critical Angles for a Plane Liquid–Solid Interface," *J. Appl. Phys.*, **44**, 48–55 (1973).

7. W. G. Mayer, "Energy Partition of Ultrasonic Waves at Flat Boundaries," *Ultrason.*, **1965** (Apr.-Jun), 62–68.

8. A. Schoch, *Acustica*, **2**, 1 (1952); *Ergeb. Exact. Naturwiss.*, **23**, 127 (1950).

9. F. Goos and H. Hänchen, *Ann. Physik*, **1**, 333 (1947).

10. H. K. Lotsch, "Reflection and Refraction of a Beam of Light at a Plane Interface," *J. Opt. Soc. Am.*, **58**, 551–561 (1968).

11. B. R. Horowitz and T. Tamir, "Lateral Displacement of a Light Beam at a Dielectric Interface," *J. Opt. Soc. Am.*, **61**, 586–594 (1971).

12. H. L. Bertoni and T. Tamir, "Reflection Phenomena for Acoustic Beams Incident on a Solid at the Rayleigh Angle," *1973 Ultrason. Symp. Proc.*, 226–229 (1973).

13. O. I. Diochok and W. G. Mayer, "Ultrasonic Scattering from Polycrystalline Solids and Plates," *J. Acoust. Soc. Am.*, **53**, 946–947 (1973).

14. F. L. Becker, "Phase Measurements of Reflected Ultrasonic Waves Near the Rayleigh Critical Angle," *J. Appl. Phys.*, **42**, 199–202 (1971).

15. A. L. Van Buren and M. A. Breazeale, "Reflection of Finite-Amplitude Ultrasonic Waves. I. Phase Shifts," *J. Acoust. Soc. Am.*, **44**, 1014–1020 (1968).

16. B. A. Auld, *Acoustic Fields and Waves in Solids*, Vol. II, Wiley, New York, 1973, Chap. X, p. 89.

17. H. L. Bertoni and T. Tamir, "Characteristics of Wedge Transducers for Acoustic Surface Waves," *IEEE Trans. Sonics Ultrason.*, **SU-22**, 415–420 (1975).

18. H. L. Bertoni, "Coupling Layers for Efficient Wedge Transducers," *IEEE Trans. Sonics Ultrason.*, **SU-22**, 421–430 (1975).

19. R. Stonely, *Proc. Roy. Soc. (London)*, *Ser. A*, **116**, 416 (1924).

20. R. O. Claus and C. H. Palmer, "Optical Measurements of Ultrasonic Waves in Interfaces Between Bounded Solids," *IEEE Trans. Sonics Ultrason.*, **SU-27**(3), 97–102 (1980).

21. W. C. Wang, P. Staecker, and R. C. M. Li, "Elastic Waves Guided by a Fluid Layer: A New Type of Ultrasonic Wave Propagation," *Appl. Phys. Lett.*, **16**, 291–293 (1970).

22. R. Ulrich, "Theory of the Prism-Film Coupler by Plane-Wave Analysis," *J. Opt. Soc. Am.*, **60**, 1337–1350 (1970).

23. D. Sarid, "High Efficiency Input-Output Prism Waveguide Coupler: An Analysis," *Appl. Opt.*, **18**, 2921–2926 (1979).

24. Ref. 16, Chap. IX.

EXERCISES

1. (a) Show that a one-dimensional wave equation in liquids for wave propagation in the z direction can be obtained from

$$\frac{\partial p}{\partial z} = -\rho_m \frac{\partial v}{\partial t}, \qquad \kappa \frac{\partial p}{\partial t} = -\frac{\partial v}{\partial z}$$

with the change in mass density δ given by $\delta = \rho_m \Delta$.

(b) Using the velocity potential defined by Eq. (9.9), show that $p(v_l t - z) = \rho_m v_l \varphi' = \rho_m v_l v(v_l t - z)$, where $\varphi = \varphi(v_l t - z)$.

(c) Show that the extension to three dimensions of the above equations is given by

$$\nabla p = -\rho_m \frac{\partial \vec{v}}{\partial t}, \qquad \kappa \frac{\partial p}{\partial t} = -\nabla \cdot \vec{v}$$

2. By numerical evaluation of Eq. (9.61), find the two critical angles for a water–stainless steel interface. For stainless steel, use $\rho_m = 8090$ kg/m^3, $v_L = 5610$ m/s, and $v_S = 3180$ m/s.

3. In the text, Eq. (9.61) has been discussed under the assumption that $v_L > v_S > v_l$, and it has been pointed out that two critical angles of incidence exist. Consider now $v_L > v_l > v_S$, which is the case of water ($v_l = 1480$ m/s, $\rho_m = 1000$ kg/m^3) and Lucite ($v_L = 2592$ m/s, $v_S = 1287$ m/s, $\rho_{m2} = 1240$ kg/m^3). Plot the magnitude of the reflection coefficient, and show that $\theta_{cL} \approx 35°$, while θ_{cS} does not exist, and consequently no real value of the Rayleigh angle exists.

4. Consider the problem of determining the reflection coefficient of a lossy liquid–lossless isotropic solid interface. Formally the lossy liquid can be represented by substituting $\beta - j\alpha$, $\alpha > 0$, for β in Eqs. (9.45). Show that Eqs. (9.61) to (9.64) remain in force with the exception of Z_0, which must be defined for this case as

$$Z_0 = \frac{\rho_{m1} v_l}{[1 - j(\alpha v_l/\omega)]\cos \theta_i}$$

Discuss how the presence of losses affects Snell's law.

5. Find the Rayleigh wave velocity on (a) aluminum oxide, Al$_2$O$_3$, having $\rho_m = 4000$ kg/m^3, $v_L = 10{,}460$ m/s, $v_S = 6010$ m/s and (b) stainless steel, $\rho_m = 8090$ kg/m^3, $v_L = 5610$ m/s, $v_S = 3180$ m/s, each immersed in water, using the dispersion equation[5]

$$4[v_s/v]\big[1 - [v_S/v]^2\big]^{1/2}\big[(v_S/v_L)^2 - (v_S/v)^2\big]^{1/2} + \big[1 - 2(v_S/v)^2\big]^2$$

$$= -\frac{\rho_{m1}}{\rho_{m2}}\left\{\frac{(v_S/v_L)^2 - (v_S/v)^2}{(v_S/v_l)^2 - (v_S/v)^2}\right\}^{1/2}$$

where v_S, v_L, and ρ_{m2} are the shear velocity, longitudinal velocity, and mass density in the solid, and v_l and ρ_{m1} are the acoustic velocity and mass density of the liquid. Show that $v = v_R \approx 2930$ m/s for stainless steel and $v = v_R \approx 5530$ m/s for Al$_2$O$_3$. Find the Rayleigh angle in each case.

Chapter 10

Radiation and Diffraction of Acoustic Waves, Planar Radiators, and Acoustic Lenses

Radiation, diffraction, and scattering problems in acoustics, as in optics or electromagnetics, consist of solving the field equations in the presence of various sources and/or obstacles under prescribed boundary and initial conditions. Historically, the advancements in acoustic radiation and diffraction have taken different aspects.

In 1880, Rayleigh[1] formulated exactly the required radiation integral for acoustic waves in fluids. Another aspect of acoustic radiation research parallels the analogous problem in optics.[†] The latter is related to the works of Huygens, Young, Fresnel, Sommerfeld, and Kirchhoff,[2] and more recently to the works of Maggi,[3] Rubinowicz,[4] Miyomoto and Wolf,[5] and Keller.[6] These works resulted in a theory known as the Rubinowicz-Maggi theory, which is given in its most general form by the Miyomoto-Wolf formula.[7] In 1941, Schoch[8] applied the concepts of the Rubinowicz-Maggi theory to the study of acoustic radiation phenomena. This study has been further carried out by Stenzel,[9] Carter and Williams,[10] Kozina and Makarov,[11] and Chetayev.[12] Miller and Pursey,[13] in 1954, formulated exactly the required integral expression for acoustic radiation into a semiinfinite isotropic solid for certain geometries of the radiator.

According to the Rubinowicz-Maggi theory, the radiated field of an acoustic (optic) source consists of two components. The first is the geometrical wave field component, which is a wave field of spatially discontinuous wavefronts that conforms in shape to the radiating surface of the source. The second component is the boundary diffraction wave field that is radiated by the edges of the surface. This radiation, due to wave interference, is usually localized at particular points on the edges of the radiating surface of the source. Each

---

[†] For instance, many books on optics use photographs of acoustic waves in water tanks to illustrate various diffraction problems.

component of the total radiated field is spatially discontinuous, but the discontinuities are such that they exactly compensate one another, thus providing a total radiated field that is continuous everywhere.

In most of the present literature on optics, a conventional scalar wave diffraction theory based on the Fresnel-Kirchhoff or Rayleigh-Sommerfeld diffraction formula or another, such as the geometrical diffraction theory of Keller, is used. In acoustics, the use of the conventional scalar wave diffraction theory, based on Kirchhoff's mathematical formulation of Fresnel's ideas, is also widespread today. This is probably due to the fact that the application of the Rubinowicz-Maggi theory to problems of practical interest requires rather tedious mathematical evaluations.

The scalar wave theory assumes that each component of a vector (such as the vector potential in optics or acoustics) can be predicted from the scalar wave equation for that component.[2] The scalar wave theory thus applies rigorously to acoustics of liquids, where the potential is given by the scalar wave equation for longitudinal waves. In solids, where the shear waves require the formulation of a vector potential,[13, 14] and in optics where the transverse nature of light requires again a formulation in terms of a vector potential,[5] the application of scalar wave theory is questionable. However, experiments in optics indicate that the scalar theory gives excellent results in most circumstances, provided certain conditions of a geometrical nature are met. Besides Kirchhoff's boundary conditions, which must be satisfied, it is further assumed that the linear dimensions of the aperture must be large compared to wavelength. At points distant from the aperture, minor inaccuracies in assuming the wave field are smoothed out. The scalar theory is thus used mainly to predict the relative amplitude and phase (but not the polarization) at points many wavelengths away from the source. However, in the acoustics of isotropic solids we do not expect that the scalar wave theory would predict the correct phase and amplitude for each scalar component of transverse acoustic waves.[†]

10.1 PHYSICAL AND GEOMETRICAL ACOUSTICS

According to Huygens' principle, every point on the primary wavefront is considered a new source of a secondary spherical wave. Furthermore, a secondary wavefront is constructed as the envelope of these secondary spherical waves, as shown in Fig. 10.1(a). By contrast, geometrical acoustics uses ray paths rather than wavefronts to postulate a point-to-point correspondence between successive positions of the wavefront as shown in Fig. 10.1(b). Thus in geometrical acoustics as in geometrical optics,[15] Fig. 10.1(c), the power passing

---

[†] It must be remembered that in isotropic solids an obliquely incident acoustic wave generates two modes (longitudinal and shear), while in optics only one mode (TEM) is present. For further discussion, see Sect. 10.7 and 10.8.

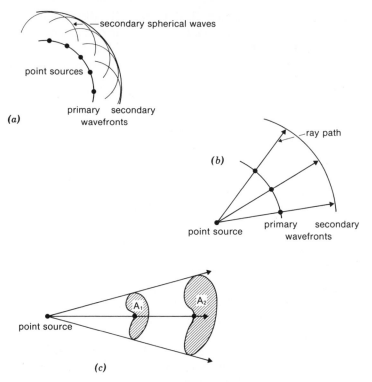

Figure 10.1. (*a*) Illustration of Huygens' principle in physical acoustics. (*b*) Ray paths in geometrical acoustics. (*c*) Conservation of power in geometrical acoustics.

through areas A_1 and A_2 must be conserved, that is,

$$I_1 A_1 = I_2 A_2 \qquad (10.1)$$

where I is given by Eq. (9.20), $I_i = |p_i|^2/2Z_l$, $i = 1, 2$.

To illustrate the fundamental difference between the two approaches, consider a uniform plane pressure wave incident on the obstacle shown in Fig. 10.2(*a*). The obstacle, also called a baffle, is assumed to have zero compliance, that is, it is ideally rigid (no energy can be coupled into it, and therefore all energy incident upon it will be reflected). All the points along the z axis, for $z \geqslant a$, are considered point sources. The pressure p at some point B is found according to Huygens' principle as

$$p(B) = \int_{z=a}^{\infty} dp \qquad (10.2)$$

where dp is the pressure at point B due to the point source at dz. Using the

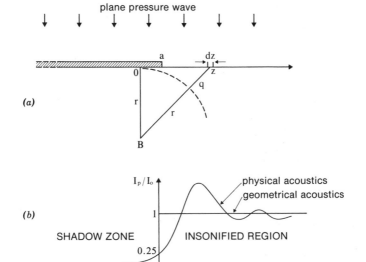

Figure 10.2. (a) A uniform pressure wave incident on an obstacle of length ($-\infty, a$). (b) The diffraction curves corresponding to the prescriptions of geometrical and physical acoustics.

solution for spherical waves, Eq. (9.17), it follows that

$$dp = \frac{p_0}{r} e^{-j\beta(r+q)} \, dz \tag{10.3}$$

or

$$p(B) = p_0 \frac{e^{-j\beta r}}{r} \int_a^\infty e^{-j\beta q} \, dz \tag{10.4}$$

For $r \gg q$ it follows that $q \approx z^2/2r$. Letting $m^2 = 2/r\lambda_a$ and $u = mz$, where $\beta = 2\pi/\lambda_a$ and λ_a is the wavelength, Eq. (10.4) is written as

$$p(B) = \frac{p_0 e^{-j\beta r}}{mr} \int_{ma}^\infty e^{-j\beta u^2/2} \, du \tag{10.5}$$

containing the Fresnel integral.[16] The latter has been defined in Eq. (3.58) and can be written as

$$p(B) = \frac{p_0}{mr} e^{-j\beta r} \left[\tilde{Z}(\infty) - \tilde{Z}(ma) \right]$$

$$= \frac{p_0 e^{-j\beta r}}{mr} \left[\frac{1 - j1}{2} - \tilde{Z}(ma) \right] \tag{10.6}$$

The intensity is found from Eq. (9.20) as

$$I_p = \frac{pp^*}{2Z_l} = \frac{I_0}{2}\{[0.5 - C(ma)]^2 + [0.5 + S(ma)]^2\} \qquad (10.7)$$

where $I_0 = |p_0|^2 \lambda_a / 2Z_l r$. The relative intensity curve, also called the diffraction curve, is shown in Fig. 10.2(b), where it is compared with the corresponding curve based on the predictions of geometrical acoustics. Thus, as we move from the insonified side to the shadow side, the power density oscillates in space, decreasing slowly to zero. This conclusion is easily verified by water-tank experiments. It is seen from Eq. (10.7) that when $ma \to \infty$ (complete obstruction), $I_p = 0$, and for $ma \to -\infty$ (no obstruction), $I_p = I_0$. On the other hand, the diffraction curve based on the predictions of geometrical acoustics can be written as

$$I_g = I_0 \chi(ma) \qquad (10.8)$$

where χ is the Heaviside step function. In Schoch's formulation of the Rubinowicz-Maggi theory, emphasis is made on discerning between I_g, which conforms is shape to the radiating surface of the source, and $I_p - I_g$, which is due to the boundary diffraction wave.

10.2   THE IMPULSE RESPONSE OF A PLANAR ACOUSTIC SOURCE

In this section, the mathematical formulation necessary for derivation of the impulse response of a planar acoustic source[17] is developed. The conclusions reached are then generalized and applied to transducers used in practice for excitation of bulk waves in liquids. Certain aspects of these transducers have been already discussed in Sect. 5.7.

    Consider a planar radiator (often called a piston) of surface A_0 located in the $z = 0$ plane of a rectangular coordinate system, as shown in Fig. 10.3(a), where the remaining portion of the $z = 0$ plane, surrounding the surface A_0, is considered to act as a rigid baffle. The initial condition is that at $t = 0$ the radiator is excited with velocity v_0, which is uniform over the area A_0 and is an impulsive function of time. If $\vec{v}(t, \vec{r}) = \vec{v}(t, x, y, z)$ is the acoustic particle velocity at an observation point $B = B(\vec{r})$, in the half-space $z > 0$, then the corresponding velocity potential $\varphi(t, \vec{r})$ satisfies the wave equation (9.11) written as

$$\nabla^2 \varphi - \frac{1}{v_l^2} \frac{\partial^2 \varphi}{\partial t^2} = 0 \qquad (10.9)$$

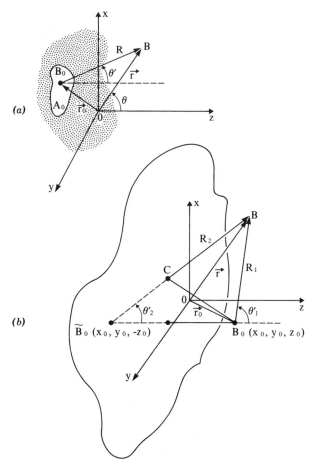

Figure 10.3. (*a*) Geometry for studying the impulse response of a planar acoustic radiator of surface A_0 located in $z = 0$ plane. (*b*) The geometrical meaning of Green's function given by Eq. (10.15). See text for more details.

subject to the initial and boundary conditions

$$\vec{v}(t, x, y, 0) = \begin{cases} \hat{z}v_0\delta(t), & \text{for } B \text{ on } A_0 \\ 0, & \text{elsewhere on the } z = 0 \text{ plane} \end{cases} \qquad (10.10)$$

Since

$$\vec{v} = -\nabla\varphi \qquad (10.11)$$

it follows from Eq. (10.10) that

$$\frac{\partial\varphi(t, \vec{r})}{\partial z} = \begin{cases} -v_0\delta(t), & \text{fo } B \text{ on } A_0 \\ 0, & \text{elsewhere on the } z = 0 \text{ plane} \end{cases} \qquad (10.12)$$

Another solution g of Eq. (10.10), valid for $z > 0$, is obtained by considering an impulsive point source at $B_0 = B_0(\vec{r}_0)$ as shown in Fig. 10.3(b). In this case, the velocity potential g of some point $B = B(\vec{r})$ in the half-space $z > 0$ must satisfy the wave equation

$$\nabla^2 g - \frac{1}{v_l^2} \frac{\partial^2 g}{\partial t^2} = -\delta(\vec{r} - \vec{r}_0)\delta(t) \tag{10.13}$$

with the boundary condition

$$\frac{\partial g}{\partial z} = 0 \quad \text{in the } z = 0 \text{ plane} \tag{10.14}$$

The function g is known as Green's function and is denoted by $g(t, \vec{r}; 0, \vec{r}_0)$, meaning that the impulse response at a point determined by the radius vector $\vec{r}$ at time t is due to the source point located at point $\vec{r}_0$ at time $t = 0$. The solution of Eq. (10.13) is well known[18] and can be written as

$$g(t, \vec{r}; 0, \vec{r}_0) = \frac{\delta(t - R_1/v_l)}{R_1} + \frac{\delta(t - R_2/v_l)}{R_2} \tag{10.15}$$

where

$$R_1^2 = (x - x_0)^2 + (y - y_0)^2 + (z - z_0)^2$$

$$R_2^2 = (x - x_0)^2 + (y - y_0)^2 + (z + z_0)^2 \tag{10.16}$$

and can be interpreted with the help of Fig. 10.3(b) to mean that the response at point B is due not only to the direct paths from the source point B_0 but also to reflections[19] occurring on the rigid baffle, that is, via B_0CB, belonging apparently to the virtual (mirror image) source point $\tilde{B}_0$ given by the second term on the right-hand side of Eq. (10.15).

Taking the Laplace transform, defined as

$$\tilde{a}(s) = \int_0^\infty a(t)\exp(-st)\, dt \tag{10.17}$$

$$a(t) = \frac{1}{2\pi j}\int_{\sigma - j\infty}^{\sigma + j\infty} \tilde{a}(s)\exp(st)\, ds \tag{10.18}$$

of Eqs. (10.9), (10.13), and (10.15), it follows that

$$\nabla^2 \tilde\varphi(s, \vec r) - \left(\frac{s}{v_l}\right)^2 \tilde\varphi(s, \vec r) = 0 \tag{10.19}$$

$$\nabla^2 \tilde g - \left(\frac{s}{v_l}\right)^2 \tilde g = -\delta(\vec r - \vec r_0) \tag{10.20}$$

$$\tilde g = \tilde g(s, \vec r; 0, \vec r_0) = \frac{\exp(-sR_1/v_l)}{R_1} + \frac{\exp(-sR_2/v_l)}{R_2} \tag{10.21}$$

If the source and the observation points exchange roles in Eq. (10.21), it is seen that

$$\tilde g(s, \vec r; 0, \vec r_0) = \tilde g(s, \vec r_0; 0, \vec r) \tag{10.22}$$

which is a reciprocity relation. If we exchange the roles of $\vec r$ and $\vec r_0$ in Eqs. (10.19) and (10.20), multiply the first by $\tilde g(s, \vec r_0; 0, \vec r)$ and the second by $\tilde\varphi(s, \vec r_0)$, and then subtract the resulting equations, an expression relating g and φ will be obtained. The latter can be integrated over a volume V that contains both B and B_0. The volume V, as shown in Fig. 10.4, is bounded by a hemispherical closed surface A_∞ of radius R. Using Green's theorem, it follows that

$$\iiint \left[\tilde g \nabla_0^2 \tilde\varphi - \tilde\varphi \nabla_0^2 \tilde g \right] dV = \iint_{A_\infty} \left[\tilde g \nabla_0 \tilde\varphi - \tilde\varphi \nabla_0 \tilde g \right] \cdot d\vec A$$

$$= \lim_{R \to \infty} \iint_{A_0} \left[\tilde g \nabla_0 \tilde\varphi - \tilde\varphi \nabla_0 \tilde g \right] \cdot d\vec A \tag{10.23}$$

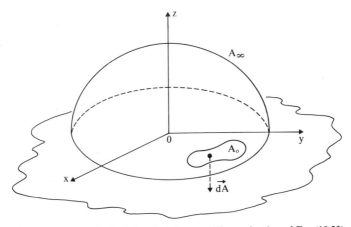

Figure 10.4. Hemispherical closed surface used for evaluation of Eq. (10.23).

where the last expression is valid for $R \to \infty$ since in this case both functions $\tilde{g}$ and $\tilde{\varphi}$ and their gradients decrease to zero (Sommerfeld's radiation condition), the integral over the hemisphere vanishes and only the integral over the area A_0 remains. The zero subscripts in the Laplacian and gradient operators indicate that they are to be taken using the components of $\vec{r}_0$, for instance, $\nabla_0^2 = \partial^2/\partial x_0^2 + \partial^2/\partial y_0^2 + \partial^2/\partial z_0^2$. Using the Helmholtz-Kirchhoff theorem,[20] the last expression in Eq. (10.23) can be written as

$$4\pi \tilde{\varphi}(s, \vec{r}) = \iint_{A_0} [\tilde{g}\nabla_0 \tilde{\varphi} - \tilde{\varphi}\nabla_0 \tilde{g}] \cdot d\vec{A} \qquad (10.24)$$

or, since $d\vec{A} = -dA\hat{z}_0$, as seen from Fig. 10.4, in the scalar form as

$$\tilde{\varphi}(s, \vec{r}) = -\frac{1}{4\pi} \iint_{A_0} \left[\tilde{g}\frac{\partial \tilde{\varphi}}{\partial z_0} - \tilde{\varphi}\frac{\partial \tilde{g}}{\partial z_0} \right] dA \qquad (10.25)$$

Since $\vec{r}$ and $\vec{r}_0$ have interchanged places, Eq. (10.14) is now written as

$$\frac{\partial \tilde{g}(t, \vec{r}_0; 0, \vec{r})}{\partial z_0} = 0, \qquad \text{on the } z_0 \text{ plane} \qquad (10.26)$$

and the corresponding Laplace transform as $\partial \tilde{g}(s, \vec{r}_0; 0, \vec{r})/\partial z_0 = 0$. Using the latter result and Eq. (10.25), it follows that

$$\tilde{\varphi}(s, \vec{r}) = -\frac{1}{4\pi} \iint_{A_0} \tilde{g}\frac{\partial \tilde{\varphi}}{\partial z_0} dA \qquad (10.27)$$

where $z_0 = 0$, g is obtained from Eqs. (10.21) and (10.22) as

$$g = \frac{2\exp(-sR/v_l)}{R}, \qquad (10.28)$$

with $R = (x - x_0)^2 + (y - y_0)^2 + z^2$. Again, since $\vec{r}$ and $\vec{r}_0$ have exchanged places, Eq. (10.12) is written as

$$\frac{\partial \tilde{\varphi}(t, \vec{r}_0)}{\partial z_0} = -v_0 \delta(t) \qquad (10.29)$$

and the corresponding Laplace transform as

$$\frac{\partial \tilde{\varphi}(s, \vec{r}_0)}{\partial z_0} = -v_0 \qquad (10.30)$$

Combining Eqs. (10.27), (10.28), and (10.30), it follows that

$$\tilde{\varphi}(s, \vec{r}) = \frac{v_0}{2\pi} \iint_{A_0} \frac{\exp(-sR/v_l)}{R} \, dA \qquad (10.31)$$

with the corresponding inverse Laplace transform

$$\varphi(t, \vec{r}) = \frac{v_0}{2\pi} \iint_{A_0} \frac{\delta(t - R/v_l)}{R} \, dA \qquad (10.32)$$

and resulting in the expression for pressure, Eq. (9.12),

$$p(t, \vec{r}) = \rho_m \frac{\partial \varphi}{\partial t} = \rho_m \frac{v_0}{2\pi} \iint_{A_0} \frac{\delta'(t - R/v_l)}{R} \, dA \qquad (10.33)$$

Using the same mathematical approach, the last two expressions can be generalized to the case of a uniform radiator velocity of arbitrary time variation; that is, instead of Eq. (10.10), the initial and boundary conditions are now

$$\vec{v}(t, x, y, 0) = \begin{cases} \hat{z}v(t)\chi(t), & \text{for } B \text{ on } A_0 \\ 0, & \text{elsewhere on the } z = 0 \text{ plane} \end{cases} \qquad (10.34)$$

where $\chi(t)$ is the Heaviside step function. This results in Rayleigh's expressions for the potential and the pressure given by

$$\varphi(t, \vec{r}) = \frac{1}{2\pi} \iint_{A_0} \frac{v(t - R/v_l)}{R} \, dA \qquad (10.35)$$

and

$$p(t, \vec{r}) = \frac{\rho_m}{2\pi} \iint_{A_0} \frac{v'(t - R/v_l)}{R} \, dA \qquad (10.36)$$

10.3 TIME-DOMAIN AND STEADY-STATE ANALYSES OF THE DISK RADIATOR

The time-domain analysis of the baffled radiator (piston) is simplified if the impulse response is first defined. Then the response for any other initial (uniform) velocity of arbitrary time variation is obtained by convolution. Using Eq. (10.33), the impulse response is defined as

$$h(t, \vec{r}) = \frac{p(t, \vec{r})}{\rho_0} = \frac{1}{2\pi} \iint_{A_0} \frac{\delta'(tv_l - R)}{R} \, dA \qquad (10.37)$$

where $p_0 \equiv \rho_m v_l v_0 = Z_l v_0$. The pressure for any other initial (uniform) radiator velocity of arbitrary time variation, Eq. (10.34), with $v(t) = v_0 f(t)$ is then given by

$$p(t, \vec{r}) = p_0 f(t) * h(t, \vec{r}) \qquad (10.38)$$

which is identical to Eq. (10.36). The particle radial velocity can be computed using Eq. (9.16).

Consider the problem of finding the impulse response at an on-axis point z_0 of a disk radiator as shown in Fig. 10.5(a). After integration over the angular coordinate ψ, Eq. (10.37) can be written as

$$h(t, z_0) = -\int_0^a d\delta\left[t v_l - \left(z_0^2 + q^2 \right)^{1/2} \right] \qquad (10.39)$$

resulting in the expression for pressure

$$p(t, z_0) = Z_l v_0 \left\{ \delta(t v_l - z_0) - \delta\left[t v_l - \left(z_0^2 + a^2 \right)^{1/2} \right] \right\} \qquad (10.40)$$

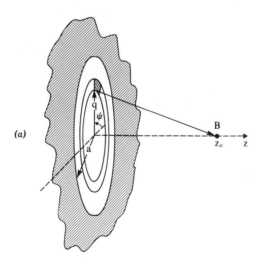

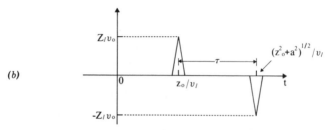

Figure 10.5. (a) Geometry for finding pressure at an on-axis point of a circular baffled radiator of radius a. (b) The impulse response of the baffled radiator.

From Eq. (10.40) it is seen that the pressure at an on-axis point consists of two impulse functions, as shown in Fig. 10.5(b). The first impulse has a delay equal to the propagation time from the center of the radiator to the observation point, and the second a delay equal to the propagation time from the edge of the radiator to the observation point z_0. For a radiator of infinite radius, Eq. (10.40) reduces to (Exercise 7)

$$p(t, z_0) = Z_l v_0 [\delta(tv_l - z_0)] \qquad (10.41)$$

which is the plane-wave approximation used in Chap. 5, Eq. (5.67), to analyze the transient response of the piezoelectric transducer of large cross-sectional dimensions.

In the more general case,[21] where the initial radiator velocity is of arbitrary time variation, $v(t) = v_0 f(t)$, the on-axis response is obtained from Eq. (10.38) as

$$p(t, z_0) = Z_l \{ v(tv_l - z_0) \chi(tv_l - z_0)$$
$$- v[tv_l - (z_0^2 + a^2)^{1/2}] \chi[tv_l - (z_0^2 + a^2)^{1/2}] \} \qquad (10.42)$$

As in Eq. (10.40), the pressure consists of two separate waveforms having the same variation as the radiator initial velocity. The mutual time separation of the two waveforms is $\tau = [(z_0^2 + a^2)^{1/2} - z_0]/v_l$. If the duration of $v(t)$ is T, then when $T < \tau$, two separate pressure waveforms will be observed at z_0. In the opposite case, when $T > \tau$, due to overlapping, a single pressure waveform is observed. Thus as the observation coordinate z_0 is changed, a number of maxima and minima will be observed. The expression for the impulse response corresponding to an off-axis observation point can also be derived.

In the steady state, the on-axis pressure resulting from sinusoidal velocity of the radiator is simply obtained by the Fourier transform of Eq. (10.40),

$$\tilde{p}(\omega, z_0) = \int_{-\infty}^{+\infty} p(t, z_0) e^{-j\omega t} \, dt = Z_l v_0 H(\omega, z_0) \qquad (10.43)$$

where $H(\omega, z_0)$ is the radiator transfer function given by

$$H(\omega, z_0) = \exp\left(-\frac{j\omega z_0}{v_l} \right) - \exp\left[-\frac{j\omega (z_0^2 + a^2)^{1/2}}{v_l} \right] \qquad (10.44)$$

On the z_0 axis and in the cylindrical coordinates implied in Fig. 10.5, it follows that $\tilde{v} = -\partial\tilde{\varphi}/\partial z_0$. From Eq. (10.33) we have $\tilde{p} = j\omega\rho_m\tilde{\varphi}$, thus giving

$$\tilde{v} = \frac{j}{\omega\rho_m} \frac{\partial\tilde{p}}{\partial z_0} = \frac{jZ_l v_0}{\omega\rho_m} \frac{\partial H}{\partial z_0} \qquad (10.45)$$

and the intensity on the z_0 axis as

$$
I_z = \text{Re}\left\{\frac{\tilde{p}\tilde{v}^*}{2}\right\} = \text{Re}\left\{\frac{Z_l^2 v_0^2}{2 j\omega\rho_m} H \frac{\partial H^*}{\partial z_0}\right\}
$$

$$
= Z_l v_0^2 \left(1 + \frac{z_0}{\left(z_0^2 + a^2\right)^{1/2}}\right)\sin^2\left\{\frac{\omega}{2v_l}\left[\left(z_0^2 + a^2\right)^{1/2} - z_0\right]\right\} \quad (10.46)
$$

$$
I_z \approx 2Z_l v_0^2 \sin^2\left\{\frac{\omega}{2v_l}\left[\left(z_0^2 + a^2\right)^{1/2} - z_0\right]\right\}, \qquad z_0 \gg a \quad (10.47)
$$

exhibits maxima and minima. As explained before, this can be regarded as the interference effect between the two waves,[21] the first originating at the edge of the disk and the second originating at the center of the disk. Defining $I_{\max} = Z_l v_0^2$ and using $\beta = 2\pi/\lambda_a = \omega/v_l$, the normalized intensity function is obtained as

$$
\frac{I}{I_{\max}} = \sin^2\left\{\frac{\pi}{\lambda_a}\left[\left(z_0^2 + a^2\right)^{1/2} - z_0\right]\right\}\left(1 + \frac{z_0}{\left(z_0^2 + a^2\right)^{1/2}}\right) \quad (10.48)
$$

The intensity at the center of the disk, $z_0 = 0$, is proportional to $\sin^2(\pi a/\lambda_a)$ and is equal to zero whenever $a = n\lambda_a$, where n is an integer. The situation here is similar to an electromagnetic circular waveguide where the radial distribution of the Poynting vector exhibits nodes and antinodes. The maxima in Eq. (10.48) occur for the value of z_0 given by

$$
z_{\max} = \frac{4a^2 - \lambda_a^2(2n + 1)^2}{4\lambda_a(2n + 1)} \quad (10.49)
$$

where $n = 0, 1, 2, \ldots$, and the minima for

$$
z_{\min} = \frac{a^2 - \lambda_a^2 m^2}{2\lambda_a m} \quad (10.50)
$$

where $m = 1, 2, 3, \ldots$. The last axial maximum occurs for $n = 0$ and is given by

$$
z_F = \frac{4a^2 - \lambda_a^2}{4\lambda_a} \quad (10.51)
$$

or for $a > \lambda_a$ by

$$
z_F \simeq \frac{a^2}{\lambda_a} \quad (10.52)
$$

defining the transition between the near field (Fresnel zone) and the far field (Fraunhofer zone). As seen from Fig. 10.6(a), in the Fresnel zone the axial intensity exhibits many maxima and minima, while in the Fraunhofer zone, $z_F > a^2/\lambda_a$, the axial intensity is slowly decreasing, $I/I_{\max} \simeq \beta^2 a^4/16z_0^2$, due to the spherical beam divergence, Eq. (9.22).

The small path difference

$$\Delta z_0 = \tau v_l = \left(z_0^2 + a^2\right)^{1/2} - z_0 \simeq \frac{a^2}{2z_0} < \frac{\lambda_a}{2} \tag{10.53}$$

causes the contributions of waves emanating from various parts of the trans-ducer to add up, this being the reason that the last extremum occurs at $z_F \simeq a^2/\lambda_a$. In the Fresnel zone, $z < z_F$, cross sections of the acoustic beam show concentric rings corresponding to intensity radial maxima and minima, with the number of rings increasing with decreasing z.[23, 22]

In order to analyze the radial distribution, consider an off-axis observation point B, as shown in Fig. 10.6(b). It is easy to show that the following

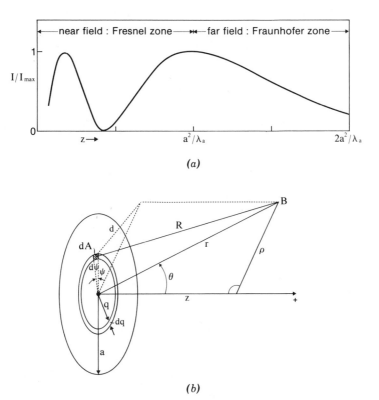

(a)

(b)

Figure 10.6. (a) The axial intensity of the circular radiator. (b) Geometry used in the analysis of acoustic beam radial variations.

coordinate relations apply:

$$\rho = r \sin \theta \tag{10.54}$$

$$d^2 = q^2 + \rho^2 - 2q\rho \cos \psi \tag{10.55}$$

$$R^2 = z^2 + d^2 \tag{10.56}$$

$$dA = q \, dq \, d\psi \tag{10.57}$$

$$r^2 = z^2 + \rho^2 \tag{10.58}$$

In the steady state, the expression for pressure is obtained from Eq. (10.37) via the Fourier transform,

$$\tilde{p}(\omega, \vec{r}) = \frac{j\omega p_0}{2\pi v_l} \iint_{A_0} \frac{\exp\left[-j(\omega/v_l)R\right]}{R} dA \tag{10.59}$$

and upon substitution of Eqs. (10.54) to (10.58) it can be shown that the resulting integral must be evaluated numerically. In the Fraunhofer zone, however, where $r \gg q$, the expression for R in the exponent of the integrand can be approximated by

$$R \simeq r - q \cos \psi \sin \theta \tag{10.60}$$

and the expression for R in the denominator of the integrand by $R \simeq r$. Using these approximations, Eq. (10.59) can be written as

$$\tilde{p}(\omega, \vec{r}) = \frac{jp_0 \beta}{2\pi} \frac{e^{-j\beta r}}{r} \int_0^{2\pi} d\psi \int_0^a e^{j\beta q \cos \psi \sin \theta} q \, dq \tag{10.61}$$

and on using the relations

$$\int_0^{2\pi} e^{j\xi \cos \psi} \, d\psi = 2\pi J_0(\xi) \tag{10.62}$$

and

$$\xi J_1(\xi) = \int \xi J_0(\xi) \, d\xi$$

it follows that

$$\tilde{p}(\omega, \vec{r}) = \frac{jp_0 \beta a^2}{2} \frac{e^{-j\beta r}}{r} \frac{2J_1(\beta a \sin \theta)}{\beta a \sin \theta} \tag{10.63}$$

Defining the strength of the radiator as $q_T = v_0 \pi a^2$, this can be written as

$$p_T = \frac{j\rho_m v_l \beta}{2\pi r} q_T e^{-j\beta r} \left[\frac{2J_1(\beta a \sin\theta)}{\beta a \sin\theta} \right] \qquad (10.64)$$

It is seen from the last equation, when compared with Eqs. (9.23) and (9.24), that the expression outside the brackets on the right represents a spherical wave, while the expression in the brackets is defined as the pattern factor, giving the variation of the pressure with angle θ. The pattern factor has the first zero for $\beta a \sin\theta = 2\pi a\rho/\lambda_a r \approx 2\pi a\rho/\lambda_a z = 3.83$, as shown in Fig. 10.7($a$). The function designated as

$$\text{jinc}(\xi) \equiv \frac{J_1(\pi\xi)}{2\xi} \qquad (10.65)$$

is often encountered in different problems of acoustics with circular symmetry. The pattern factor can then be written as

$$\frac{2J_1(\pi x)}{x} = \frac{4}{\pi}\text{jinc}(x) \approx \text{jinc}(x) \qquad (10.66)$$

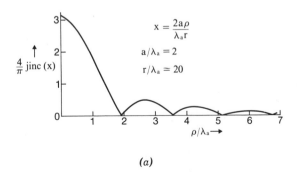

$$x = \frac{2a\rho}{\lambda_a r}$$

$$a/\lambda_a = 2$$

$$r/\lambda_a = 20$$

(a)

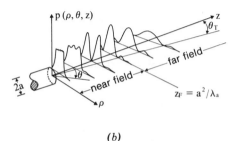

(b)

Figure 10.7. (a) The pressure pattern factor. (b) Overall behavior of pressure.

with $x = 2a\rho/\lambda_a r$. Therefore the pressure magnitude in Eq. (10.64) decreases with the polar angle θ, the phase reversal occurring at $\beta a \sin \theta_T = 3.83$, or for $\sin \theta_T \simeq 0.61\lambda_a/a$, where θ_T is defined as the theoretical beam divergence semiangle. Since $\sin \theta_T \simeq \rho/z$, for $\rho = a$ it follows that this corresponds to the point $3a^2/2\lambda_a$ on the axis of Fig. 10.6(a). In practice, however, the zeros of the pattern factor are not always discernible. The beam divergence is therefore often characterized by measuring the angle of rotation θ_M that the disk must be given relative to its initial orientation, so that it receives, by reflection in an acoustic mirror (target), an amplitude equal, for instance, to one-half the maximum amplitude received when properly oriented. Therefore from 4 jinc$(x)/\pi = 0.5$, it is found that this occurs for $\sin \theta_M \simeq 0.35\lambda_a/a$. The corresponding point obtained for $\rho = a$ on the axis of Fig. 10.6(a) is then given approximately by $3a^2/\lambda_a$.

Since the width of the acoustic beam cannot be smaller than $2a$, a practical way to model the diffraction of a circular radiator is to assume that the beam is well confined within a cylinder of radius a up to distance z_F, thereafter assuming spherical beam divergence in a cone with apex angle $2\theta_T$, and with the apex centered in the middle of the radiator.[†] In experiments with two radiators, one transmitting and the other receiving, it has been found[24] that for radiator separations less than $6a^2/\lambda_a$, any obstructions at radial distances $r > 3a$ have no influence on the pressure.

The overall behavior of the near and far fields is shown in Fig. 10.7(b).

In order to find the intensity, equations $\vec{v} = -\nabla\tilde{\varphi}$ and $\tilde{p} = j\omega\rho_m\tilde{\varphi}$ are first combined to give

$$v_r = -\frac{1}{j\omega\rho_m}\frac{\partial p}{\partial r} \quad \text{and} \quad v_\theta = -\frac{1}{j\omega\rho_m}\frac{1}{r}\frac{\partial p}{\partial \theta} \tag{10.67}$$

and on using Eq. (9.63) it follows that

$$v_r = \frac{\tilde{p}(\omega, \vec{r})}{Z_l}\left(1 + \frac{1}{j\beta r}\right) \approx \frac{\tilde{p}(\omega, \vec{r})}{Z_l} \tag{10.68}$$

and

$$v_\theta = -\frac{p_0}{2Z_l}\left(\frac{a}{r}\right)^2 e^{-j\beta r}\frac{\partial}{\partial\theta}\left[\frac{2J_1(\beta a \sin \theta)}{\beta a \sin \theta}\right] \tag{10.69}$$

Using Eq. (9.20), it is found that $I_\theta = 0$ and that

$$I_r = \frac{p_0^2}{2Z_l}\left(\frac{\beta a^2}{2r}\right)^2\left[\frac{2J_1(\beta a \sin \theta)}{\beta a \sin \theta}\right]^2 \tag{10.70}$$

[†]It can be assumed that the pressure wave is a plane wave for $z < z_F$ and a spherical wave for $z > z_F$.

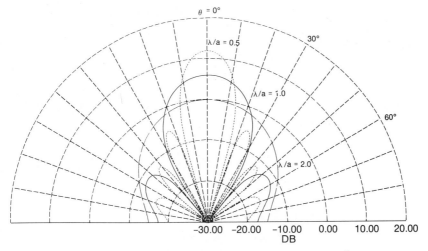

Figure 10.8. The polar diagram of the far field intensity, $10\log[2I_r Z_l/p_0^2]$, Eq. (10.70).

It is seen from Eqs. (10.63), (10.69), and (10.70) that $v_\theta = 0$ at intensity (pattern factor) maxima and the particle movement is purely radial, while at sidelobe zeros, $\text{jinc}(x) = 0$, the particle movement is purely angular. The polar plot of $\log[2Z_l I_r/p_0^2]$ is shown in Fig. 10.8.

10.4   CYLINDRICAL AND STRIP SYMMETRIES, NONUNIFORM EXCITATION VELOCITY, AND RADIATION IMPEDANCE

The Rayleigh formulas, Eq. (10.35) and (10.36), can be generalized to include the case of nonuniform velocity excitation. Instead of Eq. (10.34) for the initial and boundary conditions, we now have

$$\vec{v}(t,\rho,\psi,0) = \begin{cases} \hat{z}v_0 g(q)f(t), & \text{for } B \text{ on } A_0 \\ 0, & \text{elsewhere in the } z = 0 \text{ plane} \end{cases} \tag{10.71}$$

where g is the source shape function and (ρ, ψ, z) are the cylindrical coordinates, as shown in Fig. 10.6(b). The resulting formulas are

$$\varphi(t,\vec{r}) = \frac{v_0}{2\pi} \iint_{A_0} \frac{g(q)f(t - R/v_l)}{R} dA \tag{10.72}$$

$$p(t,\vec{r}) = \frac{\rho_m}{2\pi} \iint_{A_0} \frac{g(q)f'(t - R/v_l)}{R} dA \tag{10.73}$$

where $R^2 = \rho^2 + q^2 - 2q\rho \cos\psi + z^2$, where (r, θ, ψ) are the polar coordinates. Various field expressions will be written, for instance, as $p(\omega, \vec{\rho}) = p(\omega, \rho, \psi, z)$ in cylindrical coordinates, or $\vec{v}(t, \vec{r}) = v(t, r, \theta, \psi)$ in polar coordinates, where the absence of any argument means independence from that argument.

As usual, the impulse response is obtained for $f(t) = \delta(t)$ in preceding equations, and the corresponding steady-state formulas obtained as the Fourier transform of the impulse response are given by

$$\tilde{\varphi}(\omega, \vec{r}) = \frac{v_0}{2\pi} \iint_{A_0} \frac{g(q)\exp[-j(\omega R/v_l)]}{R} \, dA \qquad (10.74)$$

and

$$\tilde{p}(\omega, \vec{r}) = j\omega\rho_m \tilde{\varphi}(\omega, \vec{r}) \qquad (10.75)$$

Two forms of source shape function are usually studied, uniform excitation and Gaussian excitation. For a disk radiator of radius a, these are given by

$$g_U(q) = \chi(a - q) \qquad (10.76)$$

and

$$g_G(q) = \exp(-q^2/a^2) \qquad (10.77)$$

The Gaussian velocity excitation either on the disk or on a strip transducer, Fig. 10.9, produces a sidelobeless radiation pattern. Furthermore, the on-axis intensity distribution exhibits no minima or maxima. In actual design the dicing technique[25] can be used to make the radiator velocity distribution more uniform. Most practical transducers exhibit velocity distributions which are combinations of Gaussian and uniform, this being the reason that the zeros of the radiation pattern are not always discernible. All this is under the assumption that the sidelobe levels were measured using continuous wave excitation, allowing for complete cancellation of the fields to occur at nulls of the pattern factor. In the case of pulsed excitation, due to the presence of many frequencies, the nulls and the side lobes have much less distinct structure.

The Rayleigh formulas, Eqs. (10.72) and (10.73), which are written in the polar coordinates with no assumptions of symmetry, are quite general. If the radiator possesses certain symmetry, these forms can be further simplified. For instance, for the strip geometry[26] shown in Fig. 10.9, on using Eq. (10.75) with $g = g(x)$, it follows that

$$\tilde{p}(\omega, \vec{r}) = \frac{j\beta Z_l v_0}{2\pi} \int_{-b}^{+b} g(x') \, dx' \int_{-\infty}^{+\infty} \frac{e^{-j\beta R}}{R} \, dy \qquad (10.78)$$

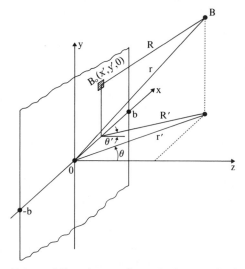

Figure 10.9. Polar and Cartesian coordinates in the case of strip symmetry.

where $R_z^2 = (x - x')^2 + (y - y')^2 + z^2$. However, since

$$\int_{-\infty}^{+\infty} \frac{e^{-j\beta R}}{R} dy = \frac{\pi}{j} H_0^{(2)}(\beta R') \tag{10.79}$$

where $R'^2 = z^2 + (x - x')^2$ and where $H_0^{(2)}$ is the zero-order Hankel function[16] of the second kind, it follows that

$$\tilde{p}(\omega, \vec{r}') = \frac{p_0 \beta}{2} \int_{-b}^{+b} g(x') H_0^{(2)}(\beta R') \, dx' \tag{10.80}$$

with $p_0 = Z_l v_0$. Using the Hankel function approximation for large arguments,

$$H_0^{(2)}(m) \simeq \left(\frac{2j}{\pi m} \right)^{1/2} e^{-jm} \tag{10.81}$$

it follows that

$$\tilde{p}(\omega, \vec{r}') = p_0 \left(\frac{j}{\lambda_a} \right)^{1/2} \int_{-b}^{+b} \frac{g(x') e^{-j\beta R'}}{\sqrt{R'}} dx' \tag{10.82}$$

where

$$\beta = \frac{2\pi}{\lambda_a}$$

Setting $R' \simeq r' - x'\sin\theta'$ and $g(x') = \chi(x' + b) - \chi(x' - b)$ in the numerator and $R' \simeq r'$ in the denominator with $\theta' \simeq \theta$, Fig. 10.9, we obtain

$$\tilde{p}(\omega, r', \theta) \simeq 2 p_0 b \left(\frac{j}{\lambda_a r'} \right)^{1/2} \frac{\sin(\beta b \sin\theta)}{\beta b \sin\theta} e^{-j\beta r'} \qquad (10.83)$$

This is seen to be similar to the diffraction pattern of an optical slit[2] characterized by the sinc function.

In cylindrical geometry, and for the case depicted in Fig. 10.6(b), it can be shown[27] that the Rayleigh formula, Eq. (10.74), can be simplified to

$$\tilde{\varphi}(\omega, \vec{\rho}) = \int_0^\infty \frac{e^{-\gamma z}}{\gamma} J_0(\rho\alpha) B(\alpha) \alpha \, d\alpha \qquad z \geqslant 0 \qquad (10.84)$$

where $\gamma^2 = \alpha^2 - \beta^2$, $\beta^2 = \omega^2/v_l^2$. In Eq. (10.84), $B(\alpha)$ is the (Fourier-Bessel integral) zero-order Hankel transform given by

$$B(\alpha) = \int_0^\infty J_0(\alpha q) v(q) q \, dq$$

$$v(q) = \int_0^\infty J_0(\alpha q) B(\alpha) \alpha \, d\alpha = - \left(\frac{\partial \tilde{\varphi}}{\partial z} \right)_{z=0} \qquad (10.85)$$

In defining impedance, it is important to note that a voltage across a piezoelectric transducer, acting as a radiator, is proportional to the pressure averaged over the radiating area of the transducer. In Eq. (9.21) the impedance has been evaluated with respect to the surface of constant phase, that is, a spherical surface. In general,[27] when the impedance is to be evaluated over a surface that is not the surface of constant phase, an averaging procedure is required. Since the surfaces of constant phase for acoustic waves emanating from the radiator are neither spherical nor planar, the impedance for the wave propagation in the $+z$ direction is defined at any plane $z =$ constant as

$$Z(\omega, z) = \frac{F(\omega, z)}{v_V(\omega, z)} \qquad (10.86)$$

where v_V is the volume-wave velocity defined as

$$v_V = \iint_{-\infty}^{+\infty} \tilde{v}(\omega, \vec{\rho}) \, dA = 2\pi \int_0^\infty v(\omega, \rho, z) \rho \, d\rho \qquad (10.87)$$

and where the last expression is valid for the case of cylindrical symmetry. The total force at any plane z = constant is given by

$$F = \iint_{-\infty}^{+\infty} \tilde{p}(\omega, \vec{\rho})\, dA = 2\pi j\omega\rho_m \int_0^{\infty} \tilde{\varphi}(\omega, \rho, z)\rho\, d\rho \qquad (10.88)$$

where again the last expression is valid in cylindrical symmetry, Fig. 10.6(b). Now consider the problem of evaluating the radiation impedance of the baffled disk radiator having a uniform velocity distribution. In this case, $v = v_0\chi(a - \rho)$ in Eq. (10.87) resulting in $v_V = \pi a^2 v_0$. In order to find the total force, z is set zero in Eq. (10.84). From Eq. (10.85) for $v(q) = v_0\chi(a - q)$, it is found that

$$B(\alpha) = v_0 \int_0^a J_0(\alpha q)q\, dq = \frac{v_0 a}{\alpha} J_1(\alpha a) \qquad (10.89)$$

Combining Eqs. (10.84), (10.85), (10.88), and (10.89), we have

$$F = 2\pi j\omega\rho_m a v_0 \int_0^a \int_0^{\infty} \frac{\rho}{\gamma} J_0(\rho\alpha) J_1(\alpha a)\, d\rho\, d\alpha$$

$$= 2\pi j\omega\rho_m a^2 v_0 \int_0^{\infty} \frac{J_1^2(\alpha a)}{\alpha\gamma}\, d\alpha \qquad (10.90)$$

The latter integral is given[27] by

$$\int_0^{\infty} \frac{J_1^2(\alpha a)}{\alpha\gamma}\, d\alpha = \frac{-j}{2\beta}\left[1 - \frac{J_1(2\beta a)}{\beta a} + j\frac{H_1(2\beta a)}{\beta a}\right] \qquad (10.91)$$

where H_1 is the Struve function.[16] Combining Eqs. (10.91), (10.90), and (10.86), it is found that

$$Z(\omega, 0) = \rho_m v_l\left[1 - \frac{J_1(2\beta a)}{\beta a} + j\frac{H_1(2\beta a)}{\beta a}\right]$$

$$\approx \rho_m v_l\begin{cases} \dfrac{(\beta a)^2}{2} + j\dfrac{8\beta a}{3\pi}, & \beta a \ll 1 \\ 1 + j0, & \beta a \gg 1 \end{cases} \qquad (10.92)$$

Thus, for $\beta a \gg 1$ it is seen that the radiation impedance is, on the average, equal to the plane-wave impedance. This is not an unexpected result, since this approximation has been used in Chap. 5 for equivalent circuit derivation of the piezoelectric transducers.

10.5 RAYLEIGH-SOMMERFELD DIFFRACTION

So far we have treated the response of the pressure field when the normal velocity (potential gradient) has been prescribed on the plane $z = 0$. The resulting integral is known as the Rayleigh-Sommerfeld diffraction formula with prescribed potential gradient or the Neumann problem in mathematics. In some situations arising in acoustical phase arrays for nondestructive testing of solids, experimental evidence[28] suggests that the boundary condition should be expressed in terms of pressure. In this case, the pressure would be nonzero over the area A_0 in Fig. 10.3(a) and zero in the rest of the $z = 0$ plane. The corresponding integral formulation is termed the Rayleigh-Sommerfeld diffraction formula with prescribed potential [in the steady state the pressure is proportional to the potential, see Eq. (10.75)], or the Dirichlet problem in mathematics. Note that what is known as the Fresnel-Kirchhoff integral formulation requires the prescription of both the potential and its gradient over the $z = 0$ plane, and as such represents an inconsistency.[20]

In what follows, the Rayleigh-Sommerfeld diffraction formula will be derived with prescribed boundary conditions in terms of pressure. For the geometry shown in Fig. 10.3(a), the boundary and initial conditions are

$$p(t, x, y, 0) = \begin{cases} p_0(\vec{r}) f(t), & \text{for } B \text{ on } A_0 \\ 0, & \text{elsewhere in the } z = 0 \text{ plane} \end{cases} \quad (10.93)$$

The required form[18] of the function $\tilde{g}$, Eq. (10.21), is now

$$\tilde{g}_- = \frac{\exp(-sR_1/v_l)}{R_1} - \frac{\exp(-sR_2/v_l)}{R_2} \quad (10.94)$$

with

$$\frac{\partial \tilde{g}_-}{\partial z_0} = -\left(\frac{s}{v_l} + \frac{1}{R_1}\right) \frac{\exp(-sR_1/v_l)}{R_1} \frac{\partial R_1}{\partial z_0} + \left(\frac{s}{v_l} + \frac{1}{R_2}\right) \frac{\exp(-sR_2/v_l)}{R_2} \frac{\partial R_2}{\partial z_0}$$

$$(10.95)$$

where, as before, R_1 and R_2 are given by Eq. (10.16), and where

$$\frac{\partial R_1}{\partial z_0} = -\frac{z - z_0}{R_1} = \cos \theta_1'$$

$$\frac{\partial R_2}{\partial z_0} = \frac{z + z_0}{R_2} = \cos \theta_2' \quad (10.96)$$

As before, R_1 and R_2 can be associated with the source point B_0 and its virtual image $\tilde{B}_0$, respectively, as depicted in Fig. 10.3(b). Angles θ_1' and θ_2' are the angles between R_1 and R_2 and the z direction, respectively, as shown in Fig. 10.3(b). For $z_0 = 0$, $R_1 = R_2 = R$, and $\cos \theta_2' = -\cos \theta_1'$, and therefore Eq. (10.95) reduces to

$$\frac{\partial \tilde{g}_-}{\partial z_0} = 2\left(\frac{s}{v_l} + \frac{1}{R}\right) \frac{\exp(-sR/v_l)}{R} \cos \theta' \qquad (10.97)$$

Since $\tilde{g} = 0$ in the $z = 0$ plane, Eq. (10.25) becomes

$$\tilde{\varphi}(s, \vec{r}) = \frac{1}{4\pi} \iint_{A_0} \tilde{\varphi} \frac{\partial \tilde{g}}{\partial z_0} dA \qquad (10.98)$$

and the Laplace transform of Eq. (10.33) yields

$$\tilde{p}(s, \vec{r}_0) = s\rho_m \tilde{\varphi}(s, \vec{r}_0) \qquad (10.99)$$

Combining the last three equations, we have

$$\tilde{\varphi}(s, \vec{r}) = \frac{1}{2\pi\rho_m} \iint_{A_0} \frac{\tilde{p}(s, \vec{r}_0)}{s} \left(\frac{s}{v_l} + \frac{1}{R}\right) \frac{\exp(-sR/v_l)}{R} \cos \theta' \, dA$$

$$(10.100)$$

Far away from the source region, $\lambda_a \ll R$, which is equivalent to $|v_l/s| \ll R$, and thus

$$\tilde{\varphi}(s, \vec{r}) = \frac{1}{2\pi\rho_m v_l} \iint_{A_0} \tilde{p}(s, \vec{r}_0) \frac{\exp(-sR/v_l)}{R} \cos \theta' \, dA \qquad (10.101)$$

or in the time domain,

$$\tilde{\varphi}(t, \vec{r}) = \frac{1}{2\pi\rho_m v_l} \iint_{A_0} \frac{p(t - R/v_l, \vec{r}_0)}{R} \cos \theta' \, dA \qquad (10.102)$$

and

$$p(t, \vec{r}) = \frac{1}{2\pi\rho_m} \iint_{A_0} \frac{p'(t - R/v_l, \vec{r}_0)}{R} \cos \theta' \, dA$$

The impulse response is obtained when $p = p_0(\vec{r}_0)\delta(t - R/v_l)$, and the steady-state equations are

$$\tilde{\varphi}(\omega, \vec{r}) = \frac{1}{2\pi Z_l} \iint_{A_0} \frac{p(\vec{r}_0)\exp[-j(\omega/v_l)R]}{R} \cos \theta' \, dA \qquad (10.103)$$

and

$$\tilde{p}(\omega, \vec{r}) = j\omega\rho_m\tilde{\varphi}(\omega, \vec{r}) = \frac{j}{\lambda_a} \iint_{A_0} \frac{p(\vec{r}_0)\exp[-j(\omega/v_l)R]}{R} \cos\theta' \, dA$$

(10.104)

where $\omega/v_l = 2\pi/\lambda_a$.

In the far field and in the strip geometry, Fig. 10.9, the required Green's function is the Hankel function for large argument, Eq. (10.81). It follows that

$$\tilde{p}(\omega, \vec{r}') = \left(\frac{j}{\lambda_a}\right)^{1/2} \int_{-b}^{+b} \frac{p_0(x')e^{-j\beta R'}}{\sqrt{R'}} \cos\theta' \, dx'$$

(10.105)

Except for the factor $\cos\theta'$, with $p_0(x') = p_0 g(x')$, this is seen to be equivalent to Eq. (10.82) derived before, but with the boundary conditions formulated in terms of the gradient of the potential. For uniform excitation, $p_0(x') = p_0[\chi(x' + b) - \chi(x' - b)]$, with $R' \simeq r' - x'\sin\theta'$ in the nominator and $R' \approx r'$ in the denominator of Eq. (10.105), and $\theta' \approx \theta$, it is found that

$$\tilde{p}(\omega, \vec{r}') \approx 2p_0 b\left(\frac{j}{\lambda_a r'}\right)^{1/2} \frac{\sin(\beta b \sin\theta)}{\beta b \sin\theta} \cos\theta e^{-j\beta r'}$$

(10.106)

Thus this type of source, sometimes called the Dirichlet source,[27] has smaller sidelobes, has two zeros in the pattern factor for $\theta = \pm\pi/2$, and is more directive than the Neumann source, Eq. (10.83).

The structure of the acoustic fields can be studied by a number of procedures.[29] The field patterns can be recorded by photographic emulsion,[23] cholesteric crystals,[30] schlieren technique,[31] thermoelectric probes,[32,33] miniature (diameter of about 1 mm) piezoelectric hydrophone probes,[34-36] and microsphere scatterers.[37,38]

10.6   FRESNEL AND FRAUNHOFER APPROXIMATIONS

Rectangular transducers[39] are also used extensively in acoustic arrays for imaging purposes.[40] With reference to Fig. 10.10, where P is a point on the radiator of area A and P_o a point in the observation plane, Eq. (10.75) can be written in cartesian coordinates as

$$\tilde{p}(\omega, \vec{r}) = \frac{j}{\lambda_a} \iint_A \frac{p(x, y)e^{-j\beta R}}{R} dA$$

(10.107)

with $\beta = \omega/v_l = 2\pi/\lambda_a$, $p(x, y) = Z_l v(x, y)$, and $R = [z^2 + (x_o - x)^2 +$

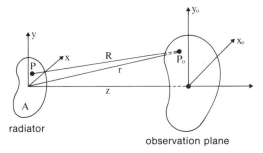

Figure 10.10. The geometry relevant to an acoustic imaging system.

$(y_o - y)^2]^{1/2}$, where $(x, y, 0)$ are the coordinates of the source points and (x_o, y_o, z) are coordinates of the observation plane. Due to complex dependence of R, Eq. (10.107) is analytically intractable. Commonly used approximations[2] are obtained by first assuming that the observation plane is far away from the transducer, that is, $z \gg x, y$, and R is written as

$$R = z\left(1 + \frac{x^2 + y^2}{z^2} + \frac{x_o^2 + y_o^2}{z^2} - 2\frac{xx_o + yy_o}{z^2}\right)^{1/2} \qquad (10.108)$$

The last three terms in the parentheses of Eq. (10.108) are, for $z \gg x, y$, much less than 1, and the Taylor expansion about z yields

$$R \simeq z + \frac{x^2 + y^2}{2z} + \frac{x_o^2 + y_o^2}{2z} - \frac{xx_o + yy_o}{z} \qquad (10.109)$$

The Fresnel approximation is obtained when $R \simeq z$ is substituted into the denominator of Eq. (10.107), and, since β can be very large, Eq. (10.109) is substituted in full in the phase term of Eq. (10.107),

$$\tilde{p}(\omega, \vec{r}) = \frac{je^{-j\beta z}}{\lambda_a z}\exp\left[-j\beta\left(\frac{x_o^2 + y_o^2}{2z}\right)\right]$$

$$\times \iint_A \exp\left[-j\beta\left(\frac{x^2 + y^2}{2z}\right)\right]p(x, y)\exp\left[j\beta\left(\frac{x_o x + y_o y}{z}\right)\right]dx\,dy$$

$$(10.110)$$

The above equation is the Fourier transform of the transducer excitation $p(x, y)$ multiplied by the quadratic phase, $\exp[-j\beta(x^2 + y^2)/2z]$ and evaluated at spatial frequencies $f_x = x_o/\lambda_a z$ and $f_y = y_o/\lambda_a z$. In most practical cases the presence of the quadratic phase term requires numerical evaluation of Eq. (10.110); however, the on-axis response $(x_o = y_o = 0)$ usually can be obtained.

If $z \gg \beta(x^2 + y^2)/2$, Eq. (10.110) can be written as

$$\tilde{p}(\omega, x_o, y_o, z) = \frac{je^{-j\beta z}}{\lambda_a z} \exp\left[-j\beta(x_o^2 + y_o^2)/2z\right]$$

$$\times \iint_A p(x, y)\exp\left[\frac{j\beta(x_o x + y_o y)}{z}\right] dx\, dy \quad (10.111)$$

which is termed the Fraunhofer approximation. It is seen that the pressure at the observation point $p_o(x_o, y_o) = \tilde{p}(\omega, x_o, y_o, z)$ is the two-dimensional Fourier transform of the transducer excitation at $p(x, y)$ multiplied by a slowly varying factor, often taken to be a constant, in front of the integral.

It has been pointed out that the voltage across the piezoelectric transducer, acting as a radiator, is proportional to pressure[†] averaged over the radiating area of the transducer. The average pressure is a function of the losses in the liquid, input voltage, and diffraction. The diffraction has an effect similar to losses and, for instance, in water, is the predominant mechanism of loss at frequencies of a few MHz.[41] These diffraction corrections have been tabulated and must be used when performing attenuation measurements.[42, 43]

10.7 ANGULAR SPECTRUM OF PLANE WAVES AND ACOUSTIC RADIATION IN SOLIDS

The Rayleigh-Sommerfeld diffraction for liquids, obtained from Eq. (10.100) and $\tilde{p}(\omega, \vec{r}) = j\omega\rho_m\tilde{\varphi}(\omega, \vec{r})$, can be written for $z > 0$, Fig. 10.3(a), as

$$p(x, y, z) = \frac{j\beta}{2\pi} \iint_{A_0} p(x_0, y_0)\left(\frac{z}{R}\right)\left(1 + \frac{1}{j\beta R}\right)\frac{\exp(-j\beta R)}{R} dx_0\, dy_0$$

$$(10.112)$$

where the notation $p(x, y, z) \equiv \tilde{p}(\omega, \vec{r})$ and $p(x_0, y_0) \equiv p(\vec{r}_0)$ was introduced and where $\cos\theta' = z/R$, $z > 0$ was used. In various acoustic radiation problems, the method called the expansion into the angular spectrum of plane waves, already stated in Sect. 9.4 for bounded acoustic beams, can be used as an alternative to Eq. (10.112). Its essence is the expansion of the pressure field $p(x, y, z)$ into the angular spectrum of plane waves.[44, 45] For instance, in the half-space $z > 0$, Fig. 10.3(a), this can be written as

$$p(x, y, z) = \int_{-\infty}^{+\infty}\int_{-\infty}^{+\infty} F(k_x, k_y)\exp(-j\vec{k}\cdot\vec{r})\frac{dk_x\, dk_y}{\beta^2} \quad (10.113)$$

[†] The "average pressure" is proportional to the total force over the area, as given by Eq. (10.88). See also Exercise 5.

where

$$\vec{k} = k_x \hat{x} + k_y \hat{y} + k_z \hat{z} = \beta(u\hat{x} + v\hat{y} + w\hat{z}) \tag{10.114}$$

and where u, v, and w are the directional cosines, $u^2 + v^2 + w^2 = 1$, of the wave vector $\vec{k}$. From Eq. (10.113) it follows that

$$p(x, y, z) = \int_{-\infty}^{+\infty} \int_{-\infty}^{+\infty} F(u, v)\exp[-j\beta(xu + yv + zw)] \, du \, dv$$

$$\tag{10.115}$$

with the value of $p(x, y, z)$ at $z = 0$ of

$$p(x, y, 0) = p(x_0, y_0) = \int_{-\infty}^{+\infty} \int_{-\infty}^{+\infty} F(u, v)\exp[-j\beta(xu + yv)] \, du \, dv$$

$$\tag{10.116}$$

Using the inverse Fourier transform, it is found that

$$F(u, v) = \left(\frac{\beta}{2\pi}\right)^2 \iint_{A_0} p(x_0, y_0)\exp[j\beta(ux_0 + vy_0)] \, dx_0 \, dy_0 \tag{10.117}$$

and combining Eqs. (10.115) and (10.117) it follows that

$$p(x, y, z) = \left(\frac{\beta}{2\pi}\right)^2 \int_{-\infty}^{+\infty} \int_{-\infty}^{+\infty} \iint_{A_0} p(x_0, y_0)$$

$$\times \exp\{-j\beta[(x - x_0)u + (y - y_0)v + zw]\} \, du \, dv \, dx_0 \, dy_0$$

$$\tag{10.118}$$

By differentiating the expression[46]

$$\int_{-\infty}^{+\infty} \int_{-\infty}^{+\infty} \frac{1}{w}\exp[-j\beta(Xu + Yv + Zw) \, du \, dv] = \frac{2\pi j}{\beta} \frac{\exp(-j\beta R)}{R}$$

$$\tag{10.119}$$

where $X = x - x_0$, $Y = y - y_0$, $Z = z$, $R^2 = X^2 + Y^2 + Z^2$, with respect to Z, it is found that

$$\int_{-\infty}^{+\infty} \int_{-\infty}^{+\infty} \exp[-j\beta(Xu + Yv + Zw)] \, du \, dv$$

$$= \frac{2\pi j}{\beta}\left(\frac{z}{R}\right)\left(1 + \frac{1}{j\beta R}\right)\frac{\exp(-j\beta R)}{R} \tag{10.120}$$

Finally, substituting Eq. (10.120) into Eq. (10.118) reduces the latter to Eq. (10.112). Therefore, in the particular case considered, the angular expansion into plane waves led to the Rayleigh-Sommerfeld diffraction formula for acoustic wave diffraction in liquids. Thus one would expect that the method is a very general one, applicable to other problems such as the acoustic wave diffraction in solids.

In isotropic solids the radiator stress fields $T_L(x_0, y_0)$, $T_{S1}(x_0, y_0)$, and $T_{S2}(x_0, y_0)$ corresponding to the longitudinal and the two shear waves, respectively, are in general coupled by the boundary condition requirements at the radiator–solid interface. It was shown by Miller and Pursey[13] and by Knopoff[14] that in isotropic solids, a purely longitudinal (or a purely shear) radiator stress field produces both longitudinal and shear wave radiation into the solid. Thus, shear and longitudinal radiation in isotropic solids cannot generally be considered separately.

Often, certain regions of space into which the acoustic power of a particular radiator is emitted contain a much higher intensity of one type of radiation (for instance, shear). For such regions, it can be assumed that $p(x_0, y_0) \approx T_{S1}(x_0, y_0)$ and Eq. (10.112) is directly applicable, where $p(x, y, z)$ is equivalent to the shear-stress radiation pattern $T_{S1}(x, y, z)$.

At high frequencies, due to much larger shear wave attenuation, and at sufficient distances from the radiator, it is possible to neglect shear wave radiation and to consider only the longitudinal wave radiation. Under these conditions, $p(x_0, y_0) \approx T_L(x_0, y_0)$ in Eq. (10.112), and $p(x, y, z)$ is equivalent to longitudinal-stress radiation pattern $T_L(x, y, z)$.

In anisotropic solids,[43–45,47] in addition to the coupling of the source fields $T_I(x_0, y_0)$, $I = S1, S2, L$ due to boundary conditions, there is, in general, coupling of the radiating fields $T_I(x, y, z)$ due to anisotropy. The decoupling of the radiating fields occurs only for propagation directions close to acoustic axes (threefold, fourfold, or sixfold), as explained in Chap. 7. For this case, assume that the acoustic axis coincides with the z axis in Fig. 10.3(a), and consider the wave-vector directions inside a cone defined by the apex semiangle θ_c, $\theta_c \ll 1$. Within the conical region the velocity surface of the particular mode[†] (L, $S1$, or $S2$) can be approximated by the elliptical surface.[48,43]

$$\left(\frac{v_x}{v_r}\right)^2 + \left(\frac{v_y}{v_r}\right)^2 + \left(\frac{v_z}{v_0}\right)^2 = 1 \tag{10.121}$$

where v_0 and v_r are the acoustic velocities along the z and x (y) directions, respectively. The components of the phase velocity, v_x, v_y, v_z, of an acoustic

[†] In what follows, unless otherwise stated, it is assumed that in the region of space considered, the radiation of one mode is of much higher intensity than the remaining two modes, which are therefore neglected.

wave propagating with velocity v at an angle $\theta \leqslant \theta_c$ to the z axis can then be written as

$$v_x^2 + v_y^2 = v^2 \sin^2\theta$$

$$v_z = v \cos\theta \tag{10.122}$$

and Eq. (10.121) is written as

$$\frac{\cos^2\theta}{v_0^2} + \frac{\sin^2\theta}{v_r^2} = \frac{1}{v^2} \tag{10.123}$$

By defining the wave numbers as $k_0 = \omega/v_0$, $k_r = \omega/v_r$, and $k = \omega/v$, this can be written as

$$k^2 = k_r^2 \sin^2\theta + k_0^2 \cos^2\theta \tag{10.124}$$

and using the approximations $\tan\theta \simeq \theta$, $\cos\theta \simeq 1 - \theta^2/2$ it follows that

$$k \simeq k_0(1 + B\theta^2) \tag{10.125}$$

where

$$\frac{k_r^2}{k_0^2} = \frac{v_0^2}{v_r^2} = \frac{1}{1 + 2B} \tag{10.126}$$

In the theory of acoustic diffraction in anisotropic solids, Eq. (10.125) is often referred to as the *parabolic approximation*.

Multiplying the numerator and denominator of each term on the left-hand side of Eq. (10.121) by t (time), the surface of constant phase is obtained as

$$R_e^2 = (x - x_0)^2 + (y - y_0)^2 + \frac{z^2}{1 + 2B} \tag{10.127}$$

where (x, y, z) are the coordinates of the observation point, (x_0, y_0) the coordinates of the source point, $x - x_0 = v_x t$, $y - y_0 = v_y t$, $z = v_z t$, $R_e = v_r t$, and Eq. (10.127) was used. By applying the angular expansion into plane waves to the problem so defined, it can be shown[44,47] that the resulting Rayleigh-Sommerfeld diffraction formula must be of the form

$$T_I(x, y, z) = \frac{jk_0}{2\pi} \iint_{A_0} T_I(x_0, y_0)\left(\frac{z}{R_e}\right)\left(1 + \frac{1}{jk_r R_e}\right)\frac{\exp(-jk_r R_e)}{R_e}\,dx_0\,dy_0$$

$$\tag{10.128}$$

where $T_I(x, y, z)$ and $T_I(x_0, y_0)$, $I = 3, 4$, or 5, are the stress fields in the observation and source planes, respectively. Using Eqs. (10.126) and (10.127), the phase term in the exponent of Eq. (10.128) is approximated by

$$k_r R_e \simeq k_0 z \left[1 + \frac{(x - x_0)^2 (1 + 2B)}{2z^2} + \frac{(y - y_0)^2 (1 + 2B)}{2z^2} \right]$$

(10.129)

and the expressions in the denominator of Eq. (10.128) by

$$k_r R_e \simeq k_0 z$$

$$R_e \simeq \frac{z}{(1 + 2B)^{1/2}}$$

(10.130)

resulting in

$$T_I(x, y, z) = \exp(-jk_0 z)\left(\frac{jk_0}{2\pi}\right)\left(\frac{1 + 2B}{z}\right) \iint_{A_0} T_I(x_0, y_0)\left(1 + \frac{1}{jk_0 z}\right)$$

$$\times \exp\left\{-jk_0 \left[\frac{(x - x_0)^2}{2} \frac{(1 + 2B)}{z} + \frac{(y - y_0)^2}{2} \frac{(1 + 2B)}{z}\right]\right\} dx_0 \, dy_0$$

(10.131)

Far away from the source and for sufficiently high frequencies, $(1 + 1/jk_0 z)$ $\simeq 1$, and Eq. (10.131) can be understood to mean that the effect of anisotropy is to decrease (increase) the distance between the source $T(x_0, y_0)$ and observed stress $T(x, y, z)$ from z to $z/(1 + 2B)$, depending on whether $B > 0$ or $B < 0$. In the isotropic case, $B = 0$ and for $B \to \infty$ the material would be self-collimating. From Eq. (10.126) it is seen that the latter condition required materials with $v_r/v_0 \to \infty$. Thus focusing and defocusing of acoustic beams is another inherent property of anisotropic materials in addition to those already described in Chap. 7. This is shown in Fig. 10.11.

As an example consider the diffraction from a rectangular aperture such that $T_I(x_0, y_0) = T_0$ for $|x_0| \leqslant W/2, |y_0| \leqslant L/2$, and zero otherwise. Introducing the variables

$$u = (x_0 - x)\left(\frac{k_0(1 + 2B)}{\pi z}\right)^{1/2}$$

$$v = (y_0 - y)\left(\frac{k_0(1 + 2B)}{\pi z}\right)^{1/2}$$

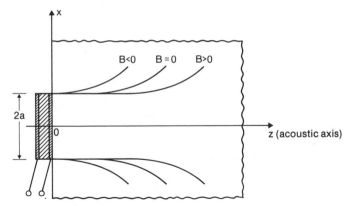

Figure 10.11. Illustration of the (de)focusing effect in anisotropic material with a strip radiator for three values of the parameter B. In isotropic materials, $B = 0$.

Eq. (10.131) can be written as

$$T(x, y, z) = j\frac{T_0\exp(-jk_0 z)}{2} \int_{u-}^{u+} \exp\left(-j\frac{\pi u^2}{2}\right) du \int_{v-}^{v+} \exp\left(j\frac{\pi u^2}{2}\right) du$$

$$(10.132)$$

with

$$u \pm = \left(x_0 \pm \frac{W}{2}\right)\left(\frac{k_0(1 + 2B)}{\pi z}\right)^{1/2}$$

$$v \pm = \left(y_0 \pm \frac{L}{2}\right)\left(\frac{k_0(1 + 2B)}{\pi z}\right)^{1/2}$$

The two-dimensional diffraction integral applicable, for instance, to a SAW device, Fig. 8.1(a), is obtained from Eq. (10.132) for $L \to \infty$. It follows that

$$T(x, y) = \frac{(-1 + j)T_0\exp(-jk_0 z)}{2} \int_{u-}^{u+} \exp\left(-j\frac{\pi u^2}{2}\right) du \quad (10.133)$$

Finally, it is important to note that the angular decomposition of plane waves has led to many useful theorems[49] for transducers operating in liquids, as well as for the transducers radiating into solid materials.[50] In the latter case the reciprocity relation[51] has proven to be of considerable importance.

In studies of acoustic radiation in solids, another approach, consisting of solving the wave equation under prescribed boundary conditions, yields more general results[13] in that the solution is valid for all values of z. To demonstrate the method, consider an infinite strip source, $|x| < a$, vibrating with frequency ω

normal to the free surface $z = 0$ of an isotropic material of semiinfinite extent, Fig. 10.11. The boundary conditions at $z = 0$ are given by

$$T_{xz} = 0 \quad \text{and} \quad T_{zz} = \begin{cases} 1, & |x| < a \\ 0, & |x| > a \end{cases} \tag{10.134}$$

Also, in this geometry $u_y = 0$, $\partial/\partial y = 0$. Combining Eq. (3.63) with $\nabla(\nabla \cdot \vec{u}) = \nabla^2 \vec{u} + \nabla \times \nabla \times \vec{u}$, it follows that

$$\rho_m \frac{\partial^2 \vec{u}}{\partial t^2} = c_{11} \nabla (\nabla \cdot \vec{u}) - c_{44} \nabla \times \nabla \times \vec{u} \tag{10.135}$$

Using auxiliary variables M and N defined as

$$M \equiv \nabla \cdot \vec{u} = \frac{\partial u_x}{\partial x} + \frac{\partial u_z}{\partial z}$$

$$N\hat{y} \equiv \nabla \times \vec{u} = \left(\frac{\partial u_x}{\partial y} - \frac{\partial u_z}{\partial x} \right) \hat{y} \tag{10.136}$$

this can be written as

$$c_{11} \frac{\partial M}{\partial z} - c_{44} \frac{\partial N}{\partial x} + \rho_m \omega^2 u_z = 0$$

$$c_{11} \frac{\partial M}{\partial x} + c_{44} \frac{\partial N}{\partial z} + \rho_m \omega^2 u_x = 0 \tag{10.137}$$

or

$$\frac{\partial^2 M}{\partial x^2} + \frac{\partial^2 M}{\partial z^2} + k_L^2 M = 0$$

$$\frac{\partial^2 N}{\partial x^2} + \frac{\partial^2 N}{\partial z^2} + k_S^2 N = 0 \tag{10.138}$$

where $k_L = \omega(\rho_m/c_{11})^{1/2}$ and $k_S = \omega(\rho_m/c_{44})^{1/2}$. The stresses are given as

$$T_{zz} = c_{12} M + 2c_{44} \frac{\partial u_z}{\partial z}$$

$$T_{xz} = c_{44} \left(\frac{\partial u_z}{\partial x} + \frac{\partial u_x}{\partial z} \right) \tag{10.139}$$

Combining Eqs. (10.137) and (10.139), it follows that

$$\frac{\rho_m \omega^2}{c_{44}^2} T_{zz} = 2 \frac{\partial^2 N}{\partial x \, \partial z} - m^2 (m^2 - 2) \frac{\partial^2 M}{\partial x^2} - m^4 \frac{\partial^2 M}{\partial z^2} \tag{10.140}$$

$$\frac{\rho_m \omega^2}{c_{44}^2} T_{xz} = \frac{\partial^2 N}{\partial x^2} - \frac{\partial^2 N}{\partial z^2} - 2m^2 \frac{\partial^2 M}{\partial x \, \partial z} \tag{10.141}$$

where $m = k_S/k_L = v_L/v_S$ is given by Eq. (3.72). The x dependence is removed by using the Fourier transform

$$\tilde{f}(\xi) = \int_{-\infty}^{+\infty} f(x)e^{-j\xi x}\, dx, \qquad f(x) = \frac{1}{2\pi}\int_{-\infty}^{+\infty} \tilde{f}(\xi)e^{j\xi x}\, d\xi \qquad (10.142)$$

Equations (10.137) to (10.141) can now be written as

$$\tilde{u}_z = -\frac{c_{44}}{\rho_m\omega^2}\left(m^2\frac{d\tilde{M}}{dz} - j\xi\tilde{N}\right), \qquad \tilde{u}_x = -\frac{c_{44}}{\rho_m\omega^2}\left(\frac{d\tilde{N}}{dz} + jm^2\xi\tilde{N}\right)$$

$$(10.143)$$

$$\frac{d^2\tilde{M}}{dz^2} - \left(\xi^2 - k_L^2\right)\tilde{M} = 0, \qquad \frac{d^2\tilde{N}}{dz^2} - \left(\xi^2 - k_S^2\right)\tilde{N} = 0 \qquad (10.144)$$

$$\frac{\rho_m\omega^2}{c_{44}^2}\tilde{T}_{zz} = -m^4\frac{d^2\tilde{M}}{dz^2} + 2j\xi\frac{d\tilde{N}}{dz} + m^2(m^2 - 2)\xi^2\tilde{M} \qquad (10.145)$$

$$\frac{\rho_m\omega^2}{c_{44}^2}\tilde{T}_{xz} = -\frac{d^2\tilde{N}}{dz^2} - 2jm^2\xi\frac{d\tilde{M}}{dz} - \xi^2\tilde{N} \qquad (10.146)$$

and the boundary conditions, Eq. (10.134),

$$\tilde{T}_{zz} = \frac{2\sin\xi a}{\xi}, \qquad \tilde{T}_{xz} = 0 \qquad (10.147)$$

The relevant solutions of Eqs. (10.144) are

$$\tilde{M} = A\exp\left[-z\left(\xi^2 - k_L^2\right)^{1/2}\right], \qquad \tilde{N} = B\exp\left[-z\left(\xi^2 - k_S^2\right)^{1/2}\right]$$

$$(10.148)$$

The constants A and B are determined by substituting Eqs. (10.147) and (10.148) in Eqs. (10.145) and (10.146) with $z = 0$. It follows that

$$\tilde{M} = \frac{2\rho_m\omega^2\left(k_S^2 - 2\xi^2\right)}{c_{44}^2 m^2\xi F(\xi)}\sin(\xi a)\exp\left[-z\left(\xi^2 - k_L^2\right)^{1/2}\right]$$

$$\tilde{N} = \frac{4j\rho_m\omega^2\left(\xi^2 - k_L^2\right)^{1/2}}{c_{44}^2 F(\xi)}\sin(\xi a)\exp\left[-z\left(\xi^2 - k_S^2\right)^{1/2}\right]$$

where

$$F(\xi) = \left(2\xi^2 - k_S^2\right)^2 - 4\xi^2\left[\left(\xi^2 - k_L^2\right)\left(\xi^2 - k_S^2\right)\right]^{1/2}$$

Using the last three equations and Eqs. (10.143) it follows

$$\tilde{u}_z = \frac{2\left(\xi^2 - k_L^2\right)^{1/2}\sin(\xi a)}{c_{44}\xi F(\xi)}\left\{2\xi^2\exp\left[-z\left(\xi^2 - k_S^2\right)^{1/2}\right]\right.$$

$$\left. + \left(k_S^2 - 2\xi^2\right)\exp\left[-z\left(\xi^2 - k_L^2\right)^{1/2}\right]\right\}$$

$$\tilde{u}_x = \frac{2\sin(\xi a)}{jc_{44}F(\xi)}\left\{2\left[\left(\xi^2 - k_L^2\right)\left(\xi^2 - k_S^2\right)\right]^{1/2}\exp\left[-z\left(\xi^2 - k_S^2\right)^{1/2}\right]\right.$$

$$\left. + \left(k_S^2 - 2\xi^2\right)\exp\left[-z\left(\xi^2 - k_L^2\right)^{1/2}\right]\right\}$$

Miller and Pursey[13] where able to derive asymptotic expressions from the Fourier inversion of the last two equations. However, in all practical cases, numerical methods are required for integral evaluation. This approach also yields a surface wave solution in the plane $z = 0$. The results of an asymptotic numerical evaluation, $a/\lambda_a \rightarrow 0$ in Fig. 10.11, are shown in Figs. 10.12 to 10.14 for three values of Poisson's ratio (0.05, 0.33, and 0.45). The radiation pattern shown are those associated with the longitudinal displacement u_z and the shear displacement u_x.

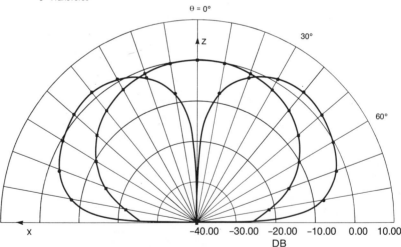

Figure 10.12. Radiation pattern of the strip radiator shown in Fig. 10.11 in the limit of $a/\lambda_a \rightarrow 0$ for the value of Poisson's ratio $\sigma = 0.05$. The radiation patterns shown are those associated with the displacements u_z and u_x.

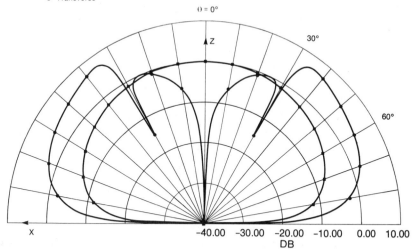

Figure 10.13. Radiation pattern of the strip radiator shown in Fig. 10.11 in the limit of $a/\lambda_a \to 0$ for the value of Poisson's ratio $\sigma = 0.33$. The radiation patterns shown are those associated with the displacements u_z and u_x.

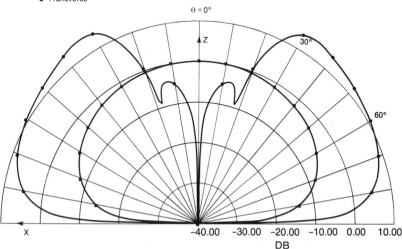

Figure 10.14. Radiation pattern of the strip radiator shown in Fig. 10.11 in the limit of $a/\lambda_a \to 0$ for the value of Poisson's ratio $\sigma = 0.45$. The radiation patterns shown are those associated with the displacements u_z and u_x.

10.8   ACOUSTIC LENSES

In this chapter, the analogy between optics and acoustics has been emphasized. Therefore one would expect that optical devices such as lenses and reflectors would have their acoustical counterparts.[52] One such device, the acoustical prism coupler, has already been considered in Chap. 9.

If the acoustic wavelength is short enough, a natural approach to solving certain problems is geometrical acoustics would be to use the ray technique, a simple method of geometrical optics. However, it must be remembered that there is no perfect analogy between geometrical optics and geometrical acoustics.[52] For instance, in isotropic solids an obliquely incident acoustic wave generates two modes (longitudinal and shear), while in optics only one mode (TEM) is present. The existence of two modes in an acoustic lens may produce a double focus. Furthermore, the ratios of the geometrical dimensions to light wavelength (~ 0.6 μm) for optical lenses are on the order of 10^4, while in acoustics this ratio is on the order of 20. The latter result has been computed by assuming a 5-MHz acoustic wave in Lucite of 0.5-mm wavelength and taking the thickness of the lens to be about 1 cm. Also, interference phenomena are much more pronounced in geometrical acoustics. For instance, acoustic waves emitted from parts of a lens separated by few millimeters interfere strongly at the focal point. In optics there may be several hundred wavelengths difference in phase between parts of an optical beam a few millimeters apart. The imperfections of the lens, either on the surface or in the volume, may be on the order of a fraction of the wavelength, thus causing no interference at the focal point.

As in optics, concave reflectors, zone plates, prisms, and lenses are used in acoustics[52,53] as well. Acoustic lenses have far wider applications than other acoustic energy concentrators. Two types of acoustic lenses[54] can be distinguished: immersion-type lenses, where a liquid surrounds a lens made of solid material, Fig. 10.15(a), and direct contact lenses also called angled lenses, in which only solid materials are employed, Fig. 10.15(b). Acoustic lenses can also be made with liquid materials. In this case a shell made of acoustically thin walls is used to define the shape of the lens.

From the point of view of energy losses, if $Z_L = \rho_{ml} v_L$ is the acoustic impedance of the lens and $Z_l = \rho_m v_l$ is the acoustic impedance of surrounding material, it is desirable to have $Z_L = Z_l$. However, the focusing process requires that $v_l \neq v_L$, since the index of refraction is defined as the ratio of velocities. Solid lenses are usually made of Lucite, sapphire, quartz, or some piezoelectric or synthetic acoustic material (Chap. 1).

For immersion lenses, the acoustic velocity in the lens is generally higher than the acoustic velocity of the surrounding liquid. This is the opposite of what happens in optics for the corresponding light velocities. Therefore, acoustic converging lenses are concave. Consider a plano-concave lens, shown in Fig. 10.16, where the concave side is made with a curvature corresponding to the spherical radius R. For plane acoustic waves incident from the left, the focal distance f is determined by the requirement that at the focus F all rays

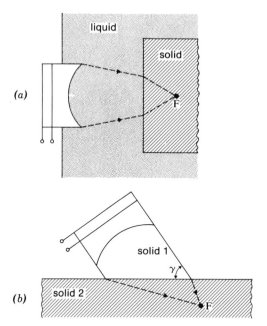

Figure 10.15. (*a*) Immersion type lens. (*b*) Angled lens.

must have equal phase. Requiring that rays I and II have zero phase difference at F, it follows that

$$\frac{f}{v_l} = \frac{d-s}{v_L} + \frac{1}{v_l}\left[(f-d+s)^2 + 2R(d-s) + (d-s)^2\right]^{1/2} \qquad (10.149)$$

or

$$f = \frac{R}{1 - v_l/v_L} - \frac{(d-s)(v_l/v_L)^2}{2(1 - v_l/v_L)} \qquad (10.150)$$

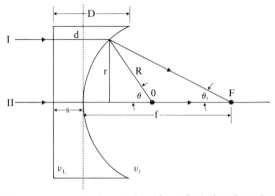

Figure 10.16. Plano-concave lens made with a spherical surface of radius R.

For small apertures, $(d - s) \approx 0$, and

$$f \approx \frac{R}{1 - v_l/v_L} \tag{10.151}$$

As the aperture increases, as seen from Eq. (10.150), the focal point moves toward the lens and the focal area is axially widened.

Various formulas derived for plane radiators in the preceding sections can be used here as a good approximation, provided that the radiating surface is only slightly curved.[55]

The function of the lens is to increase the pressure at the focal point and to bring the far-field diffraction pattern to a plane much closer to the lens. It has been shown[55] that the relative pressure in the focal plane at angle θ from the axis is proportional to the jinc function, Eq. (10.65), which is the pattern factor of a plane circular disk at large distances from the disk. This can be formulated as follows. If the pressure in front of the lens is given by Eq. (10.110), then the pressure behind the lens is also given by Eq. (10.110) provided that the multiplicative term

$$\exp\left[j\beta\left(\frac{x^2 + y^2}{2f} \right) \right] \tag{10.152}$$

is included in the integrand, that is,

$$\iint_A \exp\left[-\frac{j\beta}{2}(x^2 + y^2)\left(\frac{1}{z} - \frac{1}{f} \right) \right] p(x, y)\exp\left[j\beta\left(\frac{xx_0 + yy_0}{2} \right) \right] dx\, dy \tag{10.153}$$

In the above expression the quadratic term vanishes in the focal plane, $z = f$, reducing Eq. (10.153) to the far-field approximation given by Eq. (10.111).

In acoustics, due to a high mismatch ratio v_l/v_L, a single concave spherical surface such as that shown in Fig. 10.16 is capable of achieving a diffraction-limited focal point. As in optics, the numerical aperture[56] is given by

$$\text{N.A.} = \sin \theta_1 = \sin\left[\left(1 - \frac{v_l}{v_L} \right)\theta \right] \tag{10.154}$$

where θ and θ_1 are shown in Fig. 10.16, and the f-number is given approximately by f-number $= 0.5$ N.A. The resolution[56] corresponding to the Rayleigh criterion is given by

$$\epsilon = \frac{0.61\lambda_l}{\text{N.A.}} \tag{10.155}$$

where λ_l is the acoustic wavelength in the liquid. An ideal lens transforms a plane wave into a spherical wave. In practice there is always a certain amount of deviation between the spherical wave and a wave produced by a lens. The departure from the spherical wavefront, called the *spherical aberration*,[56] is defined as the variation of focus with aperture. For instance, for the single concave spherical surface lens shown in Fig. 10.16, the spherical aberration is given approximately by[57]

$$W(\theta, c) = 2c^2(1 - c)\left[\sin^4\frac{\theta}{2} + 2c(1 - c)\sin^6\frac{\theta}{2}\right]R \qquad (10.156)$$

where θ defines the aperture and $c = v_l/v_L$. If $W(\theta, c)$ is less than $\lambda_l/4$, the lens is nearly diffraction limited.

It is important to note that for angled lenses, Fig. 10.15(b), the radius of curvature of the lens must satisfy certain conditions[54] in order to avoid the total reflection of a part of the emitted beam at the interface between solid 1 and solid 2.

In anisotropic materials and along the acoustic axis, Fig. 10.17, the focal point of an acoustic lens will be dependent on the value of B. For this case, Eq. (10.151) is written as

$$f \simeq \frac{R(1 + 2B)}{1 - v_{L2}/v_{L1}}. \qquad (10.157)$$

However, the stress intensity at the focal point remains unaffected[47] by anisotropy and is equal to that of the corresponding isotropic material ($B = 0$).

In many fields, such as in medical ultrasonic imaging and in NDT, it is often necessary to scan rapidly a volume or a surface of a specimen while maintaining an adequate resolution in both lateral dimensions. Acoustic lenses with mechanical movements often cannot meet these requirements. Acoustic array transducers,[58] consisting of a large number of small transducers and allowing the composite acoustic beam to be controlled electronically, are used in these

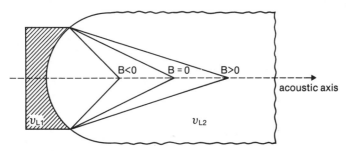

Figure 10.17. The (de)focusing effect in anisotropic materials with a cylindrical lens for three values of the parameter B. In isotropic materials, $B = 0$.

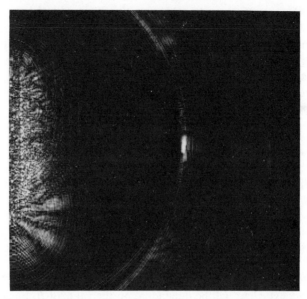

Figure 10.18. Image of a fatigue crack (bright spot close to center of the photograph) of 10 mm width and 3.5 mm depth in aluminum produced by a 32-element shear wave linear array,[59] using synthetic aperture imagining system.[60] The fringes at the left are due to transducer and amplifier ring-down. The center frequency of the transducers was 3.3 MHz. The fringes near the center top and bottom of the photograph are due to image sidelobes. (*Courtesy of G. S. Kino, Stanford University.*)

applications. By changing electronically the delays of the individual transducers across the aperture, it is possible to steer the composite acoustic beam of the array in a desired direction and to focus the composite acoustic beam at a desired point. Figure 10.18 shows the image of a fatigue crack (bright spot close to the center of the photograph) produced by a 32-element shear-wave linear array.[59] The image was produced using a synthetic aperture imaging system.[60]

10.9   DIFFRACTION OF LIGHT BY ACOUSTIC WAVES

In 1906, Pockels[61] showed that when a transparent material is mechanically deformed its index of refraction is affected. Some time later, Brillouin[62] predicted the diffraction of light by acoustic waves, which was subsequently demonstrated by Debye and Sears.[63] Raman and Nath[64,65] gave the description of the acousto-optic interaction in which the acoustic wave was considered as a phase grating.[2] Von Laue and coworkers[66] demonstrated in 1913 the diffraction pattern using X-rays and derived equations (Laue equations) for X-ray diffraction in crystals. Bragg[67] observed that beams of X-rays diffracted from a crystal could be linked to reflections from the crystal lattice plane and developed an equation (Bragg equation) useful in practical work.

In current literature on acousto-optics, Bragg diffraction is referred to as the interaction of acoustic and light waves when there is only a single order of diffracted light, Fig. 10.19(a). Raman-Nath diffraction, Fig. 10.19(b), results in the many orders of a diffracted light beam. Brillouin scattering is presently referred to as light scattering from thermal fluctuations (phonons). In certain materials the acousto-optic interaction can cause an intense light wave and an intense acoustic wave,[68] which is called stimulated Brillouin scattering.

In 1967 Klein and Cook[69] presented a unified theory treating both Raman-Nath and Bragg diffraction and used the quantity Q, defined as

$$Q = \frac{2\pi}{n_0} \frac{\lambda_0 L}{\lambda_a^2} \qquad (10.158)$$

where n_0 is the unperturbed index of refraction, λ_0 is the free-space wavelength

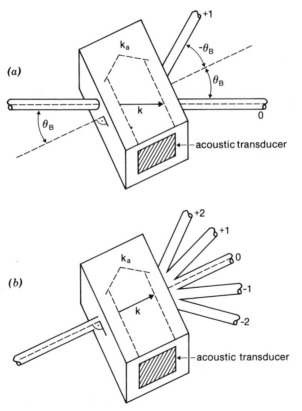

Figure 10.19. (a) The Bragg type of acousto-optic interaction. k_a and k are acoustical and optical wavenumbers, respectively, and θ_B is the Bragg angle. (b) The Raman-Nath type of acousto-optic interaction.

of the incident light, $\lambda_0 = n_0\lambda$, $k = 2\pi/\lambda$, λ_a is the acoustic wavelength, $k_a = 2\pi/\lambda_a$, and L is the interaction length. The interaction length is a distance measured along the direction of light propagation, over which the light and the acoustic waves overlap. For $Q < 1$, Raman-Nath diffraction occurs, and for $Q > 10$ Bragg diffraction takes place. The acousto-optic diffraction of light has been observed in both liquids and solids belonging to all crystal classes. Acousto-optic devices with Bragg-type diffraction developed for optical beam control are now widely used.[70-72]

Bragg diffraction can be understood by using the following model.[73] The index of refraction perturbation due to a homogeneous plane longitudinal wave propagating in the z direction can be written as

$$n(t, z) = n_0 + \Delta n \sin(\omega_a t - k_a z) \tag{10.159}$$

where Δn is the change in the index of refraction arising from the presence of the acoustic wave. Because the acoustic wave velocity is much smaller than the light wave velocity, the light beam encounters a phase grating of spatial periodicity given by the acoustic wavelength λ_a. This is due to the fact that since the index of refraction is higher in the compressed regions of the material, the light reflection coefficient will be higher in these regions also and lower in rarefied regions. Therefore the compressed regions can be approximated by partially reflecting mirrors. Consider a light beam incident at an angle θ_i on such a structure, Fig. 10.20(a). In order for the light diffraction to occur in a given direction, all the points on a mirror must contribute to the diffraction in this direction. Consider the diffraction from points R and Q in the "mirror" (M) in Fig. 10.20(a). For constructive interference of the light wave to occur in the θ_r direction, the optical path difference $NQ - RP$ must be an integral multiple of the free-space optical wavelength, λ_0. This can be written as

$$b(\cos\theta_i - \cos\theta_r) = m\lambda_0 \tag{10.160}$$

where $m = 0, \pm 1, \pm 2, \ldots$. By considering other points located between points P and Q on the mirror (M), it is found that Eq. (10.160) is satisfied only for $m = 0$, that is, for $\theta_i = \theta_r$. Since there are a number of mirrors, a requirement must be imposed that the reflections from two adjacent mirrors, Fig. 10.20(b), must add up in phase. The phase difference of the two optic wavefronts AC and BC, $AD + DB$, must be equal to the free-space optical wavelength λ_0.[73] Therefore it follows that

$$\sin\theta_B \simeq \theta_B = \frac{\lambda_0}{2\lambda_a} \tag{10.161}$$

where θ_B is the Bragg angle. Furthermore, the conservation of energy[72] requires that the frequency of the diffracted light ω_1 be upshifted with respect to the

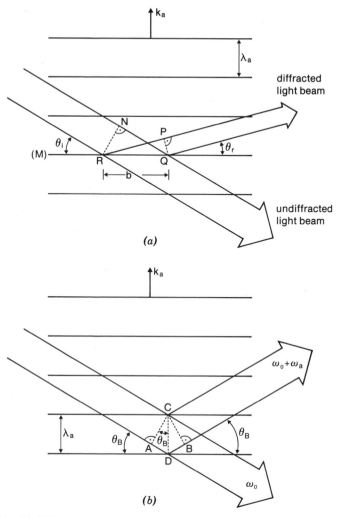

Figure 10.20. Modeling of the Bragg interaction with partially reflecting mirrors. (*a*) Effect of reflection from points *R* and *Q*. (*b*) Reflections from two adjacent mirrors.

incident light, ω_0, by the acoustic frequency ω_a,

$$\omega_1 = \omega_0 + \omega_a \tag{10.162}$$

In Raman-Nath diffraction, Fig. 10.19(*b*), the light beam is near normal incidence, $\theta_i \approx 90°$. It can be shown[2] that when a coherent light beam of frequency ω_0 is normally incident on a phase grating, the light is diffracted into several orders, Fig. 10.19(*b*). The angle of diffraction θ_n measured from the light beam axis is given by

$$\sin \theta_n \simeq \theta_n = \frac{n\lambda_0}{\lambda_a} \tag{10.163}$$

$n = 0, \pm 1, \pm 2, \ldots$, and the frequency of the diffracted light orders ω_n is given by

$$\omega_n = \omega_0 + n\omega_a. \tag{10.164}$$

Although the model presented was developed with longitudinal acoustic waves, shear acoustic waves also diffract light beams. The intensity of the diffracted light beam in Bragg diffraction and the intensity of the diffracted light beams of order $n = \pm 1$ in Raman-Nath diffraction are to a good approximation all proportional to acoustic power integrated over the interaction length. The latter property is used extensively in various diagnostic visualization techniques[74,75] and various optical modulators and light beam deflectors.[70–72]

REFERENCES

1. J. W. Strutt, Lord Rayleigh, *The Theory of Sound*, Vol. 2, Dover, New York, 1945.
2. M. Born and E. Wolf, *Principles of Optics*, Pergamon, London, 1959.
3. G. A. Maggi, *Ann. di Mat. Ser. 2a*, **16**, 21 (1888).
4. A. Rubinowicz, *Ann. Phys. Ser. 4*, **53**, 257 (1917).
5. K. Miyamoto and E. Wolf, "Generalization of the Maggi-Rubinowicz Theory of the Boundary Diffraction Wave—Part I," *J. Opt. Soc. Am.*, **52**, 615 (1962); *J. Opt. Soc. Am.*, **52**, 626 (1962).
6. J. B. Keller, "Geometrical Theory of Diffraction," *J. Opt. Soc. Am.*, **52**, 116 (1962).
7. A. Rubinowicz, "The Miyamoto-Wolf Diffraction Wave," in *Progress in Optics* (E. Wolf, Ed.), Vol. IV, Wiley, New York, 1965, Chap. 5.
8. A. Schoch, *Akust. Z.*, **6**, 318 (1942).
9. H. Stenzel, *Ann. Phys.* 5, **41**, 242 (1942).
10. A. H. Carter and A. O. Williams, "A New Expansion for the Velocity Potential of a Piston Source," *J. Acoust. Soc. Am.*, **23**, 179–184 (1951).
11. O. G. Kozina and G. I. Makarov, "Transient Processes in the Acoustic Fields of Special Piston Membranes," *Sov. Phys.-Acoust.*, **8**, 49 (1962).
12. D. N. Chetayev, *Dokl. Akad. Nauk SSSR*, **76**, 813 (1951).
13. G. F. Miller and H. Pursey, "The Field and Radiation Impedance of Mechanical Radiators on the Free Surface of Semi-Infinite Isotropic Solids," *Proc. Roy. Soc. (London) Ser. A*, **223**, 521–541 (1954).
14. L. Knopoff, "Diffraction of Elastic Waves," *J. Acoust. Soc. Am.*, **28**, 217–229 (1956).
15. J. D. Kraus and K. R. Carver, *Electromagnetics*, McGraw-Hill, New York, 1973, Chap. 12.
16. See, for instance, M. Abramowitz and I. A. Stegun, Eds., *Handbook of Mathematical Functions*, Dover, New York, 1970.
17. A. J. Rudgers, "Spatial Impulse Response of an Acoustic Line Radiator—A Study of Boundary-Diffraction-Wave Phenomena and their Experimental Detection," NRL Rep. 8191, 1978.

18. E. Wolf, "Some Recent Research on Diffraction of Light," *Proc. Symp. Mod. Opt., Polyt. Inst. Brooklyn*, March 23–24, 1967, Wiley, New York, 1967, pp. 435–452.

19. P. M. Morse and K. V. Ingard, *Theoretical Acoustics*, McGraw-Hill, New York, 1968, Chap. 7.

20. See, for instance, J. W. Goodman, *Introduction to Fourier Optics*, McGraw-Hill, New York, 1968, Chap. 3.

21. P. R. Stephanishen, "Transient Radiation from Pistons in an Infinite Baffle," *J. Acoust. Soc. Am.*, **49**(5) (Part 2), 1629–1638 (1971).

22. Y. Torikai and K. Negishi, "On the Ultrasonic Field Near the Circular Piston-Like Source: I. Visualization of the Ultrasonic Fields," *J. Acoust. Soc. Japan*, **13**, 117 (1957).

23. J. T. Dehn, "Interference Patterns in the Near Field of a Circular Piston," *J. Acoust. Soc. Am.*, **32**(12), 1692–1696 (1960).

24. E. F. Carome, J. M. Witting, and P. A. Fleury, *J. Acoust. Soc. Am.*, **33**(10), 1417–1425 (1961).

25. D. E. Hann, "Experimental Evaluation of Diced-Ceramic Elements for Sonar Transducers," *1972 Ultrason. Symp. Proc.*, 95–98 (1972).

26. C. J. Drost, "Near and Far Field of Strip-Shaped Acoustic Radiators," *J. Acoust. Soc. Am.*, **65**(3), 565–572 (1979).

27. M. Greenspan, "Piston Radiator: Some Extension of the Theory," *J. Acoust. Soc. Am.*, **65**(3), 608–621 (1979).

28. A. R. Selfridge, G. S. Kino, and B. T. Khuri-Yakub, "A Theory for the Radiation Pattern of a Narrow-Strip Acoustic Transducer," *Appl. Phys. Lett.*, **37**(1) 35–36 (1980).

29. E. P. Papadakis, "Theoretical and Experimental Methods to Evaluate Ultrasonic Transducers for Inspection and Diagnostic Applications," *IEEE Trans. Sonics Ultrason.*, **SU-26**, 14–27 (1979).

30. B. D. Cook and R. E. Werchen, "Mapping Ultrasonics Fields with Cholesteric Liquid Crystals," *Ultrason.*, **9**, 101–102 (1971).

31. L. R. Dragonette, "Schlieren Visualization of Radiation Caused by Illumination of Plates with Short Acoustic Pulses," *J. Acoust. Soc. Am.*, **51**, 920–935 (1971).

32. F. Dunn and W. J. Fry, "Precision Calibration of Ultrasonic Fields by Thermoelectric Probes," *IRE Trans. Ultrason. Eng.*, **UE-5**, 59–65 (1957).

33. P. E. Woolley, R. J. Barnett, and J. B. Pond, "The Use of Probes to Measure Megahertz Ultrasonic Fields in Liquids," *Ultrason.*, **13**, 68–72 (1975).

34. J. Saneyoshi, M. Okujima, and M. Ide, "Wide Frequency Calibrated Probe Microphones for Ultrasound in Liquid," *Ultrason.*, **4**, 64–66 (1966).

35. E. V. Romaneko, "Miniature Piezoelectric Ultrasonic Receivers," *Soviet Phys.-Acoust.*, **3**, 364–370 (1957).

36. P. A. Lewin, "Miniature Piezoelectric Polymer Ultrasonic Hydrophone Probes," *Ultrason.*, **19**, 213–216 (1981).

37. P. L. Edwards and J. Jarzynski, "Use of Microsphere Probe for Pressure Field Measurements in the Megahertz Frequency Range," *J. Acoust. Soc. Am.*, **68**, 356–359 (1980).

38. P. L. Edwards, B. D. Cook, and H. D. Darby, "Comparison of the Experimental and Theoretical Pressure Fields in the Near Field of Ultrasonic Transducer-Lens Systems," *J. Acoust. Soc. Am.*, **68**, 1528–1530 (1980).

39. A. Freeman, "Sound Field of a Rectangular Piston," *J. Acoust. Soc. Am.*, **32**(2), 197–209 (1960).

40. A. Macovski, "Theory of Imaging with Arrays," in *Acoustic Imaging* (G. Wade, Ed.), Plenum, New York, 1976.

41. H. Seki, A. Granato, and R. Truell, *J. Acoust. Soc. Am.*, **28**, 230 (1956).

42. R. Truell, C. Elbaum, and B. B. Chick, *Ultrasonics Methods in Solid State Physics*, Academic, New York, 1969.

43. E. P. Papadakis, "Ultrasonic Diffraction Loss and Phase Change in Anisotropic Materials," *J. Acoust. Soc. Am.*, **40**(4), 863–876 (1966).

44. L. Bergstein and T. Zachos, "A Huygens Principle for Uniaxially Anistropic Media," *J. Opt. Soc. Am.*, **56**(7), 931–937 (1966).

45. N. R. Ogg, "A Huygens' Principle for Anisotropic Media," *J. Phys. A: Gen. Phys.*, **4**, 382–388 (1971).

46. L. Brekhovskih, *Waves in Layered Media*, Academic, New York, 1960.

47. M. G. Cohen, "Optical Study of Ultrasonic Diffraction and Focusing in Anisotropic Media," *J. Appl. Phys.*, **38**(10), 3821–3828 (1967).

48. P. C. Watermann, Phys. Rev., **113**, 1240 (1959).

49. E. B. Miller and A. D. Yaghijan, "Two Theoretical Results Suggesting a Method for Calibrating Ultrasonic Transducers by Measuring the Total Near-Field Force," *J. Acoust. Soc. Am.*, **66**(6), 1601–1608 (1979).

50. A. Atalar, "A Backscattering Formula for Acoustic Transducers," *J. Appl. Phys.*, **51**(6), 3093–3098 (1980).

51. B. A. Auld, "General Electromechanical Reciprocity Relations Applied to the Calculation of Elastic Wave Scattering Coefficient," *Wave Motion*, **1**, 3–10 (1979).

52. T. Tarnoczy, "Sound Focusing Lenses and Waveguides," *Ultrason.*, July–Sept., 115–127 (1965).

53. L. D. Rozenberg, Ed., *Sources of High-Intensity Ultrasound*, Vol. 2, Plenum, New York, 1969.

54. U. Schlengermann, "Criteria for the Selection of Focusing Ultrasonic Probes," *Materialprüf.*, **19**(10), 416–420 (1977).

55. H. T. O'Neill, "Theory of Focusing Radiators," *J. Acoust. Soc. Am.*, **21**(5), 516–526 (1949).

56. W. J. Smith, *Modern Optical Engineering*, McGraw-Hill, New York, 1966, Chaps. 3 and 6.

57. R. A. Lemons and C. F. Quate, "Acoustic Microscopy," in *Physical Acoustics* (W. P. Mason and R. N. Thurston, Eds.), Vol. XIV, Academic, New York, 1979.

58. A. Macovski, "Ultrasonic Imaging Using Arrays," *Proc. IEEE*, **67**, 484–495 (1979).

59. R. L. Baer, A. R. Selfridge, B. T. Khuri-Yakub, and G. S. Kino, "Contacting Transducers and Transducer Arrays for NDE," *1981 Ultrason. Symp. Proc.*, 969–973 (1981).

60. G. S. Kino, D. Corl, S. Bennett, and K. Peterson, "Real Time Synthetic Aperture Imaging System," *1980 Ultrason. Symp. Proc.*, 722–731 (1980).

61. F. Pockels, *Lehrbuch der Kristalloptik*, Teubner, Leipzig, 1906.

62. L. Brillouin, "Diffusion de la lumière et des rayons X par un corps transparent homogène," *Ann. Physique*, **17**, 88 (1922).

63. P. Debye and F. W. Sears, "On the Scattering of Light by Supersonic Waves," *Proc. Nat. Acad. Sci. U.S.*, **18**, 409 (1932).

64. C. V. Raman and N. S. N. Nath, "The Diffraction of Light by High Frequency Sound," *Proc. Indian Acad. Sci.*, Part I, **2A**, 406 (1935); Part II, **2A**, 413 (1935); Part III, **3A**, 75 (1936); Part IV, **3A**, 119 (1936); Part V, **3A**, 459 (1936).

65. N. S. N. Nath, "Generalized Theory," *Proc. Indian Acad. Sci.*, **4A**, 222 (1937).

66. W. Friendrich, P. Knipping, and M. von Laue, *Ann. Phys.*, **41**, 971 (1913).

67. W. L. Bragg, *Proc. Cambridge Phil. Soc.*, **17**, 43 (1913).

68. R. Y. Chiao, C. H. Townes, and B. P. Stoicheff, *Phys. Rev. Lett.*, **12**, 592 (1964).

69. W. R. Klein and B. D. Cook, "Unified Approach to Ultrasonic Light Diffraction," *IEEE Trans. Sonics Ultrason.*, **SU-14**, 123–134 (1967).

70. R. W. Damon, W. T. Maloney, and D. H. McMahon, "Interaction of Light and Ultrasound and Applications," in *Physical Acoustics* (W. P. Mason and R. N. Thuston, Eds.), Vol. 7, Academic, New York, 1970, Chap. 5.

71. E. G. Lean, "Interaction of Light and Acoustic Surface Waves," in *Progress in Optics* (E. Wolf, Ed.), Vol. XI, North-Holland, New York, 1973, Chap. 3.

72. D. Mayden, "Acousto-Optical Pulse Modulators," *IEEE J. Quantum Electron.*, **QE-6**, 15–24 (1970).

73. A. Yariv, *Introduction to Optical Electronics*, Holt, Rinehart and Winston, New York, 1971.

74. E. Dieulesaint and D. Royer, *Elastic Waves in Solids*, Wiley, New York, 1980, Chaps. 8 and 9.

75. V. M. Ristic, M. Zuliani, G. Stegeman, and P. Vella, "Probing of Acoustic Shear Wave Radiation in Surface Wave Devices," *Appl. Phys. Lett.*, **21**(3), 85–87 (1972).

EXERCISES

1. Using Huygens' principle, find the intensity for a system consisting of a slit $z \leqslant (a/2)$ in an infinite baffle when a plane pressure wave is incident on the system.

2. Consider a plane piston radiator with initial velocity $v = v_0 \, \text{tri}(t/T)$, where the function tri is defined by Eq. (3.75). Sketch the radiator on-axis response when $\tau > T$, $\tau = T$, and $\tau < T$, where τ is defined by Eq. (10.53).

3. In the study of acoustic lenses, the so-called paraxial approximation is often used. With reference to Fig. 10.6(b), with B close to the z axis ($\rho \approx 0$), this approximation is obtained from Eqs. (10.54) to (10.58) as

$$R \approx z\left[1 + \frac{q^2 + \rho^2 - q\rho\cos\psi}{2z^2}\right]$$

for the numerator and $R \approx z$ for the denominator of Eq. (10.59). Show that the latter equation can then be written as

$$p(\rho, z) = \frac{p_0 j\beta r^{-j\beta z}}{2\pi z} \int_0^{2\pi}\int_0^a q \exp\left[\frac{j\beta}{2z}(\rho^2 + q^2 - 2q\rho\cos\psi)\right] d\psi \, dq$$

4. (a) From Eqs. (10.61) and (10.62), show that for a circular radiator of radius a the pressure in the Fraunhofer zone can be written as

$$\tilde{p}(\omega, \vec{r}) = \frac{j p_0 \beta}{r} e^{-j\beta r} \int_0^a J_0(\beta q \sin\theta) q \, dq$$

(b) Using the formula derived in (a), find the pressure of an annular radiator consisting of an area bounded by two circles of radii a_1 and a_2.

5. The diffraction corrections can be derived from Eq. (10.84) by assuming a radiator of uniform velocity distribution. Show that at distance $z \neq 0$ from the radiator, the total force, Eq. (10.88), is given approximately by

$$F = 2F_0 e^{-j\beta z} \int_0^\infty \frac{J_1(y)\exp(jSy^2/4\pi)}{y} \, dy$$

where $F_0 = \rho_m v_l v_0 \pi a^2 = p_0 \pi a^2$, $y = \alpha a$, and $S = \lambda_a z/a^2$, and the paraxial approximation $\gamma \approx j\beta - j\alpha^2/2\beta$ has been used in the exponential term of the integrand and $\gamma \approx j\beta$ in the denominator of the integrand. The diffraction corrections to phase and amplitude for a wave traveling the normalized distance S are then found from $(F/F_0)\exp j\beta z$. The diffraction losses in dB are given by

$$\alpha_d [\mathrm{dB}] = 20 \log |F/F_0|$$

Note: It can be shown that a good approximation to the above equation is α_d [dB] $\simeq \lambda_a z/a^2$, giving an average loss of 1 dB per path length a^2/λ_a. In many practical situations this approximation provides adequate correction. Also, a square transducer and a circular transducer of equal areas display approximately the same diffraction loss.

6. Find the surfaces of constant phase for the waves emanating from the source region using (a) Fresnel approximation and (b) Fraunhofer approximation.

7. The apparent discrepancy between Eq. (10.41) with $a \to \infty$ and Eq. (5.67) stems from the property of the δ function, $\delta(\alpha\beta) = \delta(\beta)/\alpha$. To clarify this point, use the stated property to find the dimension of the Green's function, g in Eq. (10.13). Furthermore, keep in mind that if $g(x - vt)$ is a solution of the wave equation $\partial^2 g/\partial x^2 - \partial^2 g/v^2 \partial t^2 = 0$, then $g[(x - vt)/C]$, where C is a constant, is also a solution of the wave equation.

8. With reference to Fig. 10.16, find the spherical aberrations and the resolution of (a) an f2 sapphire/water lens with $f = 2$ mm (b) an f5 fused quartz/water lens with $f = 5$ mm.

9. Show that the paraxial approximation derived in Exercise 3 in the case when z coincides with an acoustic axis can be written as

$$T(\rho, z) = \frac{jT_0 k_0 (1 + 2B)\exp(-jk_0 z)}{2\pi z}$$

$$\times \int_0^{2\pi} \int_0^a q \exp\left[\frac{jk_0(1 + 2B)}{2z}(\rho^2 + q^2 - 2q\rho \cos \psi)\right] d\psi \, dq$$

Author Index

Subject Index

DATE DUE

DEC 14 1990			
NOV 19 1990			
OCT 14 1992			
NOV 30 1992			

DEMCO 38-297